College Calculus

A One-Term Course for Students with Previous Calculus Experience

About the cover

The Mandelbrot set is a set of points in the complex plane with a remarkable boundary (see Section 8.1.1). The image on the cover is a view of a portion of the boundary centered at the complex number $c = -0.74364085 + 0.13182733i$ that resembles an integral sign. It is adapted from one created by Wolfgang Beyer using the software program *Ultra Fractal 3*. Beyer's image is licensed under the Creative Commons Attribution-Share Alike 1.0 Generic license, which may be found at creativecommons.org/licenses/by-sa/1.0/legalcode.

© 2015 by

The Mathematical Association of America (Incorporated)

Library of Congress Control Number: 2014954340

Print ISBN: 978-1-93951-206-2

Electronic ISBN: 978-1-61444-616-3

Printed in the United States of America

Current Printing (last digit):
10 9 8 7 6 5 4 3 2 1

College Calculus

A One-Term Course for Students with Previous Calculus Experience

Michael E. Boardman
Pacific University

and

Roger B. Nelsen
Lewis & Clark College

Published and distributed by
The Mathematical Association of America

MAA TEXTBOOKS

A Radical Approach to Lebesgue's Theory of Integration, David M. Bressoud
A Radical Approach to Real Analysis, 2nd edition, David M. Bressoud
Real Infinite Series, Daniel D. Bonar and Michael Khoury, Jr.
Topology Now!, Robert Messer and Philip Straffin
Understanding our Quantitative World, Janet Andersen and Todd Swanson

MAA Service Center
P.O. Box 91112
Washington, DC 20090-1112
1-800-331-1MAA   FAX: 1-301-206-9789

Contents

Preface

The calculus is one of the grandest edifices constructed by mankind.

Cambridge Conference on
School Mathematics (1963)

In a joint position statement on calculus dated March 2012, the Mathematical Association of America and the National Council of Teachers of Mathematics wrote "The college curriculum should acknowledge the ubiquity of calculus in secondary school [and] shape the college calculus curriculum so that it is appropriate for those who have experienced introductory calculus in high school." It is in that spirit that we have written this textbook.

College Calculus: A One-Term Course for Students with Previous Calculus Experience is, as its subtitle indicates, a textbook for students who have successfully experienced a course in introductory calculus in high school such as the Advanced Placement® Calculus AB course (or its equivalent). By "successfully experienced" a course in calculus, we mean that the student understands the concepts of calculus and its methods and applications at the level of earning a score of 4 or 5 on the AP® Calculus AB Exam. The College Board® describes students earning a 4 or 5 on the Exam as "well qualified" or "extremely well qualified" for placement into the appropriate next course at their chosen college or university.[1]

But what is the appropriate next course? At many colleges and universities it is the second course in the single variable calculus sequence, which we shall call "Calc II." While Calc II varies from institution to institution, it frequently begins with the definition of the definite integral and the fundamental theorem, and ends with infinite series. The AP Calculus AB course definitely does not cover all the content of Calc II, but there is significant overlap in the topics covered in the AP course and in Calc II, as the following table indicates.

---

[1] Advanced Placement®, AP®, and College Board® are trademarks registered and/or owned by the College Board, which was not involved in the production of, and does not endorse, this product.

AP Students	
have studied these topics	but may not have seen these
The fundamental theorem	Integration by parts
Substitution technique	Trigonometric substitution
Area using integrals	Partial fractions
Volume by known cross-sections	Volume by cylindrical shells
Volume by disks and washers	Work and hydrostatic force
Trapezoidal rule	Arc length
Midpoint rule	Simpson's rule
Separable differential equations	Euler's method
Slope fields	L'Hôpital's rule
Average value of a function	Parametric equations
	Polar coordinates
	Logistic models
	Infinite series

College Calculus begins with a gentle review of some of the content in the AP course, and proceeds to give students a thorough grounding in the remaining topics in single variable calculus. Our aim is to give students a solid foundation in single-variable calculus and prepare them for their next course in college level mathematics, be it multivariable calculus, linear algebra, a course in discrete mathematics, statistics, etc.

To the student

The fact that you are reading this indicates that you were successful in your previous calculus experience (which we will abbreviate PCE in this book)—congratulations! The content of this course will provide you with a firm foundation for your college-level mathematics courses and courses in other disciplines that use concepts and methods from calculus. To get off to a good start in this course, we recommend that you consult *A Description of the AP Calculus AB Course* in Appendix A for a review of the AP course, which serves as the prerequisite for College Calculus, and a careful reading of Chapter 0 *Preparation for College Calculus*.

To the instructor

The students enrolled in this course may be different than you have experienced in traditional single variable calculus courses. They have successfully studied calculus at the level of the AP Calculus AB course, now taught in over 18,000 high schools worldwide. We have written this book with just such a class of well prepared students in mind, and, if you are not familiar with the AP Calculus program, we also encourage you to consult *A Description of the AP Calculus AB Course* in Appendix A to better understand the background of the students enrolled in your course.

Note: In addition to the Calculus AB course and exam, there is the AP Calculus BC course and exam. Calculus BC includes the same topics in differential and integral calculus as Calculus

AB, plus additional material usually found in a two-semester calculus sequence in college such as polynomial approximations and infinite series. While this book has been written with the student who scores a 4 or 5 on the AP Calculus AB exam in mind, it is also appropriate for students who took the Calculus BC exam, and received a Calculus AB subscore (the score for the approximately 60% of the BC exam that covers AB topics) of 4 or 5.

About the authors

Michael Boardman and Roger Nelsen both have long affiliations with the AP Calculus program. Nelsen has participated in the annual summer AP readings for 25 years, most of these in AP Calculus, (with a stint in AP Statistics). For much of that time, Nelsen served as a Table Leader, and more recently as a member of a question team, responsible for working out the details of scoring a particular free-response question. Nelsen is the author or co-author of eight books including *Proofs Without Words I*, *Proofs Without Words II*, *Math Made Visual*, *The Calculus Collection*, *Charming Proofs*, and *Icons of Mathematics*. Boardman's affiliation with AP Calculus began with the 1994 reading. He was the moderator of the AP Calculus listserv for 10 years, and served four years as Chief Reader for AP Calculus. In this role, he was in charge of all aspects of the scoring of approximately 300,000 exams per year including selecting and supervising 800 readers, finalizing scoring rubrics, overseeing the logistics of the summer reading, and working with College Board personnel to set final cut scores. Boardman also served on the Development Committee for AP Calculus (2007–2011) whose members are responsible for updating the course syllabus, writing the exams, and providing outreach to high school teachers and college faculty. Boardman is currently involved in professional development of AP Calculus teachers, instructing summer and school-year workshops. Boardman serves on several MAA committees including the Committee on the Undergraduate Program in Mathematics.

Acknowledgments

Special thanks to Zaven Karian, Stan Seltzer, and the members of the editorial board of the MAA Textbook series for their careful reading of an earlier draft of the book and their many helpful suggestions. We would also like to thank Carol Baxter, Beverly Ruedi, and Samantha Webb of the MAA's book publication staff for their expertise in preparing this book for publication. Finally, special thanks to Don Albers, the MAA's former editorial director for books, who encouraged us to pursue this project and guided its production.

Michael E. Boardman
Pacific University
Forest Grove, Oregon

Roger B. Nelsen
Lewis & Clark College
Portland, Oregon

0

Preparation for College Calculus

Contrary to common belief, the calculus is not the height of the so-called "higher mathematics." It is, in fact, only the beginning.

Morris Kline
Mathematics in Western Culture

The definitions, theorems, and techniques in this review chapter should be familiar to you from your previous calculus experience (which we will abbreviate PCE in the rest of this book) and will be used throughout this course. Most are referred to sufficiently often in calculus and later courses that they have acquired names. For many we include a picture. The picture is, of course, not a part of the definition or theorem, but may help you see what the definition means or why the theorem is true, and also help you remember it. You will notice that this is not a complete and thorough review of your PCE, but rather a review of several of the topics from your PCE that are important in *College Calculus*.

0.1 Limits, continuity, and the derivative

The three primary objects studied in calculus—the derivative, the definite integral, and infinite series—are defined in terms of *limits*. We begin by reviewing some definitions and theorems from your PCE pertaining to continuity and the derivative. We assume you are familiar with limits and their properties from your PCE (see Appendix A for an outline of topics in the Calculus AB course). The first really important use of limits is to define *continuity*. Let f be a function with domain D (often an interval of real numbers or all real numbers). Then f is *continuous at a in D* if

$$\lim_{x \to a} f(x) = f(a).$$

The simple limit statement actually says three things:

(i) a is in D so that $f(a)$ exists.
(ii) $\lim_{x \to a} f(x)$ exists.
(iii) $\lim_{x \to a} f(x) = f(a)$.

The second really important use of limits is to define the derivative. For any x in D the derivative of f is the function f' defined by

$$f'(x) = \lim_{h \to 0} \frac{f(x+h) - f(x)}{h}$$

for all x in D for which the limit exists. Often "Δx" replaces h in the definition of f'. When $f'(x)$ exists, we say that f *is differentiable at x*.

Continuity and differentiability are related by the following theorem.

Theorem 0.1 *If f is differentiable at x, then f is continuous at x.*

In your PCE you encountered a number of theorems with *names*. We give names to theorems to indicate their importance and to help us remember them. When the "limit laws" from your PCE aren't applicable (as will occur in this course), the following named theorem can be useful.

Theorem 0.2 (The squeeze theorem) *If f, g, and h are functions such that $f(x) \le g(x) \le h(x)$ on an open interval containing a, except perhaps at a itself, and if*

$$\lim_{x \to a} f(x) = L = \lim_{x \to a} h(x),$$

then $\lim_{x \to a} g(x) = L$. See Figure 0.1.

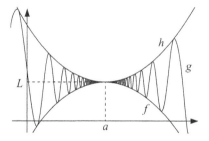

Figure 0.1. The squeeze theorem.

Three named theorems from your PCE have the word *value* in the title. Each is what mathematicians call an *existence theorem*, in that it tells us that a number exists with a certain property, but does not tell us how to find that number. The first two express important properties of *continuous* functions on *closed* intervals.

Theorem 0.3 (The intermediate value theorem) *Let f be a continuous function on the closed interval $[a, b]$ where $f(a) \ne f(b)$, and let N be any number between $f(a)$ and $f(b)$. Then there exists at least one number c in (a, b) such that $f(c) = N$. See Figure 0.2.*

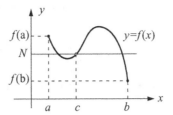

Figure 0.2. The intermediate value theorem.

Theorem 0.4 (The extreme value theorem) *Let f be a continuous function on the closed interval [a, b]. Then f has an absolute maximum value f(M) at some number M in [a, b] and an absolute minimum value f(m) at some number m in [a, b]. See Figure 0.3.*

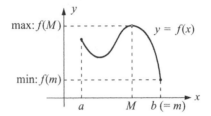

Figure 0.3. The extreme value theorem.

The third value theorem is the mean value theorem, used throughout calculus to prove other theorems. It also has practical applications. In Chapter 2 we will use the theorem to find the length of a curve using an integral.

Theorem 0.5 (The mean value theorem) *Let f be a function continuous on the closed interval [a, b] and differentiable on the open interval (a, b). Then there exists at least one number c in (a, b) such that*

$$f'(c) = \frac{f(b) - f(a)}{b - a}.$$

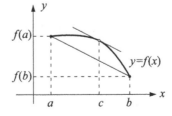

Figure 0.4. The mean value theorem.

In Figure 0.4 we recall from your PCE an illustration of the mean value theorem showing that, when the hypotheses of the theorem hold, there is a point $(c, f(c))$ on the graph of $y = f(x)$ where the slope $f'(c)$ of the tangent line equals the slope of the secant line joining the points

Table 0.1. Basic Differentiation Rules

$\dfrac{d}{dx}c = 0$	$\dfrac{d}{dx}cf(x) = cf'(x)$
$\dfrac{d}{dx}[f(x) \pm g(x)] = f'(x) \pm g'(x)$	$\dfrac{d}{dx}[f(x)g(x)] = f'(x)g(x) + f(x)g'(x)$
$\dfrac{d}{dx}\dfrac{f(x)}{g(x)} = \dfrac{f'(x)g(x) - f(x)g'(x)}{[g(x)]^2}$	$\dfrac{d}{dx}f(g(x)) = f'(g(x))g'(x)$
$\dfrac{d}{dx}x^r = rx^{r-1}$	$\dfrac{d}{dx}|x| = \dfrac{x}{|x|}$
$\dfrac{d}{dx}e^x = e^x$	$\dfrac{d}{dx}a^x = a^x \ln a$
$\dfrac{d}{dx}\ln x = \dfrac{1}{x}$	$\dfrac{d}{dx}\log_a x = \dfrac{1}{x \ln a}$
$\dfrac{d}{dx}\sin x = \cos x$	$\dfrac{d}{dx}\cos x = -\sin x$
$\dfrac{d}{dx}\tan x = \sec^2 x$	$\dfrac{d}{dx}\sec x = \sec x \tan x$
$\dfrac{d}{dx}\cot x = -\csc^2 x$	$\dfrac{d}{dx}\csc x = -\csc x \cot x$
$\dfrac{d}{dx}\arcsin x = \dfrac{1}{\sqrt{1 - x^2}}$	$\dfrac{d}{dx}\arccos x = \dfrac{-1}{\sqrt{1 - x^2}}$
$\dfrac{d}{dx}\arctan x = \dfrac{1}{x^2 + 1}$	$\dfrac{d}{dx}\text{arccot}x = \dfrac{-1}{x^2 + 1}$
$\dfrac{d}{dx}\text{arcsec}x = \dfrac{1}{x\sqrt{x^2 - 1}}$	$\dfrac{d}{dx}\text{arccsc}x = \dfrac{-1}{x\sqrt{x^2 - 1}}$

$(a, f(a))$ and $(b, f(b))$. The word *mean* in the title of the theorem means *average* in the following sense: when the hypotheses of the theorem hold, there is a number in the interval (a, b) where the *instantaneous* rate of change of f equals the *average* rate of change of f over the interval $[a, b]$.

In Table 0.1 we list some of the differentiation rules that should be familiar to you. We list them here for you to refer to as needed.

0.2 The integral and integration

Among the major topics in this book are applications of the definite integral and techniques of integration. Consequently we want to be sure that you recall exactly what a definite integral is, what an indefinite integral is, and what the fundamental theorem of calculus says about integrals, their evaluation, and the relationship between integration and differentiation.

The definite integral is defined in terms of the limit of *Riemann sums*, named for the German mathematician Georg Friedrich Bernhard Riemann (1826–1866). Rather than give a formal definition, we'll describe a construction procedure that we will use repeatedly in applications of the definite integral.

We begin with a function f defined on an interval $[a, b]$, and construct a Riemann sum for f on $[a, b]$ in three steps:

1. Let n be a positive integer and let P denote the *partition* of $[a, b]$ into n subintervals of equal width $\Delta x = (b - a)/n$, and let $x_0 < x_1 < x_2 < \cdots < x_{n-1} < x_n$ be the endpoints of the subintervals with $a = x_0$ and $b = x_n$. For each i between 1 and n inclusive the ith subinterval is $[x_{i-1}, x_i]$. Note that $x_i = a + i \Delta x$.
2. Choose a *sample point* x_i^* in each subinterval $[x_{i-1}, x_i]$ and evaluate $f(x_i^*)$.
3. Multiply each $f(x_i^*)$ by Δx and form a *Riemann sum* for f and the partition P:

$$\sum_{i=1}^{n} f(x_i^*)\Delta x = f(x_1^*)\Delta x + f(x_2^*)\Delta x + \cdots + f(x_n^*)\Delta x.$$

In Figure 0.5 we have an illustration, familiar to you from your PCE, of a Riemann sum approximation (with $n = 4$) for the area under the graph of a positive function. In this application the Riemann sum represents the sum of the areas of the four rectangles, and approximates the area under the graph of $y = f(x)$ over the interval $[a, b]$.

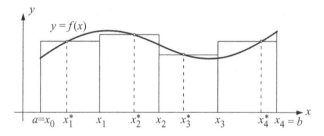

Figure 0.5. A Riemann sum with $n = 4$.

We will apply this three-step process to construct Riemann sums for a variety of geometry problems such as finding the length of a curve and the surface area of a solid, and physics problems such as calculating work and hydrostatic force.

In the first step of the three-step process, $[a, b]$ can be partitioned into subintervals of unequal widths. However, equal-width subintervals will suffice in *College Calculus*.

The definition of a definite integral and the fundamental theorem

Let f be defined on the interval $[a, b]$, and suppose that

$$\lim_{n \to \infty} \sum_{i=1}^{n} f(x_i^*)\Delta x$$

exists. This limit is called *the definite integral of f from a to b* and we write

$$\int_a^b f(x)\, dx = \lim_{n \to \infty} \sum_{i=1}^{n} f(x_i^*)\Delta x.$$

It can be shown that when f is continuous on $[a, b]$, the limit of the Riemann sum as $n \to \infty$ always exists and is independent of the choice of the sample points $\{x_i^*\}_{i=1}^n$ in $[a, b]$. In this case, we say that f is *integrable* on $[a, b]$ and call the process of evaluating an integral *integration*. Also recall that we call the elongated "S" symbol an *integral sign*, the a and b decorating the integral sign the *lower* and *upper limits of integration*, and the function following the integral sign the *integrand*.

While the definite integral is defined in terms of Riemann sums, evaluating the limit of the sums is usually not possible. For that we have *the fundamental theorem of calculus* (the name hints at its importance!) that connects integration to differentiation. It has two parts.

Theorem 0.6 (The fundamental theorem of calculus, part 1) *Let f be continuous on $[a, b]$, and x be such that $a \le x \le b$. If $g(x)$ is the function defined by the definite integral*

$$g(x) = \int_a^x f(t)\,dt,$$

then g is continuous on $[a, b]$ and differentiable on (a, b) with $g'(x) = f(x)$.

The second part of the fundamental theorem relates a definite integral to an *antiderivative*. Recall that F is an *antiderivative* of f if $F' = f$, that is, f is the derivative of F.

Theorem 0.7 (The fundamental theorem of calculus, part 2) *Let f be continuous on $[a, b]$, and let F be any antiderivative of f. Then*

$$\int_a^b f(x)\,dx = F(b) - F(a). \tag{0.1}$$

Recall from your PCE that the symbol $\int f(x)\,dx$ represents the general antiderivative $F(x)$ of $f(x)$ (e.g., $\int 2x\,dx = x^2 + C$ since $\frac{d}{dx}\left(x^2 + C\right) = 2x$ for any value of C). In practice, we almost always choose the antiderivative with $C = 0$ when applying part 2 of the fundamental theorem. Also recall that we use the symbol $F(x)\big|_a^b$ for $F(b) - F(a)$. Combining these notations permits us to write the conclusion of part 2 of the fundamental theorem as

$$\int_a^b f(x)\,dx = \int f(x)\,dx \bigg|_a^b = F(x)\bigg|_a^b = F(b) - F(a),$$

expressing the three step procedure to evaluate a definite integral:

1. find an antiderivative of the integrand,
2. evaluate it at a and b, and
3. subtract.

Since $f = F'$, (0.1) can be rewritten as

$$F(b) = F(a) + \int_a^b F'(x)\,dx,$$

that is, for any b in the domain of F, $F(b)$ can be recovered if we know the value of F at a single point a in its domain and its derivative at every point.

To make efficient use of the second part of the fundamental theorem, you need to be able to find an indefinite integral of the integrand. One technique for being able to do so is the *substitution* (of variables) *technique* or *rule* (also called the *chain rule for antidifferentiation* or *u-substitution*).

The substitution technique

Let $u = g(x)$ be a differentiable function whose range is an interval in the domain of a continuous function f. Then

$$\int f(g(x))\,g'(x)\,dx = \int f(u)\,du.$$

In applying the substitution technique we hope to go further, finding an antiderivative F of f, reversing the substitution, and writing

$$\int f(g(x))\,g'(x)\,dx = \int f(u)\,du = F(u) + C = F(g(x)) + C.$$

In this course you will learn several additional ways to evaluate indefinite and definite integrals. Many techniques will be even more useful when combined with substitution, so substitution is arguably the most important technique for integration.

In your PCE you studied a variety of applications of integration, including finding the area of a region, the volumes of certain solids, and the distance traveled by a particle. You will learn many more in this course. Another application of the integral in your PCE was finding the *average value* of a continuous function f, the number f_{ave} given by

$$f_{\text{ave}} = \frac{1}{b-a} \int_a^b f(x)\,dx.$$

Is the average value of f equal to a value of f at some point in $[a, b]$, that is, is there a number c in $[a, b]$ where $f_{\text{ave}} = f(c)$? The following theorem, which you may not have seen in your PCE, answers the question.

> **Theorem 0.8 (The mean value theorem for integrals)** *If f is continuous on $[a, b]$, then there exists at least one number c in $[a, b]$ such that $f_{\text{ave}} = f(c)$, or equivalently,*
>
> $$\int_a^b f(x)\,dx = f(c)(b - a).$$

When f is positive on $[a, b]$ there is a nice geometric interpretation of the theorem illustrated in Figure 0.6: the area under the graph of f over $[a, b]$ is exactly equal to the area of a rectangle with base $[a, b]$ and height $f_{\text{ave}} = f(c)$ for some c in $[a, b]$.

Theorem 0.8 can be proven from theorems reviewed previously in two different ways:

1. apply the ordinary mean value theorem (Theorem 0.5) and the first part of the fundamental theorem (Theorem 0.6) to the function $g(x) = \int_a^x f(t)\,dt$, and then replace x by b, or
2. apply the extreme value theorem (Theorem 0.4) to f to find m and M such that $f(m) \le f(x) \le f(M)$ on $[a, b]$, integrate to show that $f(m) \le f_{\text{ave}} \le f(M)$ and then apply the intermediate value theorem (Theorem 0.3) to f.

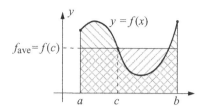

Figure 0.6. The mean value theorem for integrals.

In Table 0.2 we have some of the integration formulas familiar from your PCE. We will learn many more in this course.

Table 0.2. Basic integration formulas

$$\int cf(x)\,dx = c\int f(x)\,dx \qquad\qquad \int [f(x)\pm g(x)]\,dx = \int f(x)\,dx \pm \int g(x)\,dx$$

$$\int x^n\,dx = \frac{x^{n+1}}{n+1} + C\ (n \neq -1) \qquad \int \frac{1}{x}\,dx = \ln|x| + C \qquad \int e^x\,dx = e^x + C$$

$$\int a^x\,dx = \frac{a^x}{\ln a} + C\ (a > 0) \qquad\qquad \int \sin x\,dx = -\cos x + C$$

$$\int \cos x\,dx = \sin x + C \qquad\qquad \int \tan x\,dx = \ln|\sec x| + C$$

$$\int \cot x\,dx = -\ln|\csc x| + C \qquad\qquad \int \sec^2 x\,dx = \tan x + C$$

$$\int \csc^2 x\,dx = -\cot x + C \qquad\qquad \int \sec x \tan x\,dx = \sec x + C$$

$$\int \csc x \cot x\,dx = -\csc x + C \qquad\qquad \int \frac{1}{1+x^2}\,dx = \arctan x + C$$

$$\int \frac{1}{\sqrt{1-x^2}}\,dx = \arcsin x + C \qquad\qquad \int \frac{1}{x\sqrt{x^2-1}}\,dx = \operatorname{arcsec} x + C$$

$$\int \sec x\,dx = \ln|\sec x + \tan x| + C \qquad\qquad \int \csc x\,dx = -\ln|\csc x + \cot x| + C$$

0.3 Further review

In Appendix A near the end of this book you will find the topic outline for the AP Calculus AB course. It lists the calculus topics covered on the AP Calculus AB exam. We have written this book for students who scored well on that exam, so all, or most all, of the topics in Appendix A should be familiar to you. We encourage you to read that Appendix carefully. If any of the topics seem unfamiliar, or if you feel that you've become a bit rusty since the AP exam, we encourage you to review them. Resources for such a review include your school library, online web sites (search for "calculus help"), a math help center (check to see if there is one at your school), your instructor and fellow students, and the calculus textbook used at your school for students who did not take AP. We encourage you to do this review as soon as possible in the semester to prepare yourself for *College Calculus*!

I

Volume Integrals and Integration by Parts

A basic problem-solving technique is the construction of an approximate solution to a problem, and then do some math to improve the approximation so that the approximate solution becomes closer and closer to the exact solution. In your PCE you learned how to solve area and volume problems this way using approximations by Riemann sums, and the math that improved the approximations led to definite integrals.

In this chapter you will learn a second method for computing Riemann sum approximations to volumes of solids of revolution, which leads to the *cylindrical shell method*. Many of the integrals that result from this technique involve products of functions, and so we will explore an integration technique, called *integration by parts*, useful for evaluating many such integrals. In the final section we will combine two applications of definite integrals from your PCE—finding the area of a plane region and finding the volume of a solid with a known cross-sectional area—to finding volumes of more general solids.

1.1   The cylindrical shell technique

In your PCE you encountered methods for finding the volume of certain solids of revolution. They were obtained by rotating a region in the plane (usually bounded by graphs of functions and lines) about a line in the same plane (the axis of revolution) to either the x- or y-axis. In the disk and washer methods, you constructed Riemann sums to approximate the volume by partitioning the coordinate axis *parallel* to the axis of revolution. This method works in many instances, but fails in many others.

Example 1.1 Find the volume V of the solid obtained by revolving about the y-axis the region R in the first quadrant bounded by the graph of $y = x - x^3$ and the x-axis. See Figure 1.1.

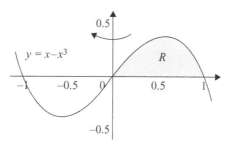

Figure 1.1. The region R in Example 1.1.

To use the washer method we must partition an interval on the y-axis and express the inside and outside radii of the washer in terms of functions of y. But it is difficult to solve $y = x - x^3$ for the two functions of y representing the two radii. We will return to this example after developing a method that begins with partitioning an interval $[0, 1]$ on the x-axis (the axis *perpendicular* to the axis of revolution). ∎

Now consider the general case—revolving about the y-axis a region R bounded by the graph $y = f(x) \geq 0$ of a continuous function, the x-axis, and the lines $x = a$ and $x = b$, $0 < a < b$, as illustrated in Figure 1.2. We partition the interval $[a, b]$ on the x-axis, and consider the object we obtain by revolving the slices of R generated by the partition $P = \{x_0, x_1, x_2, \ldots, x_{n-1}, x_n\}$ of $[a, b]$ where $a = x_0$ and $b = x_n$. In Figure 1.2 we show an approximation to one of the slices of R, a rectangle with base $[x_{i-1}, x_i]$ and height $f(x_i^*)$ where x_i^* is an arbitrary point in $[x_{i-1}, x_i]$.

When this rectangle is revolved about the y-axis, the object we obtain is called a *cylindrical shell* (only the back half of the shell is shown in Figure 1.2). We can find its volume the same way we found the volume of a washer in the washer method, as the difference of the volumes of two cylinders, an outer cylinder and an inner cylinder. Of course there will be n cylindrical shells in our approximation to the volume of the entire solid.

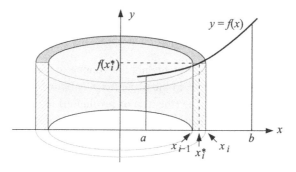

Figure 1.2. A cylindrical shell

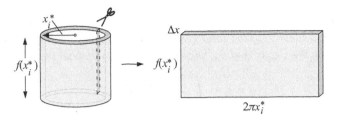

Figure 1.3. Computing the volume of a shell

Let V_i denote the volume of the ith shell. The outer cylinder has radius x_i and height $f(x_i^*)$, and the inner cylinder has radius x_{i-1} and height $f(x_i^*)$, hence

$$V_i = \pi x_i^2 f(x_i^*) - \pi x_{i-1}^2 f(x_i^*) = \pi \left(x_i^2 - x_{i-1}^2 \right) f(x_i^*).$$

The sum $\sum_{i=1}^n V_i$ of the volumes of the n shells represents an approximation to the true volume V of the solid. As n increases without bound, the approximation gets closer and closer to V and thus

$$V = \lim_{n \to \infty} \sum_{n=1}^n \pi \left(x_i^2 - x_{i-1}^2 \right) f(x_i^*). \tag{1.1}$$

If the sum in (1.1) were a Riemann sum, its limit would be a definite integral. But a Riemann sum has the form $\sum_{i=1}^n g(x_i^*) \Delta x$ where Δx is the width of each subinterval in the partition and $g(x_i^*)$ is a function evaluated at a single point x_i^* in the interval $[x_{i-1}, x_i]$. And (1.1) does not have that form—Δx is missing, and the summand is evaluated at the three points (x_{i-1}, x_i, x_i^*) in the interval.

But Δx isn't really missing in (1.1), it is one of the factors of $x_i^2 - x_{i-1}^2$. The other factor is $x_i + x_{i+1}$, which is twice the value of the midpoint $(x_i + x_{i-1})/2$ of the interval $[x_{i-1}, x_i]$. Recalling our choice of the point x_i^* was arbitrary, we now specify that we will use midpoints in each subinterval in the partition, that is, we set $x_i^* = \frac{x_i + x_{i-1}}{2}$ for $i = 1, 2, \ldots, n$. So x_i^* is the *average radius of revolution* of the ith cylindrical shell. Thus $x_i^2 - x_{i-1}^2 = 2x_i^* \Delta x$ and the volume V_i of the ith shell is $V_i = 2\pi x_i^* f(x_i^*) \Delta x$. To obtain this volume geometrically, think of the ith shell as having been cut as shown and opened up flat as in Figure 1.3. Then the volume of the shell results from multiplying (length) × (height) × (depth) to obtain $2\pi x_i^* \times f(x_i^*) \times \Delta x$.

Since the volume V_i of the ith shell is $V_i = 2\pi x_i^* f\left(x_i^*\right) \Delta x$, we can write (1.1) as

$$V = \lim_{n \to \infty} \sum_{i=1}^n 2\pi x_i^* f\left(x_i^*\right) \Delta x. \tag{1.2}$$

The sum in (1.2) is a Riemann sum, and hence we have a procedure for finding the volume V of the solid obtained by revolving R about the y-axis.

Theorem 1.1 *If a function f is continuous and positive on the interval $[a, b]$ where a is positive, then the volume V of the solid generated by revolving the region bounded by the graph of $y = f(x)$, the x-axis, and the lines $x = a$ and $x = b$ about the y-axis is given by*

$$V = \int_a^b 2\pi x f(x) \, dx. \qquad (1.3)$$

Back to Example 1.1 When the region R in Figure 1.1 is revolved about the y-axis, then the volume of the resulting solid is

$$V = \int_0^1 2\pi x \left(x - x^3\right) dx = 2\pi \left[\frac{x^3}{3} - \frac{x^5}{5}\right]_0^1 = \frac{4\pi}{15}. \quad \blacksquare$$

Example 1.2 (The volume of a sphere) In your PCE, you may have used the disk method to find the volume V of a sphere of radius r. If not, we will do it now.

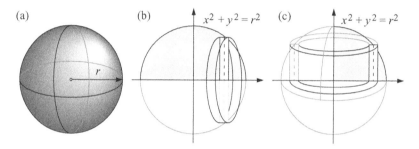

Figure 1.4. Computing the volume of a sphere

To obtain a sphere of radius r we revolve the gray semicircle bounded by the graph of $y = \sqrt{r^2 - x^2}$ and the x-axis for x in $[-r, r]$ about the x-axis, as shown in Figure 1.4(b). Then the volume, V, is

$$V = \int_{-r}^r \pi \left(\sqrt{r^2 - x^2}\right)^2 dx = \pi \left[r^2 x - \frac{x^3}{3}\right]_{-r}^r = \frac{4}{3}\pi r^3.$$

The volume can also be computed by shells, as twice the volume of the solid obtained by revolving the quarter circle shown in Figure 1.4(c) about the y-axis. The height of the shell is $f(x) = \sqrt{r^2 - x^2}$ and (1.3) yields

$$V = 2 \int_0^r 2\pi x \sqrt{r^2 - x^2} \, dx.$$

To evaluate the integral we use the substitution $u = r^2 - x^2$, $du = -2x \, dx$ to obtain

$$V = 2 \int_{r^2}^0 \pi u^{1/2}(-du) = 2\pi \int_0^{r^2} u^{1/2} \, du = 2\pi \cdot \frac{2}{3} u^{3/2} \Big|_0^{r^2} = \frac{4}{3}\pi r^3. \quad \blacksquare$$

We doubt that you are surprised that we obtain the same answer for the volume of a sphere using shells as when the disk method is used. In Exploration 1.4.1 at the end of the chapter you will show that in cases like this the two methods *must* give you the same answer.

☺ Whenever you can solve a problem in two different ways, you have a good way to check your work!

In solving volume problems by the shell method, keep in mind that the x_i^* in formula (1.2) is the average radius of revolution of the shell (so that $2\pi x_i^*$ is the *average circumference* of the circle of revolution), and $f\left(x_i^*\right)$ represents the *height* of the shell, and Δx the *thickness* of the shell. Hence the same is true of x, $2\pi x$, $f(x)$, and dx in (1.3).

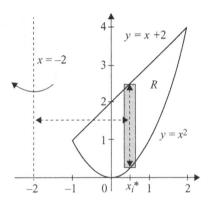

Figure 1.5. Region bounded by $y = x^2$ and $y = x + 2$

Example 1.3 Find the volume V of the solid obtained by revolving about the line $x = -2$ the region R bounded by the graphs of $y = x^2$ and $y = x + 2$. See Figure 1.5.

The line and the parabola intersect at $(-1, 1)$ and $(2, 4)$, and so we partition the interval $[-1, 2]$ on the x-axis. When the ith rectangle in the partition of R is revolved about the line $x = -2$, it generates a shell of height $(x_i^* + 2) - (x_i^*)^2$ (the length of the vertical dashed arrow in Figure 1.5) and radius of revolution $x_i^* - (-2) = x_i^* + 2$ (the length of the horizontal dashed arrow in Figure 1.5). In the integral, each x_i^* is replaced by x, and hence

$$V = 2\pi \int_{-1}^{2} (x + 2)(x + 2 - x^2)\, dx = \cdots = \frac{45\pi}{2}.$$

☺ You should be able to easily fill in the gap indicated by the ellipsis ($\cdots$) ∎

Example 1.4 Find the volume V of the solid obtained by revolving about the x-axis the region bounded by the graph of $y = \ln x$, the x-axis, and the line $x = e$. See Figure 1.6.

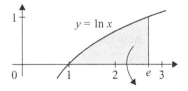

Figure 1.6. Region in first quadrant bounded by $y = \ln x$ and $x = e$

To use the disk method we partition the interval $[1, e]$ on the x-axis, and obtain

$$V = \pi \int_1^e (\ln x)^2 \, dx.$$

Since we do not (yet) have a technique for evaluating this integral, we try the shell method. Writing the function as $x = e^y$ and partitioning the interval $[0, 1]$ on the y-axis yields

$$V = 2\pi \int_0^1 y(e - e^y) \, dy, \tag{1.4}$$

another integral we cannot yet evaluate. While we could resort to a numeric method (e.g., the trapezoidal rule) at this point, what we really need is a technique for evaluating integrals of products, as this often occurs when using the shell method. The method, a product rule for integration, is called *integration by parts*. After presenting the method in the next section we will return to this example. ∎

1.1.1 Exercises

In Exercises 1 through 8 we consider some volume problems that can be solved using either the cylindrical shell method or the circular disk (or washer) method from your PCE. Compute the volume of each solid of revolution both ways, to review the disk technique while you practice the shell technique. In the statement of each problem we give the bounds of the region being revolved followed by the axis of revolution.

1. $x = \sqrt{4 - y}$, $x = 0$, $y = 0$; about the y-axis.

2. $2y + x = 4$, $x = 0$, $y = 0$; about the x-axis.

3. $y = x^3$, $x = 0$, $y = 8$; about the y-axis.

4. $y = \sqrt{9 - x}$, $x = 0$, $y = 0$; about the x-axis.

5. $y = x^2$, $y = 4x - x^2$; about the y-axis.

6. $y = \sqrt{x}$, $y = x/2$; about the x-axis.

7. $y = x^2$, $x = 0$, $y = 4$; about the line $x = 2$.

8. $y = x^2$, $y = 2x$; about the line $y = 4$.

In Figure 1.7, R is the region bounded by the y-axis, the line $y = 1$, and the curve $x = y^2$; S is the region bounded by the curves $x = y^2$ and $y = x^3$; and T is the region bounded by the x-axis, the line $x = 1$, and the curve $y = x^3$.

In Exercises 9 through 20, use the cylindrical shell method to find the volume of the solid obtained when the indicated region is revolved about the indicated axis:

9. R about the y-axis   10. S about the y-axis   11. T about the y-axis

12. R about the x-axis   13. S about the x-axis   14. T about the x-axis

15. R about $x = 1$ 16. S about $x = 1$ 17. T about $x = 1$

18. R about $y = 1$ 19. S about $y = 1$ 20. T about $y = 1$.

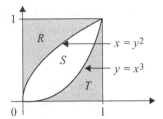

Figure 1.7. For Exercises 9–20

21. Find the volume of the solid obtained by revolving the region bounded by the parabola $y = 2x - x^2$ and the x-axis about
    (a) the y-axis (b) the line $x = 4$.

22. Find the volume of the solid obtained by revolving the region bounded by the parabola $x = y^2$ and the lines $x = 1$ and $x = 2$ about
    (a) the y-axis (b) the line $x = 2$.

23. Find the volume of the solid obtained by revolving the region bounded by the parabola $x = y^2 - 1$ and the line $y = x - 1$ about the line $y = -1$.

24. Find the volume of the solid obtained by revolving the region bounded by the curves $y = \sin\left(x^2\right)$ and $y = \cos\left(x^2\right)$ for x between $\sqrt{\pi}/2$ and $\sqrt{5\pi}/2$ about the y-axis.

25. A *torus* (perhaps from the Latin word for "knot") is a doughnut-shaped object (*doughnut* was originally spelled *dough knot*) obtained by revolving a circle about a line outside the circle, as shown in Figure 1.8.

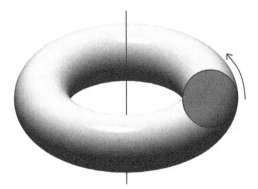

Figure 1.8. A torus

If the circle has radius r and the distance from the center of the circle to the axis of revolution is R where $R > r$, set up integrals for the volume of a torus using (a) washers and (b) shells. Then (c) evaluate whichever you think is easier to do. (Hint: the equation of the circle is simple if you place its center at the origin. Then the axis of revolution can be either $x = R$ or $y = R$.)

26. A *bead* results from drilling a hole of radius a through a sphere of radius r, $0 < a < r$, as shown in Figure 1.9. Show that to find the volume of the bead you need only measure the height h of the hole.

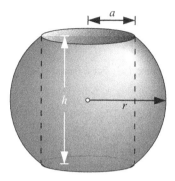

Figure 1.9. A bead

1.2 Integration by parts

Integration by parts is the integration analog of the product rule for differentiation, and is useful in evaluating integrals whose integrands contain products of functions, and for integrals of functions such as the natural logarithm, inverse trigonometric functions, etc.

Let f and g be differentiable functions of x. Then

$$\frac{d}{dx} f(x)g(x) = f(x)g'(x) + g(x)f'(x).$$

We now find the indefinite integral (i.e., the antiderivative) of each side. Since the antiderivative of the derivative of $f(x)g(x)$ is $f(x)g(x)\pm$ some constant, we have

$$f(x)g(x) = \int f(x)g'(x)\,dx + \int g(x)f'(x)\,dx,$$

where the constant on the left side has been absorbed into one of the constants of integration on the right side. Rearranging the terms yields

Theorem 1.2 (Integration by parts) *If f and g are functions with continuous derivatives, then*

$$\int f(x)g'(x)\,dx = f(x)g(x) - \int g(x)f'(x)\,dx. \tag{1.5}$$

This is the integration by parts formula. Its purpose is to replace one integral by another that hopefully is easier to evaluate. Before presenting examples, a change of variables in the formula makes it easier to apply. If we let $u = f(x)$ and $v = g(x)$, then $du = f'(x)dx$ and $dv = g'(x)dx$, and the integration by parts formula can be written as

$$\boxed{\int u \, dv = uv - \int v \, du.}$$ (1.6)

The key to successfully using (1.6) to evaluate an integral (in the variable x) lies in finding an auspicious choice of functions of x for the parts u and dv. Two suggestions:

1. The function v should be easy to find from dv,
2. The integral $\int v \, du$ should be (in some sense) simpler (or at least not more complicated) than $\int u \, dv$.

Example 1.5 Evaluate $\int x \cos x \, dx$.
Which part of the integrand should appear as u and which part as dv? Since we differentiate u and integrate dv, an auspicious choice of u is a function that simplifies (or becomes no more complicated) by differentiation, and an auspicious choice of dv is a function that simplifies (or becomes no more complicated) by integration. Let's try $u = x$ and $dv = \cos x \, dx$. Then $du = dx$ and $v = \int \cos x \, dx = \sin x$, and thus

$$\int x \cos x \, dx = x \sin x - \int \sin x \, dx,$$ (1.7)

so that the original integral has been replaced by a simpler one that can be readily evaluated to yield

$$\int x \cos x \, dx = x \sin x - \int \sin x \, dx = x \sin x + \cos x + C.$$

The result can be checked by differentiation, which you should do. Note that you use the product rule, which is to be expected since the parts technique comes from the product rule.

But wait—shouldn't we have introduced a constant of integration when we evaluated v from dv? If we let $v = \sin x + c$, then (1.7) becomes

$$\int x \cos x \, dx = x (\sin x + c) - \int (\sin x + c) \, dx,$$

and c drops out after the next integration. So we need not introduce a constant of integration when finding v from dv.

What would have happened with a different choice of parts? If we choose $u = \cos x$ and $dv = x \, dx$, then $du = -\sin x \, dx$ and $v = x^2/2$, and hence

$$\int x \cos x \, dx = \frac{x^2}{2} \cos x + \int \frac{x^2}{2} \sin x \, dx,$$

which is a true statement—it's just not a very useful true statement since the integral on the right side is not simpler than the one we began with on the left. ◼

Example 1.6 Evaluate $\int x^2 \sin x \, dx$.
We let $u = x^2$ and $dv = \sin x \, dx$ so that $du = 2x \, dx$ and $v = -\cos x$, hence

$$\int x^2 \sin x \, dx = -x^2 \cos x + 2 \int x \cos x \, dx.$$

The new integral is simpler; in fact, it is the one we did in the previous example, and so

$$\int x^2 \sin x \, dx = -x^2 \cos x + 2x \sin x + 2 \cos x + C. \quad \blacksquare$$

☺ As always, it is good practice to check the result by differentiation!

Integration by parts is a technique that may need to be applied more than once in a single integration problem. Furthermore, integration by parts can be used to integrate functions like $\ln x$ and $\arctan x$ for which we do not yet know an antiderivative (but that we do know how to differentiate). In addition, integration by parts may need to be combined with another integration technique, as the following example shows.

Example 1.7 Evaluate $\int \arctan x \, dx$.
We have no real choice of parts here. We must set $u = \arctan x$ and $dv = dx$, so that $du = dx/(x^2 + 1)$ and $v = x$. Hence

$$\int \arctan x \, dx = x \arctan x - \int \frac{x}{x^2 + 1} \, dx.$$

To finish this problem, we evaluate the new integral with an ordinary substitution, $t = x^2 + 1$ so that

$$\int \frac{x}{x^2 + 1} = \frac{1}{2} \int \frac{1}{t} \, dt = \frac{1}{2} \ln |t| + C = \frac{1}{2} \ln \left(x^2 + 1\right) + C.$$

We may omit the absolute value inside the logarithm since $x^2 + 1 > 0$. Thus we have

$$\int \arctan(x) \, dx = x \arctan x - \frac{1}{2} \ln \left(x^2 + 1\right) + C. \quad \blacksquare$$

If the derivatives $f'(x)$ and $g'(x)$ are continuous on an interval $[a, b]$, then the definite integral version of (1.5) is

Theorem 1.3 (Definite integration by parts) *If f and g are functions with continuous derivatives on the interval $[a, b]$, then*

$$\int_a^b f(x)g'(x) \, dx = f(x)g(x) \Big|_a^b - \int_a^b g(x)f'(x) \, dx. \tag{1.8}$$

Back to Example 1.4 We can now evaluate the volume integral from (1.4). Let $u = y$ and $dv = (e - e^y) \, dy$ so that $du = dy$ and $v = ey - e^y$. Then

$$V = 2\pi \int_0^1 y(e - e^y) dy = 2\pi \left[y(ey - e^y) \Big|_{y=0}^1 - \int_0^1 (ey - e^y) dy \right]$$

$$= 2\pi \left[ey^2 - ye^y - \frac{e}{2}y^2 + e^y \right]_0^1 = \pi (e - 2). \quad \blacksquare$$

⊘ While Theorem 1.3 is more efficient than Theorem 1.2 when using integration by parts to evaluate a definite integral, it does have one drawback: You lose the opportunity to use differentiation to check your work.

Example 1.8 Evaluate $\int e^{-x} \cos x \, dx$.

Our heuristic advice about how to choose the parts doesn't help much here, since the derivative and indefinite integral of e^{-x} are identical (both are $-e^{-x}$), and similarly the derivative and indefinite integral of $\sin x$ differ only in sign. Nevertheless, let's try integration by parts anyway, with $u = e^{-x}$ and $dv = \cos x \, dx$, so that $du = -e^{-x} \, dx$ and $v = \sin x$. Setting $I = \int e^{-x} \cos x \, dx$ will simplify our work a bit. Thus,

$$I = e^{-x} \sin x + \int e^{-x} \sin x \, dx.$$

For the new integral, let's keep u the same and use $dv = \sin x \, dx$:

$$I = e^{-x} \sin x + \left(-e^{-x} \cos x - \int e^{-x} \cos x \, dx \right),$$

that is,

$$I = e^{-x} \sin x - e^{-x} \cos x - I.$$

The original integral I on the left has reappeared on the right, but the two Is do not cancel since I on the right is preceded by a negative sign. Furthermore, its constant of integration is *arbitrary*, and hence not necessarily the same as the constant of integration in the original integral. Moving the integral on the right side to the left (but leaving an arbitrary constant C on the right) yields

$$2I = e^{-x} (\sin x - \cos x) + C$$

and hence

$$\int e^{-x} \cos x \, dx = \frac{e^{-x}}{2} (\sin x - \cos x) + C.$$

The last two Cs are not the same but this is fine since C represents an *arbitrary* constant. ∎

☺ Don't forget to check the result by differentiation!

Example 1.9 Evaluate $\int \sec^3 x \, dx$.

There are many ways to split $\sec^3 x \, dx$ into the u and dv parts for integration, but the need to integrate dv leads to the choice $u = \sec x$ and $dv = \sec^2 x \, dx$, so that $du = \sec x \tan x \, dx$ and $v = \tan x$. Letting $I = \int \sec^3 x \, dx$ we have

$$I = \sec x \tan x - \int \sec x \tan^2 x \, dx.$$

Employing the identity $\tan^2 x = \sec^2 x - 1$ yields

$$I = \sec x \tan x - I + \int \sec x \, dx$$

so that

$$2I = \sec x \tan x + \int \sec x \, dx$$

or

$$\int \sec^3 x \, dx = \frac{1}{2} \sec x \tan x + \frac{1}{2} \int \sec x \, dx. \qquad (1.9)$$

(The constants of integration from the two appearances of the integral I have been absorbed into the remaining integral.) To finish this example, we need to evaluate $\int \sec x \, dx$. This is easily accomplished using what you know about the derivatives of the tangent and secant from your PCE:

$$\frac{d}{dx} \sec x = \sec x \tan x \quad \text{and} \quad \frac{d}{dx} \tan x = \sec^2 x.$$

Adding the two together and factoring yields

$$\frac{d}{dx}(\sec x + \tan x) = (\sec x \tan x + \sec^2 x) = (\sec x + \tan x) \sec x$$

and hence

$$\sec x = \frac{1}{\sec x + \tan x} \frac{d}{dx}(\sec x + \tan x).$$

Integrating both sides with respect to x and using the substitution $t = \sec x + \tan x$ yields

$$\boxed{\int \sec x \, dx = \ln |\sec x + \tan x| + C.} \qquad (1.10)$$

Returning to $\int \sec^3 x \, dx$ in (1.9), we have

$$\boxed{\int \sec^3 x \, dx = \frac{1}{2}\left(\sec x \tan x + \ln |\sec x + \tan x| \right) + C.} \qquad (1.11)$$

An easy way to remember (1.11) is to observe that the antiderivative of $\sec^3 x$ is the average of the derivative and the antiderivative of $\sec x$. ■

1.2.1 Exercises

In Exercises 1 through 21, evaluate the indefinite integral. Be sure to check your work by differentiation.

1. $\int x \sin x \, dx$

2. $\int xe^{-2x} \, dx$

3. $\int \theta \sec^2 \theta \, d\theta$

4. $\int \ln x \, dx$

5. $\int t \ln t \, dt$

6. $\int x \tan^{-1} x \, dx$

7. $\int \dfrac{\ln y}{y^2} \, dy$

8. $\int \theta^2 \cos \theta \, d\theta$

9. $\int x^3 e^{-x} \, dx$

10. $\int \arcsin z \, dz$

11. $\int x \arcsin x \, dx$

12. $\int (\ln x)^2 \, dx$

13. $\int e^{2x} \cos 3x \, dx$

14. $\int \cos (\ln t) \, dt$

15. $\int \arctan \sqrt{x} \, dx$

16. $\int e^{\sqrt{x}} \, dx$

17. $\int \sin 2x \cos 3x \, dx$

18. $\int x^n \ln x \, dx$

19. $\int x \sin x \cos x \, dx$

20. $\int xe^x \sin x \, dx$

21. $\int \sec^5 x \, dx$

In Exercises 22 through 27, evaluate the definite integral.

22. $\int_1^e x^3 \ln x \, dx$

23. $\int_0^{\pi/4} \theta \sec \theta \tan \theta \, d\theta$

24. $\int_0^2 x 2^x \, dx$

25. $\int_{\pi/6}^{\pi/2} \cos x \ln (\sin x) \, dx$

26. $\int_0^1 \ln \sqrt{x^2 + 1} \, dx$

27. $\int_{-1}^1 \dfrac{\tan x}{1 + x^2} \, dx$

28. In your PCE you may have learned to integrate $\int \sin^2 x \, dx$ and $\int \cos^2 x \, dx$ using the double-angle formulas $\sin^2 x = (1 - \cos 2x)/2$ and $\cos^2 x = (1 + \cos 2x)/2$ followed by the substitution $u = 2x$. If you have (temporarily) forgotten these formulas you will be glad to hear you can use integration by parts to evaluate the integrals.
   (a) Integrate $\int \sin^2 x \, dx$ by parts. (Hints: set $u = \sin x$ and $dv = \sin x \, dx$. When the integral $\int \cos^2 x \, dx$ appears, replace $\cos^2 x$ by $1 - \sin^2 x$.)
   (b) Integrate $\int \cos^2 \, dx$.
   (c) Part (b) can be done without integration by parts. Do you know a function whose derivative involves $\cos^2 x$? Yes, using the product rule, the derivative of $\sin x \cos x$ is $\cos^2 x - \sin^2 x$. Write the integrand as $\frac{1}{2}(\cos^2 x + \cos^2 x)$, and use the Pythagorean identity to obtain $\frac{1}{2}(1 + \frac{d}{dx}(\sin x \cos x))$ and integrate.

29. Find the volume of the solid obtained when the region bounded by the curve $y = \sin x$ and the x-axis between $x = 0$ and $x = \pi$ is revolved about the y-axis.

30. Find the volume of the solid obtained when the region bounded by the curve $y = \ln x$, the x-axis, and the line $x = e$ is revolved about (a) the y-axis, (b) the line $x = e$.

31. (a) Show that if one evaluates $\int 1/(x \ln x)\, dx$ by parts with $u = 1/(\ln x)$ and $dv = (1/x)\, dx$, the result is

$$\int \frac{1}{x \ln x}\, dx = 1 + \int \frac{1}{x \ln x}\, dx$$

and hence $0 = 1$.

(b) Where is the error?

(c) Evaluate $\int \frac{1}{x \ln x}\, dx$.

32. There are instances when it is advantageous in using integration by parts to include a nonzero constant of integration when finding v from dv. Here is an example.

(a) Evaluate $\int x \arctan x\, dx$ with $u = \arctan x$ and $dv = x\, dx$, using $v = x^2/2$.

(b) Repeat (a), but with $v = (x^2 + 1)/2$.

(c) Which is easier? Explain.

33. Formula (1.10) for the integral of the secant often appears in other forms. Show that

$$\int \sec x\, dx = \ln \left| \frac{1 + \tan(x/2)}{1 - \tan(x/2)} \right| + C = \ln \left| \tan \left(\frac{x}{2} + \frac{\pi}{4} \right) \right| + C.$$

34. Evaluate $\int \csc x\, dx$ and $\int \csc^3 x\, dx$. (Hint: the techniques for these integrals parallel those in Example 1.9.)

35. (a) Show that for an integer $n \geq 2$,

$$\int \sin^n x\, dx = -\frac{1}{n} \sin^{n-1} x \cos x + \frac{n-1}{n} \int \sin^{n-2} x\, dx.$$

(b) Use (a) to show that for an integer $k \geq 1$,

$$\int_0^{\pi/2} \sin^{2k} x\, dx = \frac{2k-1}{2k} \cdot \frac{2k-3}{2k-2} \cdots \frac{1}{2} \cdot \frac{\pi}{2} \quad \text{and}$$

$$\int_0^{\pi/2} \sin^{2k+1} x\, dx = \frac{2k}{2k+1} \cdot \frac{2k-2}{2k-1} \cdots \frac{2}{3} \cdot 1.$$

1.3 The volume of a solid with computable cross-sectional areas

After learning how to use integrals to find areas of plane regions in your PCE you learned several methods for finding volumes of solids. Perhaps the first method you learned was for a solid with a known cross-sectional area. If the solid and the x-axis could be oriented in space in such a way that for each x in an interval $[a, b]$ the area of the cross-section at x is given by a continuous function $A(x)$, then the volume V of the solid is $V = \int_a^b A(x)\, dx$. See Figure 1.10

In this section we combine these two applications of the integral. We use integration to first find the areas of the cross-sections of a given solid, and then use integration a second time to find the volume of the solid.

To accomplish this, we need to introduce some three-dimensional geometry to describe the solids. In area problems, a region in the xy−plane is often described as the set of points below the

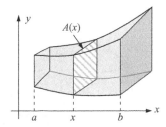

Figure 1.10. Solid with known cross-sectional area

graph of a non-negative function $y = f(x)$ over an interval $[a, b]$ of the x-axis. For our volume problems, a region in xyz-space is the set of points below the graph of a non-negative function $z = h(x, y)$ over a rectangle $[a, b] \times [c, d] = \{(x, y) | a \le x \le b, c \le y \le d\}$ in the xy-plane. For example, Figure 1.11 illustrates the region under the graph of $z = \sin x + \cos y + 2$ over the rectangle $[-2\pi, 2\pi] \times [-2\pi, 2\pi]$.

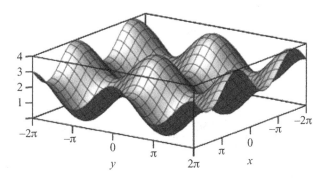

Figure 1.11. Graph of $z = \sin x + \cos y + 2$

For such a solid we will first use integration to find the areas of the cross-sections obtained by slicing the solid, with slices perpendicular to either the x- or y-axis, and then integrate this cross-sectional area function to obtain the volume of the solid. The next example illustrates the procedure.

Example 1.10 Find the volume of the solid given by the region in space under the graph of $z = h(x, y) = 6 - 2x^2 - y^2$ over the rectangle $[-1, 1] \times [-2, 2]$.

First we graph the function over the rectangle in Figure 1.12.

For a fixed x_0 in the interval $[-1, 1]$, slice the solid with a plane perpendicular to the x-axis at x_0. This plane intersects the graph of $h(x, y)$ in the curve $z = g(y) = h(x_0, y)$, and the cross-section of the solid is the region under the graph of $z = g(y)$ over the interval $[-2, 2]$. Since g is continuous, the area $A(x_0)$ of the cross-section is

$$A(x_0) = \int_{-2}^{2} g(y)\,dy = \int_{-2}^{2} (6 - 2x_0^2 - y^2)\,dy = \left[6y - 2x_0^2 y - \frac{y^3}{3} \right]_{y=-2}^{2} = \frac{56}{3} - 8x_0^2.$$

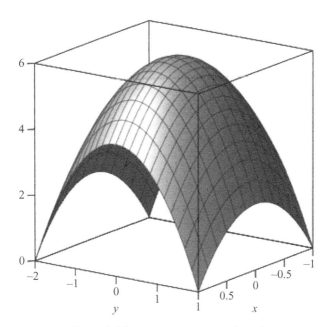

Figure 1.12. Graph of $z = 6 - 2x^2 - y^2$

Replacing x_0 by x yields the cross-sectional area function $A(x) = (56/3) - 8x^2$, which we now integrate to obtain the volume of the solid:

$$V = \int_{-1}^{1} A(x)\, dx = \int_{-1}^{1} \left(\frac{56}{3} - 8x^2 \right) dx = \left[\frac{56}{3}x - \frac{8}{3}x^3 \right]_{-1}^{1} = 32,$$

and thus the volume of the solid is $V = 32$.

Or is it? What if we sliced the solid with planes perpendicular to the y-axis? Would we obtain the same result? Intuition tells us that the answer must be yes, but let's see if calculus supports our intuition.

For a fixed y_0 in the interval $[-2, 2]$, we now slice the solid with a plane perpendicular to the y-axis at y_0. This plane intersects the graph of $h(x, y)$ in the curve $z = f(x) = h(x, y_0)$, and this cross-section of the solid is the region under the graph of $z = f(x)$ over the interval $[-1, 1]$. Since f is continuous, the area $B(y_0)$ of the cross-section is

$$B(y_0) = \int_{-1}^{1} f(x)\, dx = \int_{-1}^{1} \left(6 - 2x^2 - y_0^2 \right) dx = \left[6x - \frac{2}{3}x^3 - xy_0^2 \right]_{-1}^{1} = \frac{32}{3} - 2y_0^2.$$

Replacing y_0 by y yields another cross-sectional area function $B(y) = (32/3) - 2y^2$, which we now integrate:

$$V = \int_{-2}^{2} B(y)\, dy = \int_{-2}^{2} \left(\frac{32}{3} - 2y^2 \right) dy = \left[\frac{32}{3}y - \frac{2}{3}y^3 \right]_{-2}^{2} = 32,$$

and thus the volume of the solid is once again $V = 32$. ■

We can rewrite these results as

$$V = \int_{-1}^{1} A(x)\, dx = \int_{-1}^{1} \left(\int_{-2}^{2} g(y)\, dy \right) dx$$

$$= \int_{-1}^{1} \left(\int_{-2}^{2} h(x, y)\, dy \right) dx = \int_{-1}^{1} \int_{-2}^{2} h(x, y)\, dy\, dx$$

and

$$V = \int_{-2}^{2} B(y)\, dy = \int_{-2}^{2} \left(\int_{-1}^{1} f(x)\, dx \right) dy$$

$$= \int_{-2}^{2} \left(\int_{-1}^{1} h(x, y)\, dx \right) dy = \int_{-2}^{2} \int_{-1}^{1} h(x, y)\, dx\, dy.$$

These integral expressions with one integral inside another are examples of *iterated integrals*, which you will study in greater detail in a course in multivariable calculus. Conditions under which the two iterated integrals are equal are given in *Fubini's theorem* named for the Italian mathematician Guido Fubini (1879–1943). In its simplest form, this theorem asserts that if $h(x, y)$ is continuous on the rectangle $[a, b] \times [c, d]$, then

$$\int_{c}^{d} \int_{a}^{b} h(x, y)\, dx\, dy = \int_{a}^{b} \int_{c}^{d} h(x, y)\, dy\, dx.$$

You may have noticed that we haven't yet defined continuity for functions of two variables like $h(x, y)$, but for now you can think of it intuitively as you do for functions of a single variable: the graph has no breaks, gaps, or tears in it, and doesn't approach $\pm\infty$ anywhere in $[a, b] \times [c, d]$. In the examples (and exercises) in this section, Fubini's theorem holds, so we may compute the volume using either iterated integral.

We recommend the following two-step process for finding the volume under the graph of a function $z = h(x, y)$ over a rectangle $[a, b] \times [c, d]$:

1. Find the area of a cross-section, either $A(x) = \int_{c}^{d} h(x, y)\, dy$ or $B(y) = \int_{a}^{b} h(x, y)\, dx$.
2. Integrate the cross-sectional area to find the volume, either $V = \int_{a}^{b} A(x)\, dx$ or $V = \int_{c}^{d} B(y)\, dy$.

Example 1.11 Find the volume of the solid given by the region in space under the graph of $z = h(x, y) = x^3 - 3xy^2 + 2$ over the rectangle $[-1, 1] \times [-1, 1]$. This surface is known as a *monkey saddle*, and is graphed in Figure 1.13. A chair with this surface for the seat would be ideal for a monkey to sit on, as there are places for the monkey's two legs and its tail.

For a fixed x in the interval $[-1, 1]$, we slice the solid with a plane perpendicular to the x-axis at that point, and integrate the resulting function with respect to y (keeping in mind that x is now fixed in $z = h(x, y)$ and that h is a continuous function of y) to obtain the cross-sectional area $A(x)$:

$$A(x) = \int_{-1}^{1} h(x, y)\, dy = \int_{-1}^{1} \left(x^3 - 3xy^2 + 2 \right) dy$$

$$= \left[x^3 y - xy^3 + 2y \right]_{y=-1}^{y=1} = 2x^3 - 2x + 4.$$

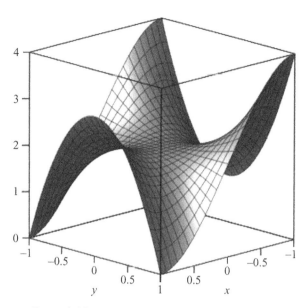

Figure 1.13. Graph of $z = h(x, y) = x^3 - 3xy^2 + 2$

We now integrate the continuous cross-sectional area function $A(x)$ to obtain the volume V

$$V = \int_{-1}^{1} A(x)\,dx = \int_{-1}^{1} \left(2x^3 - 2x + 4\right)\,dx = \left[\frac{x^4}{2} - x^2 + 4x\right]_{-1}^{1} = 8.$$

As noted earlier, we should get the same value for the volume if we first slice the solid with planes perpendicular to the y-axis. For a fixed y in $[-1, 1]$, we have the cross-sectional area (keeping y fixed in the integral):

$$B(y) = \int_{-1}^{1} h(x, y)\,dx = \int_{-1}^{1} \left(x^3 - 3xy^2 + 2\right)\,dx = \left[\frac{x^4}{4} - \frac{3}{2}x^2 y^2 + 2x\right]_{x=-1}^{x=1} = 4.$$

Thus

$$V = \int_{-1}^{1} B(y)\,dy = \int_{-1}^{1} 4\,dx = 8. \quad \blacksquare$$

☺ Computing the volume both ways yields a way to check your work!

In the final example of this section, we will see that with some solids finding the volume with cross-sections perpendicular to one axis is much easier than with cross-sections perpendicular to the other axis.

Example 1.12 Find the volume of the solid given by the region in space under the graph of $z = h(x, y) = ye^{-xy}$ over the rectangle $[1, 2] \times [0, 2]$. See Figure 1.14.

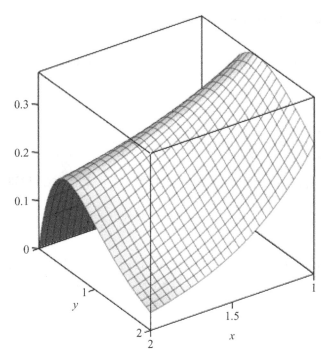

Figure 1.14. Graph of $z = h(x, y) = ye^{-xy}$

If we slice the solid with planes perpendicular to the x-axis, the cross-sectional area $A(x)$ is given by

$$A(x) = \int_0^2 ye^{-xy}\, dy.$$

Evaluating this integral requires integration by parts: setting $u = y$ and $dv = e^{-xy}\, dy$ yields $du = dy$ and $v = -\frac{1}{x}e^{-xy}$ (recall that x is constant in this integration), so that

$$A(x) = -\frac{y}{x}e^{-xy}\Big]_{y=0}^{y=2} + \frac{1}{x}\int_0^2 e^{-xy}\, dy = -\frac{2}{x}e^{-2x} - \left[\frac{1}{x^2}e^{-xy}\right]_{y=0}^{y=2}$$

$$= \frac{-2xe^{-2x} - (e^{-2x} - 1)}{x^2}.$$

To integrate $A(x)$, first note that $A(x) = \dfrac{d}{dx}\dfrac{e^{-2x} - 1}{x}$, so that

$$V = \int_1^2 A(x)\, dx = \int_1^2 \frac{d}{dx}\frac{e^{-2x} - 1}{x}\, dx = \frac{e^{-2x} - 1}{x}\Big|_1^2 = \frac{1}{2}\left(1 - e^{-2}\right)^2 \approx 0.3738.$$

That integration was difficult, using integration by parts and the not-very-obvious recognition of $A(x)$ as the derivative of a quotient. Compare it to the integration resulting from slicing the solid with planes perpendicular to the y-axis:

$$B(y) = \int_1^2 ye^{-xy}\, dx = -e^{xy}\Big|_{x=1}^{x=2} = e^{-y} - e^{-2y}.$$

Thus

$$V = \int_0^2 B(y)\, dy = \int_0^2 \left(e^{-y} - e^{-2y}\right) dy = \frac{1}{2}e^{-2y} - e^{-y}\Big|_0^2$$

$$= \frac{1}{2}\left(1 - e^{-2}\right)^2 \approx 0.3738. \quad \blacksquare$$

1.3.1 Exercises

1. Find the volume of the solid given by region in space under the graph of $z = h(x, y) = 0.5 - xye^{-x^2-y^2}$ over the rectangle $[-1, 1] \times [-1, 1]$. See Figure 1.15.

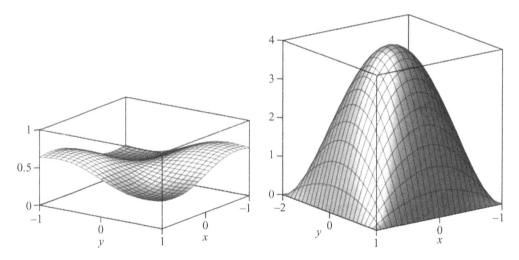

Figure 1.15. For Exercise 1 (left) and Exercise 2 (right)

2. Find the volume of the solid given by region in space under the graph of $z = h(x, y) = 4 - y^2 - 4x^2 + x^2y^2$ over the rectangle $[-1, 1] \times [-2, 2]$. See Figure 1.15.

3. Find the volume of the solid given by the region in space under the graph of $z = h(x, y) = 1 - \sin(x + y)$ over the rectangle $[-\pi, \pi] \times [-\pi, \pi]$. See Figure 1.16.

4. Find the volume of the solid given by region in space under the graph of $z = h(x, y) = 1 - 2xy(x^2 - y^2)$ over the rectangle $[-1, 1] \times [-1, 1]$. See Figure 1.16.

5. The Sagrada Familia School, built in 1908 and 1909, was designed by the Catalan architect Antoni Gaudí for the sons of the men working on the Sagrada Familia basilica in Barcelona.

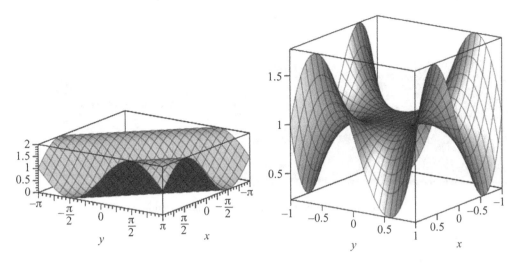

Figure 1.16. For Exercise 3 (left) and Exercise 4 (right)

It features an unusual undulating roof, which channels rain equally to opposite sides of the building, as seen in Figure 1.17.

The school measures 12m by 24m and is 6m tall at its highest point. We can model the school's roof by the function $z = h(x, y) = 5 - (x/6)\cos(\pi y/3)$ over the rectangle $[-6, 6] \times [-12, 12]$. See the right side of Figure 1.17. Find the volume of the school building.

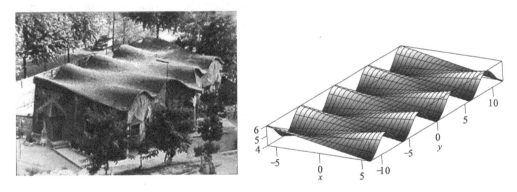

Figure 1.17. The Sagrada Familia school

6. The Warszawa-Ochota railway station was built in the early 1960s in Warsaw, Poland. Like the school building in the preceding exercise, it features an unusual roof, as seen in in Figure 1.18.

If the station measures 10m by 10m and the roof is 6m tall at its highest point, we can model the roof with the function $z = h(x, y) = 3 + (3xy/25)$ over the rectangle $[-5, 5] \times [-5, 5]$. See the right side of Figure 1.18. Find the volume of the airspace below the station roof.

 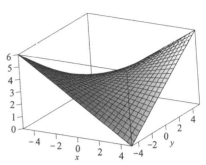

Figure 1.18. The Warszawa-Ochota railway station

1.4 Explorations

Each chapter in this book concludes with several explorations. An exploration is an extended or discovery exercise, presenting a topic or idea not often encountered in a course such as this. We encourage you to read through each one, whether or not your instructor assigns it. Answers to most of the exercises in each exploration are provided in its text.

1.4.1 The equivalence of the disk and shell methods

The goal of this exploration is to show that for a large class of solids of revolution, computing the volume by the disk method and by the shell method always yields the same value. But along the way we'll find a second way to integrate inverse functions. We begin with an easy exercise:

Exercise 1. Show that $\int_0^1 \arcsin x \, dx = \frac{\pi}{2} - \int_0^1 \frac{x}{\sqrt{1-x^2}} \, dx$.

This is an easy integration by parts problem. But we can't go any further, since the derivative of the arcsine is not continuous on $[0, 1]$ (it doesn't exist at 1). As a result, the integral on the right is not defined (it is an example of what is called an *improper integral*, which we discuss in Chapter 9). But here is a theorem we can use:

Theorem 1.4 *Suppose that $h(x)$ is continuous and monotone on the interval $[a, b]$. Then*

$$\int_{h(a)}^{h(b)} h^{-1}(y) \, dy = bh(b) - ah(a) - \int_a^b h(x) \, dx.$$

Exercise 2. Use Theorem 1.4 to evaluate $\int_0^1 \arcsin x \, dx$.

Did you get $\pi/2 - 1$, using $h(x) = \sin x$ on the interval $[0, \pi/2]$? Theorem 1.4 is a useful theorem! While we won't ask you to prove it, we will ask you to show that it is plausible, by looking at the graphs in Figure 1.19.

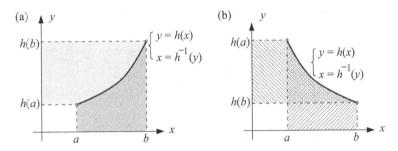

Figure 1.19. Graphs for Theorem 1.4

Exercise 3. Argue that Theorem 1.4 is reasonable, using Figure 1.19 and the area interpretation of integrals.

A rigorous proof of Theorem 1.4 uses Riemann sums for the integrals of h and h^{-1}. With Theorem 1.4, we can show that the disk and shell methods agree. We'll do the case for an increasing function; the case for a decreasing function is similar.

Theorem 1.5 *Let f be continuous, increasing, and positive on $[a, b]$, and let R denote the region bounded by the graph of $y = f(x)$, the x-axis, and the lines $x = a$ and $x = b$. Let S denote the solid obtained by revolving R about the x-axis. Then the disk and shell methods for computing the volume of S yield the same result.*

Exercise 4. Draw a sketch of R and S.

Exerices 5. Let V_{shell} denote the volume of S obtained by the shell method. Show that

$$V_{\text{shell}} = \pi \left[b[f(b)]^2 - a[f(a)]^2 - \int_{f(a)}^{f(b)} 2y f^{-1}(y)\, dy \right].$$

Let V_{disk} denote the volume of S obtained by the disk method. We know $V_{\text{disk}} = \pi \int_a^b [f(x)]^2\, dx$. The final step is

Exercise 6. Show that $V_{\text{disk}} = V_{\text{shell}}$. (Hint: Apply Theorem 1.4 to the function $y = h(x) = [f(x)]^2$ on the interval $[a, b]$, show that $h^{-1}(x) = f^{-1}\left(\sqrt{y}\right)$, and use the substitution $t = \sqrt{y}$.)

1.4.2 Is there an integration analog of the quotient rule?

In Section 1.2 you saw how the integration by parts formula is derived from the product rule for differentiation. Can a new integration by parts formula be derived from the quotient rule for differentiation? The goal of this Exploration is to answer that question.

Exercise 1. Differentiate $f(x)/g(x)$ and then integrate both sides to obtain

$$\frac{f(x)}{g(x)} = \int \frac{f'(x)}{g(x)}\, dx - \int \frac{f(x)g'(x)}{[g(x)]^2}\, dx.$$

Exercise 2. Make the change of variables $u = f(x)$ and $v = g(x)$ to obtain a quotient rule integration by parts formula:

$$\boxed{\int \frac{u}{v^2}\, dv = \int \frac{1}{v}\, du - \frac{u}{v}.}$$

Exercise 3. Is this useful? Consider the integral

$$\int \frac{x^2}{(1-x^2)^{3/2}}\, dx.$$

Let $v = (1 - x^2)^{1/2}$ and $u = -x$ to show that

$$\int \frac{x^2}{(1-x^2)^{3/2}}\, dx = \frac{x}{(1-x^2)^{1/2}} - \arcsin x + C.$$

Exercise 4. Repeat Exercise 3 using the usual integration by parts formula, and compare the amount of work to that in Exercise 3.

You probably noticed that the work in Exercise 3 was simpler: once you've chosen v, you need only compute dv, u is then the rest of the integrand, requiring only a differentiation. Exercise 4 required both a differentiation and an integration (finding v from dv).

Exercise 5. Check some other calculus texts. Can you find the quotient rule integration by parts formula in Exercise 2 in any of them?

Exercise 6. To understand why you were unsuccessful in Exercise 5, evaluate the integral $\int \frac{u}{v^2}\, dv$ with the old integration by parts formula $\int U\, dV = UV - \int V\, dU$ with $U = u$ and $dV = dv/v^2$. Is the quotient rule integration by parts formula really different from the product rule one?

1.4.3 Integrating the product of a polynomial and an exponential function

We begin with an exercise.

Exercise 1. Evaluate $\int e^{-x} \left(x^4 + 2x^3 + 3x^2 + 4x + 5\right)\, dx$.

How did you proceed? Did you multiply out the terms in the integrand to obtain five integrals and then work out each one individually, using integration by parts a total of ten times? Or did you do it more efficiently, treating the entire polynomial as the "u" part, requiring the use of integration by parts only four times? In either case, you will be happy to learn (in this Exploration) that only one integration by parts is needed! That single integration by parts is all we need in order to prove the nice formula in the following

Theorem 1.6 *If k is a nonzero constant and P is a polynomial, then*

$$\int e^{kx} P(x)\, dx = \frac{e^{kx}}{k} \left[P(x) - \frac{P'(x)}{k} + \frac{P''(x)}{k^2} - \frac{P'''(x)}{k^3} + \cdots \right] + C.$$

(The expression in brackets has only finitely many nonzero terms—do you see why?)

To prove the theorem, we first establish the following lemma (a *lemma*, from the Greek $\lambda\acute{\eta}\mu\mu\alpha$, "anything received, gain, or profit," is often a short or simple theorem used to prove a more important one).

Lemma 1.7 *Under the hypothesis of Theorem 1.6,*

$$\int e^{kx} P(x)\,dx = \frac{e^{kx}}{k} P(x) - \frac{1}{k}\int e^{kx} P'(x)\,dx.$$

Exercise 2. Prove the lemma. This is easy. The result in this lemma is an example of *recursion*, where an object (here the integral on the left) can be evaluated in terms of a similar but simpler object (the integral on the right is similar, but simpler since the degree of P' is one less than the degree of P).

Exercise 3. Prove Theorem 1.6. (Hint: use the lemma repeatedly.)

Exercise 4. Repeat Exercise 1.

Exercise 5. How would you evaluate $\int (\ln x)^5\,dx$? Integration by parts five times would be tedious. Can you make a substitution and use Theorem 1.6?

Exercise 6. How about $\int \sqrt{x}(\ln x)^5\,dx$?

1.4.4 Solids of revolution and the Cauchy-Schwarz inequality

Let f be a continuous nonnegative function on the interval $[a, b]$ where $0 < a < b$, and let R denote the region bounded by the graph of $y = f(x)$, the x-axis, $x = a$, and $x = b$. Now generate two solids, one by revolving R about the x-axis, the other by revolving R about the y-axis. See Figure 1.20.

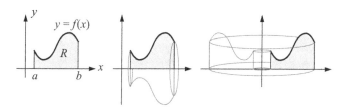

Figure 1.20. R revolved around x-axis and y-axis

Using the disk method, the volume of the solid obtained when R is revolved about the x-axis is $V_{x\text{-axis}} = \int_a^b \pi [f(x)]^2\,dx$, and using the shell method, the volume of the solid obtained when R is revolved about the y-axis is $V_{y\text{-axis}} = \int_a^b 2\pi x f(x)\,dx$. How are the volumes of the solids related? In this exploration you will show that

$$V_{y\text{-axis}}^2 \leq \frac{4}{3}\pi (b^3 - a^3) \cdot V_{x\text{-axis}}. \tag{1.12}$$

The fact that an inequality exists between the volumes is a bit surprising. The inequality itself is a bit strange, as the first term $\frac{4}{3}\pi(b^3 - a^3)$ in the product on the right doesn't depend on f, and it can be interpreted as the volume of a spherical shell, a sphere of radius b with an interior spherical hole of radius a.

Exercise 1. Under the assumptions in the first paragraph, show that for any t,

$$\int_a^b \pi [f(x) + tx]^2 \, dx \geq 0.$$

That was easy—a definite integral with a nonnegative integrand must be nonnegative.

Exercise 2. Show that the integral in Exercise 1 can be written as

$$\int_a^b \pi [f(x) + tx]^2 \, dx = At^2 + Bt + C,$$

where $A = \frac{1}{3}\pi (b^3 - a^3)$, $B = V_{y\text{-axis}}$, and $C = V_{x\text{-axis}}$.

Exercise 3. How are the coefficients A, B, and C related when $At^2 + Bt + C \geq 0$? (*Hint*: the quadratic equation $At^2 + Bt + C = 0$ has at most one real root.)

Exercise 4. Show that $B^2 - 4AC \leq 0$ is equivalent to (1.12).

Exercise 5. Show that (1.12) is a best possible inequality—that is, there are functions f that yield equality in (1.12). (Hint: consider $f(x) = mx$ for $m > 0$.)

The same procedure can be used to prove the *Cauchy-Schwarz inequality* for definite integrals: If f and g are continuous on the interval $[a, b]$, then

$$\left[\int_a^b f(x)g(x) \, dx \right]^2 \leq \int_a^b [f(x)]^2 \, dx \cdot \int_a^b [g(x)]^2 \, dx. \tag{1.13}$$

Exercise 6. Prove the Cauchy-Schwarz inequality. (Hint: use

$$\int_a^b [f(x) + tg(x)]^2 \, dx \geq 0$$

and proceed as in Exercises 2 through 4.)

The Cauchy-Schwarz inequality

The inequality in (1.13)—also known as the *Cauchy-Bunyakovsky-Schwarz* inequality—is one of the most important in mathematics, and finds applications in a variety of areas including linear algebra, probability, and analysis. In 1821 Augustin-Louis Cauchy (1789–1857) published a version for sums, which you will see in Exercise 3 in Section 3.4. The integral version of (1.13) was published in 1859 by Viktor Yakovlevich Bunyakovski (1804–1889), and rediscovered in 1885 by Hermann Amandus Schwarz (1843–1921), whose proof was essentially the same as Exercise 6.

Exercise 7. Use the Cauchy-Schwarz inequality to show that if the region R in Figure 1.20 has area A_R and the average value of f on $[a, b]$ is f_{ave}, then

$$V_{x\text{-axis}} \geq \pi A_R f_{\text{ave}}.$$

(Hint: let $g(x) = \sqrt{\pi}$, and recall that $f_{\text{ave}} = A_R/(b - a)$.) Is this inequality best-possible? (Hint: let $f(x) = k > 0$.)

1.4.5 Integrating products of sines, cosines, and exponentials

In several examples and exercises in this chapter you evaluated integrals where the integrand was the product of an exponential function and a sine or cosine, or a product of a sine and a cosine. Such integrals have the form $\int f(x)g(x)\,dx$ where $f''(x) = hf(x)$ and $g''(x) = kg(x)$ for constants h and k. In this Exploration you will develop a general formula for these integrals when $h \neq k$.

Exercise 1. Show that if we ignore the constant of integration, then $k \int g(x)\,dx = g'(x)$.

Exercise 2. Show that $k \int f(x)g(x)\,dx = f(x)g'(x) - \int f'(x)g'(x)\,dx$.

Exercise 3. Show that $k \int f(x)g(x)\,dx = f(x)g'(x) - f'(x)g(x) + h \int f(x)g(x)\,dx$.

Exercise 4. Show that

$$\int f(x)g(x)\,dx = \frac{1}{h - k}\left[f'(x)g(x) - f(x)g'(x)\right] + C.$$

Exercise 5. Use the result in Exercise 4 to show that

$$\int e^{ax} \sin bx\,dx = \frac{1}{a^2 + b^2}\left[ae^{ax}\sin bx - be^{ax}\cos bx\right] + C$$

and if $a^2 \neq b^2$,

$$\int \sin ax \cos bx\,dx = \frac{1}{b^2 - a^2}\left[a\cos ax \cos bx + b\sin ax \sin bx\right] + C.$$

1.5 Acknowledgments

1. Our approach to $\int \sec x\,dx$ is from N. Schaumberger, $\int \sec\theta\,d\theta$, *American Mathematical Monthly*, **68** (1961), p. 565.
2. Exploration 1.4.1 is adapted from E. Key, Disks, shells, and integrals of inverse functions, *College Mathematics Journal*, **25** (1994), pp. 136–138.
3. Exploration 1.4.2 is adapted from J. Switkes, A quotient rule integration by parts formula, *College Mathematics Journal*, **36** (2005), pp. 58–60, and M. Deveau and R. Hennigar, Quotient-rule-integration-by-parts, *College Mathematics Journal*, **43** (2012), pp. 254–256.
4. Exploration 1.4.3 is adapted from L. F. Meyers, Four crotchets on elementary integration, *College Mathematics Journal*, **22** (1991), pp. 410–413.
5. Exploration 1.4.5 is adapted from D. K. Pease, A useful integral formula, *American Mathematical Monthly*, **66** (1959), p. 908.

2

Arc Length, Trigonometric Substitution, and Surface Area

An idea which can only be used once is a trick. If you can use it more than once it becomes a method.

George Pólya and Gábor Szegö
Problems and Theorems in Analysis (1972)

In this chapter we continue with our method of using Riemann sums to construct definite integrals that solve certain geometric problems. You know how to use integrals to solve some two-dimensional problems (areas of certain regions in the plane) and some three-dimensional problems (volumes of certain solids). In them the graph of a function, such as $y = f(x)$, plays an important role. Now we use Riemann sums to set up integrals for the length of the graph, called the *arc length* of the curve. The resulting integral in an arc length problem often has an integrand that is in the form of the square root of a sum of two squares, and so we develop a technique for use in some of these integrals, *trigonometric substitution*. Finally, we extend the technique used to find the arc length of a curve to find the area of a surface of revolution.

2.1 Arc length

In your PCE you learned how to construct an approximation to the area of a region by slicing the region into narrow strips, using rectangles to approximate the area of each strip, and improving the approximation by using larger numbers of thinner rectangles. Later you learned you could approximate the volume of a solid in a similar fashion by slicing the solid into objects that resembled disks or washers. Let's review that procedure in a different setting: finding the length of a curve that is the graph of a function—usually referred to as finding the *arc length* of the curve.

In Figure 2.1(a) we see a graph of a continuous function $y = f(x)$ over the interval $[a, b]$. We begin by constructing a regular partition $P = \{x_0, x_1, \ldots, x_{n-1}, x_n\}$ of $[a, b]$, where $a = x_0$ and $b = x_n$, n is a positive integer, and the distance between adjacent points in P is $\Delta x = (b - a)/n$. The partition P induces a partition of the graph of f into n pieces. Letting $y_i = f(x_i)$, the points on the graph of f are $P_i = (x_i, y_i)$ for $i = 0, 1, 2, \ldots, n$.

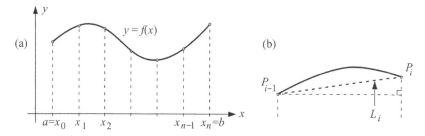

Figure 2.1. Determining arc length

In Figure 2.1(b) we see a close-up of the ith portion of the graph of f, and a line segment joining its endpoints $P_{i-1} = (x_{i-1}, y_{i-1})$ and $P_i = (x_i, y_i)$. To approximate the length of the ith portion of the graph, we use the Pythagorean theorem to find the length L_i of the line segment joining P_{i-1} to P_i:

$$L_i = \sqrt{(x_i - x_{i-1})^2 + (y_i - y_{i-1})^2} = \sqrt{(\Delta x)^2 + [f(x_i) - f(x_{i-1})]^2}.$$

The sum $\sum_{i=1}^{n} L_i$ of the lengths of the n line segments represents a piecewise linear approximation to the true length L of the curve. As n increases without bound, the approximation gets closer and closer to L and thus

$$L = \lim_{n \to \infty} \sum_{i=1}^{n} \sqrt{(\Delta x)^2 + [f(x_i) - f(x_{i-1})]^2}. \tag{2.1}$$

If the sum in (2.1) were a Riemann sum, its limit would be a definite integral. But a Riemann sum has the form $\sum_{i=1}^{n} g(x_i^*) \Delta x$ where Δx is the width of each subinterval in the partition and $g(x_i^*)$ is a function evaluated at a point x_i^* in the interval $[x_{i-1}, x_i]$. And (2.1) does not have the form—Δx is trapped under the square root sign, and the square root is evaluated at two points (x_{i-1} and x_i) in the interval. So we can use the sum in (2.1) with a large value of n to approximate L, but we cannot yet write a definite integral to compute L exactly.

To approximate L by a Riemann sum, we need a Δx in the correct place in the sum in (2.1). That is easily accomplished by factoring. Since $\sqrt{(\Delta x)^2} = \Delta x$ we have

$$\sqrt{(\Delta x)^2 + [f(x_i) - f(x_{i-1})]^2} = \sqrt{1 + \frac{[f(x_i) - f(x_{i-1})]^2}{(\Delta x)^2}} \, \Delta x$$

$$= \sqrt{1 + \left(\frac{y_i - y_{i-1}}{x_i - x_{i-1}}\right)^2} \, \Delta x. \tag{2.2}$$

Now that we have Δx in the correct place, we see that the quantity under the square root sign involves only the slope of the line segment joining P_{i-1} and P_i. But since f is differentiable, that slope is the same as the slope of the tangent line at an intermediate point, as shown in Figure 2.2.

Indeed, the mean value theorem, Theorem 0.5, tells us that if f is differentiable in addition to being continuous, then at some point x_i^* in the interval (x_{i-1}, x_i), the slope of the tangent line to f at x_i^* is the same as the slope of the approximating line segment, i.e.,

$$\frac{y_i - y_{i-1}}{x_i - x_{i-1}} = \frac{f(x_i) - f(x_{i-1})}{x_i - x_{i-1}} = f'(x_i^*).$$

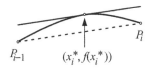

Figure 2.2. Determining arc length using the mean value theorem

Hence $L_i = \sqrt{1 + [f'(x_i^*)]^2}\,\Delta x$, and if f is differentiable on (a, b), then (2.1) can be written as

$$L = \lim_{n \to \infty} \sum_{i=1}^{n} \sqrt{1 + [f'(x_i^*)]^2}\,\Delta x. \tag{2.3}$$

In (2.3) we *do* have a Riemann sum, and if the derivative f' is continuous, then the Riemann sum's limit as $n \to \infty$ is a definite integral, and we have proved

Theorem 2.1 *If a function f has a continuous derivative on the interval $[a, b]$, then the arc length L of the graph of $y = f(x)$ from $(a, f(a))$ to $(b, f(b))$ is given by*

$$L = \int_a^b \sqrt{1 + [f'(x)]^2}\,dx. \tag{2.4}$$

Switching the roles of x and y in Theorem 2.1 yields

Corollary 2.2 *If a function g has a continuous derivative on the interval $[c, d]$, then the arc length L of the graph $x = g(y)$ from $(g(c), c)$ to $(g(d), d)$ is given by*

$$L = \int_c^d \sqrt{1 + [g'(y)]^2}\,dy. \tag{2.5}$$

Example 2.1 (The semicubical parabola) The semicubical parabola resembles a parabola, but as its name suggests its exponent is half that of the cubic. It has the distinction in the history of mathematics of having been the first algebraic curve (a plane curve is *algebraic* if it can be expressed in the form $P(x, y) = 0$ where P is a polynomial in x and y) other than the straight line and the circle to be *rectified*, i.e., to have its length measured. This was first accomplished by the English mathematician William Neile (1637–1670) when he was a 20-year-old student at Oxford. So it is appropriate that we too begin with this curve.

In Cartesian coordinates the equation of the semicubical parabola is $y^2 = a^2x^3$, where a is a positive constant. The graph of a semicubical parabola is symmetric with respect to the x-axis, and the function $y = ax^{3/2}$ represents the branch in the first quadrant. We now set $a = 1$ and find the length L_{semi} of $y = x^{3/2}$ from $(0, 0)$ to $(1, 1)$. See Figure 2.3 for graphs of (a) the semicubical parabola $y = x^{3/2}$ and for comparison (b) the parabola $y = x^2$.

Before we begin, let's approximate L_{semi}. Since the semicubical parabola is concave up between $(0, 0)$ and $(1, 1)$, it appears that L_{semi} should lie between the straight-line distance from

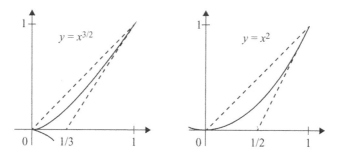

Figure 2.3. Graphs of the semicubical parabola $y = x^{3/2}$ and the parabola $y = x^2$

(0, 0) to (1, 1) and the sum of the lengths of segments of two tangent lines, one at (0, 0), the other at (1, 1). Thus we expect L_{semi} to satisfy $\sqrt{2} < L_{semi} < (1 + \sqrt{13})/3$, or approximately $1.4142 < L_{semi} < 1.5352$—and it appears that L_{semi} may actually be closer to 1.4142 than to 1.5252. ■

Since the function for the semicubical parabola is continuously differentiable (i.e., its derivative is continuous), we can use (2.4) to compute the arc length. We have $y' = (3/2)\sqrt{x}$ and hence

$$L_{semi} = \int_0^1 \sqrt{1 + \frac{4}{9}x}\, dx.$$

This integral is easily evaluated using the substitution $u = 1 + (9/4)x$, which yields

$$L_{semi} = \frac{4}{9} \int_1^{13/4} u^{1/2}\, du = \frac{4}{9} \cdot \frac{2}{3} u^{3/2} \Big|_{u=1}^{13/4} = \frac{13\sqrt{13} - 8}{27} \approx 1.4397. \ \blacksquare$$

Example 2.2 (The parabola) Having succeeded in finding the length of the semicubical parabola, let's apply the same procedure to the ordinary parabola $y = x^2$ from (0, 0) to (1, 1), as illustrated in Figure 2.3. But first let's approximate its length L_{para}. As with the semicubical parabola, L_{para} should lie between the straight-line distance from (0, 0) to (1, 1), and the sum of the lengths of segments of two tangent lines, one at (0, 0), the other at (1, 1). Thus, $\sqrt{2} < L_{para} < \varphi$, where φ denotes the *golden ratio* $(1 + \sqrt{5})/2 \approx 1.618$. The appearance of $\sqrt{2}$ and φ as bounds on L_{para} is interesting in that these were the first two numbers shown to be irrational by the ancient Greek geometers.

None of the obvious substitutions seem to help much here. But there is another substitution we might try. The fact that the integrand is the square root of a sum of two squares comes from the use of the Pythagorean theorem in our original approximation in (2.1) of the arc length using line segments. In the integral, the integrand can be interpreted as the length of the hypotenuse of a right triangle where one leg has fixed length 1 and the other variable length $2x$. See Figure 2.4.

Other variables of interest in this right triangle are the acute angles, such as the one marked as θ. Using trigonometry it is possible to express ratios of the lengths of sides in terms of trigonometric functions of θ. In any substitution for x we will need to find dx, so let's use the trigonometric function of θ in Figure 2.4 with the simplest ratio: $\tan\theta = 2x/1 = 2x$. Then $dx = (1/2)\sec^2\theta\, d\theta$ and $\sqrt{1 + 4x^2} = \sec\theta$. As x goes from 0 to 1, θ goes from 0 to arctan 2

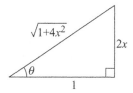

Figure 2.4. The right triangle for the substitution $\tan\theta = 2x$

and hence

$$L_{\text{para}} = \int_0^{\arctan 2} \sec\theta \cdot \frac{1}{2}\sec^2\theta\,d\theta = \frac{1}{2}\int_0^{\arctan 2} \sec^3\theta\,d\theta.$$

Before proceeding, note that the substitution is somewhat different from the ordinary u-substitution from your PCE. In the u-substitution you usually set u equal to a function of x that appears in the integral, and then find du. In the substitution above, we set a function of θ equal to a function of x, solved for x and found dx. This type of substitution is called *trigonometric substitution*, and will be explored further in the next section.

The indefinite integral $\int \sec^3\theta\,d\theta$ was evaluated by parts in Example 1.9 where we showed that $\int \sec^3\theta\,d\theta = \frac{1}{2}(\sec\theta\tan\theta + \ln|\sec\theta + \tan\theta|) + C$, and hence

$$L_{\text{para}} = \frac{1}{4}\left[\sec\theta\tan\theta + \ln|\sec\theta + \tan\theta|\right]_{\theta=0}^{\arctan 2} = \frac{1}{4}\left[2\sqrt{5} + \ln(2 + \sqrt{5})\right] \approx 1.4789. \quad \blacksquare$$

The golden ratio is omnipresent in mathematics
The appearance of the golden ratio φ in a calculus problem is not surprising. It appears in nearly every branch of mathematics. Examples from elementary mathematics include geometry (φ is the length of a diagonal in a regular pentagon with side 1), algebra ($x^5 \pm 1 = (x \pm 1)(x^2 \mp \varphi x + 1)(x^2 \pm x/\varphi + 1)$), and trigonometry ($\varphi = 2\cos(\pi/5)$).

2.1.1 Exercises

In Exercises 1 through 8, find the length of the curve over the indicated interval.

1. $y = \sqrt{x}\,(x - 1/3)$, $\quad 1 \le x \le 4$
2. $y = \frac{3}{8}x^{2/3}\left(2x^{2/3} - 1\right)$, $\quad 1 \le x \le 8$
3. $y = \ln(\sec x)$, $\quad 0 \le x \le \pi/4$
4. $y = \frac{1}{8}x^2 - \ln x$, $\quad 1 \le x \le e$
5. $3y = (x + 2)^{3/2}$, $\quad 0 \le x \le 2$
6. $3y = (x^2 + 2)^{3/2}$, $\quad 0 \le x \le 2$
7. $y = 3x^{2/3}$, $\quad 1 \le x \le 8$
8. $y = \frac{1}{3}x\sqrt{2x + 3}$, $\quad -1 \le x \le 1/2$.

9. In Figure 2.5 we see a *parabolic segment* (a region bounded by a parabola and a chord perpendicular to the axis of symmetry of the parabola) with base b and height h. Find its area and perimeter in the case where $b = 2h$ as illustrated in Figure 2.5.

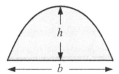

Figure 2.5. A parabolic segment.

10. *The catenary.* A catenary (from the Latin *catena*, meaning "chain") is the mathematical curve in the plane that models the shape of a uniform flexible cable, chain, or cord when suspended from two points, as illustrated in Figure 2.6a. Its equation is $y = (c/2)(e^{x/c} + e^{-x/c})$, where c is the y-intercept, as illustrated in Figure 2.6b. Find the length of the catenary from $x = a$ to $x = b$ for $a < b$.

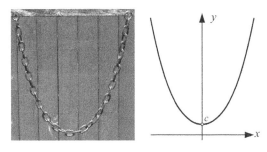

Figure 2.6. The catenary

11. In the preceding Exercise, the integration was easy because the function describing the catenary has the pleasant property that one plus the square of its derivative is again a perfect square. For each of the following three curves, find the arc length between the points $x = 1$ and $x = 2$:

(a) $f(x) = \dfrac{x^2}{4} - \dfrac{\ln x}{2}$ (b) $f(x) = \dfrac{x^3}{6} + \dfrac{x^{-1}}{2}$ (c) $f(x) = \dfrac{x^4}{8} + \dfrac{x^{-2}}{4}$

(d) Is there a pattern to these functions? What is it? (Hint: look for a pattern in the derivatives.)

12. Use the arc length formula in Theorem 2.1 to verify that the length of the line segment joining (x_1, y_1) to (x_2, y_2) (where $x_1 \neq x_2$) is the same as that given by the familiar distance formula.

In Exercises 13 through 16, find the arc length of the given function over the indicated interval. (Hint: graph the functions first!)

13. $f(x) = \cos^{-1}(\sin x), \quad -\pi/2 \leq x \leq \pi/2$ 14. $\phi(x) = x + |x|, \quad -1/2 \leq x \leq 1/2$

15. $g(x) = \sqrt{(3-x)(1+x)}, \quad 1 \leq x \leq 3$ 16. $h(x) = \sqrt{x(1-x)}, \quad 0 \leq x \leq 1$

17. (a) Set up (and simplify) an integral that represents the arc length of $f(x) = \arccos(e^{-x})$ from $(0, 0)$ to $(\ln 2, \pi/3)$. Do not evaluate the integral (yet).

(b) Set up (and simplify) an integral that represents the arc length of the inverse of $f(x)$ from $(0, 0)$ to $(\pi/3, \ln 2)$. Do not evaluate the integral (yet).

(c) Since the two integrals represent the same length, evaluate the simpler one.

18. *Why is there an "arc" in arcsine?* Show that the length of the arc of the unit circle $x^2 + y^2 = 1$ from $(1, 0)$ to $(\sqrt{1 - t^2}, t)$ is $\arcsin t$ for $t \in (0, 1)$ to conclude the arcsine is an arc length. (Hint: integrate with respect to y.)

19. The *astroid* (from the Greek $\alpha\sigma\tau\eta\rho$ "star", and the suffix *-oid*, "resembling," or "having the shape of") is a four-cusped curve with the Cartesian equation

$$x^{2/3} + y^{2/3} = a^{2/3}$$

where a is a positive constant. See Figure 2.7, where the x- and y- intercepts are $-a$ and a. Using the symmetry of the astroid, find its arc length.

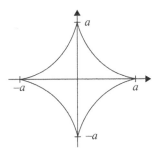

Figure 2.7. The astroid $x^{2/3} + y^{2/3} = a^{2/3}$

20. Show that the length of the curve $x^{21/20}$ between $(0, 0)$ and $(1, 1)$ is rational. (The numerator and denominator of the arc length each have 23 digits, so you may wish to use a computer algebra system to evaluate the integral!)

2.2 Trigonometric substitutions

Trigonometric substitution is a powerful integration technique. Often it can be employed in calculus problems that seem to have little to do with the trigonometric functions. It frequently can be used in integrals where a term in the integrand has one of the three forms $\sqrt{a^2 + u^2}$, $\sqrt{a^2 - u^2}$, or $\sqrt{u^2 - a^2}$, where u is a differentiable function and a is a positive constant. The integrand may contain the square root to a positive or negative integer power, and other expressions in u as well.

In Figure 2.8 we illustrate the three types of substitutions with right triangles, as we did in the last section. While the constant a can be any positive number, the value of u is restricted so that u/a is in the domain of θ, i.e., in (a) u has any real value, in (b) $|u| \le a$, and in (c) $|u| \ge a$.

Then substitute for u, du, and the square root as indicated. In each case the substitution will produce an integral with trigonometric functions that may require trigonometric identities to integrate. We illustrate the procedure with several examples.

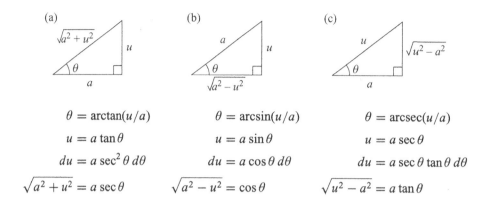

Figure 2.8. Trigonometric substitutions

Example 2.3 Evaluate $\displaystyle\int \frac{x^2}{\sqrt{4-x^2}}\,dx$.

Since the denominator has the form $\sqrt{a^2-u^2}$ for $a=2$ and $u=x$, we use the substitution illustrated in Figure 2.8(b) and set $\theta = \arcsin(x/2)$. Thus $x=2\sin\theta$, $dx = 2\cos\theta\,d\theta$ and $\sqrt{4-x^2} = 2\cos\theta$, hence

$$\int \frac{x^2}{\sqrt{4-x^2}}\,dx = \int \frac{4\sin^2\theta}{2\cos\theta}\,2\cos\theta\,d\theta = 4\int \sin^2\theta\,d\theta.$$

In your PCE you may have evaluated the integral $\int \sin^2\theta\,d\theta$ using the double-angle formula $\sin^2\theta = (1-\cos 2\theta)/2$. In Exercise 28 in Section 1.2, you evaluated the integral by parts, yielding $\int \sin^2\theta\,d\theta = (1/2)(\theta - \sin\theta\cos\theta) + C$.

In any indefinite integral substitution problem, the last step is to return to the original variable by undoing the substitution: here we have $\theta = \arcsin(x/2)$, $\sin\theta = x/2$, and $\cos\theta = \sqrt{4-x^2}/2$, hence

$$\int \frac{x^2}{\sqrt{4-x^2}}\,dx = 4\left[\frac{1}{2}\left(\arcsin\frac{x}{2} - \frac{x}{2}\cdot\frac{\sqrt{4-x^2}}{2}\right) + C\right] = 2\arcsin\frac{x}{2} - \frac{x\sqrt{4-x^2}}{2} + C. \ \blacksquare$$

☺ Don't forget to check the result by differentiation!

Example 2.4 Evaluate $\int_0^2 \dfrac{dx}{(9+4x^2)^{3/2}}$.

Since the denominator is a power of $\sqrt{a^2 + u^2}$ for $a = 3$ and $u = 2x$, we use the substitution illustrated in Figure 2.8(a) and set $\theta = \arctan(2x/3)$. Thus $x = (3/2)\tan\theta$, $dx = (3/2)\sec^2\theta\,d\theta$, and $\sqrt{9 + 4x^2} = 3\sec\theta$, hence

$$\int \frac{dx}{(9+4x^2)^{3/2}} = \int \frac{(3/2)\sec^2\theta}{(3\sec\theta)^3}\,d\theta = \frac{1}{18}\int \cos\theta\,d\theta$$

$$= \frac{1}{18}\sin\theta + C = \frac{1}{18}\frac{2x}{\sqrt{9+4x^2}} + C = \frac{x}{9\sqrt{9+4x^2}} + C.$$

Thus

$$\int_0^2 \frac{dx}{(9+4x^2)^{3/2}} = \frac{x}{9\sqrt{9+4x^2}}\bigg|_{x=0}^2 = \frac{2}{45}. \quad \blacksquare$$

Example 2.5 Evaluate $\int \dfrac{\sqrt{3x^2-5}}{x^2}\,dx$.

Here the numerator has the form $\sqrt{u^2 - a^2}$ for $a = \sqrt{5}$ and $u = x\sqrt{3}$, we use the substitution illustrated in Figure 2.8(c) and set $\theta = \operatorname{arcsec}(x\sqrt{3/5})$. Thus $x = \sqrt{5/3}\sec\theta$, $dx = \sqrt{5/3}\sec\theta\tan\theta\,d\theta$, and $\sqrt{3x^2 - 5} = \sqrt{5}\tan\theta$, hence

$$\int \frac{\sqrt{3x^2-5}}{x^2}\,dx = \int \frac{\sqrt{5}\tan\theta}{(5/3)\sec^2\theta}\sqrt{\frac{5}{3}}\sec\theta\tan\theta\,d\theta = \sqrt{3}\int \frac{\tan^2\theta}{\sec\theta}\,d\theta$$

$$= \sqrt{3}\int (\sec\theta - \cos\theta)\,d\theta = \sqrt{3}\left[\ln|\sec\theta + \tan\theta| - \sin\theta\right] + C$$

$$= \sqrt{3}\left(\ln\left|\frac{x\sqrt{3} + \sqrt{3x^2-5}}{\sqrt{5}}\right| - \frac{\sqrt{3x^2-5}}{x\sqrt{3}}\right) + C$$

$$= \sqrt{3}\ln\left|x\sqrt{3} + \sqrt{3x^2-5}\right| - \frac{\sqrt{3x^2-5}}{x} + C.$$

(Do you see where the $\sqrt{5}$ in the denominator went from inside the ln?) $\blacksquare$

Example 2.6 Evaluate $\int \dfrac{x^2}{(x^2+1)^2}\,dx$.

In this example the square root is disguised: $(x^2 + 1)^2 = \left(\sqrt{x^2+1}\right)^4$. So we use a tangent substitution with $\theta = \tan^{-1}x$ so that $x = \tan\theta$, $dx = \sec^2\theta\,d\theta$, and $\sqrt{x^2 + 1} = \sec\theta$. Thus

$$\int \frac{x^2}{(x^2+1)^2}\,dx = \int \frac{\tan^2\theta}{\sec^4\theta}\sec^2\theta\,d\theta = \int \sin^2\theta\,d\theta.$$

In Example 2.3, we learned that $\int \sin^2\theta\,d\theta = \frac{1}{2}(\theta - \sin\theta\cos\theta) + C$, and since $\theta = \tan^{-1}x$, $\sin\theta = x/\sqrt{x^2 + 1}$, and $\cos\theta = 1/\sqrt{x^2 + 1}$, we have

$$\int \frac{x^2}{(x^2+1)^2}\,dx = \frac{1}{2}\left(\tan^{-1}x - \frac{x}{x^2+1}\right) + C. \quad \blacksquare$$

Example 2.7 Evaluate $\int \dfrac{1}{\sqrt{x^2 + 4x}}\, dx$.

It is tempting, but unfortunately incorrect, to let $u = x$ and $a = \sqrt{4x}$—a must be a constant, not a function of x. But there is an algebraic technique called *completing the square*, that enables us to express any quadratic expression in x in one of the three forms $a^2 + u^2$, $a^2 - u^2$, or $u^2 - a^2$, where u is a (linear) function of x and a a positive constant. We will return to this example after presenting the technique.

You may have encountered the completing the square technique when you studied algebra in high school, as it is the foundation for the quadratic formula. But we don't need the quadratic formula here, only the following identity which you can easily verify with algebra: for any real numbers x and a,

$$x^2 + 2ax = (x^2 + 2ax + a^2) - a^2 = (x + a)^2 - a^2.$$

The term we add we add and subtract (a^2) is the square of one-half the coefficient of x (one-half of $2a$ is a). Here is a geometric illustration of the completing the square identity (for positive x and a) that may help you remember it: simply compute the area of the shaded region in Figure 2.9 in two different ways.

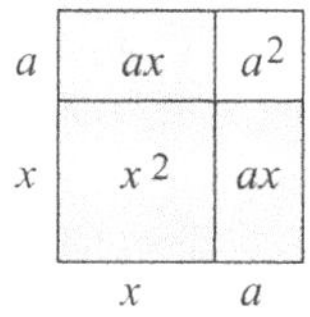

Figure 2.9. Completing the square

Back to Example 2.7 We can now evaluate $\int (1/\sqrt{x^2 + 4x})\, dx$ after completing the square in the denominator: $x^2 + 4x = (x^2 + 4x + 4) - 4 = (x + 2)^2 - 2^2$. Thus the denominator has the form $\sqrt{u^2 - a^2}$ for $u = x + 2$ and $a = 2$, so we make this substitution followed by the secant substitution illustrated in Figure 2.8(c). Hence

$$\int \frac{1}{\sqrt{x^2 + 4x}}\, dx = \int \frac{du}{\sqrt{u^2 - a^2}} = \int \frac{a \sec\theta \tan\theta\, d\theta}{a \tan\theta} = \int \sec\theta\, d\theta = \ln|\sec\theta + \tan\theta| + C$$

(recall that we evaluated $\int \sec\theta\, d\theta$ in Example 1.9). We now undo both substitutions:

$$\int \frac{1}{\sqrt{x^2 + 4x}}\, dx = \ln\left| \frac{u}{a} + \frac{\sqrt{u^2 - a^2}}{a} \right| + C = \ln\left| \frac{x + 2}{2} + \frac{\sqrt{x^2 + 4x}}{2} \right| + C.$$

Although it is not necessary to do so, the result can be simplified further since

$$\ln\left| \frac{x + 2}{2} + \frac{\sqrt{x^2 + 4x}}{2} \right| = \ln\left| \frac{x + 2 + \sqrt{x^2 + 4x}}{2} \right| = \ln\left| x + 2 + \sqrt{x^2 + 4x} \right| - \ln 2,$$

and then the constant $\ln 2$ can be absorbed into the constant C so that

$$\int \frac{1}{\sqrt{x^2 + 4x}}\, dx = \ln\left| x + 2 + \sqrt{x^2 + 4x} \right| + C. \quad \blacksquare$$

☺ Since this integration technique may be new to you, it is well worthwhile checking the work by differentiation!

Example 2.8 Evaluate $\displaystyle\int \frac{1}{\sqrt{x(1-x)}}\,dx$.

Completing the square on the quadratic expression yields

$$x(1-x) = -(x^2 - x) = -(x^2 - x + 1/4) + 1/4 = (1/2)^2 - (x - 1/2)^2.$$

Thus the denominator has the form $\sqrt{a^2 - u^2}$ for $a = 1/2$ and $u = x - 1/2$ and so we make this substitution followed by the sine substitution illustrated in Figure 2.8(b). Hence

$$\int \frac{1}{\sqrt{x(1-x)}}\,dx = \int \frac{du}{\sqrt{a^2 - u^2}} = \int \frac{a\cos\theta\,d\theta}{a\cos\theta} = \int d\theta = \theta + C.$$

Undoing both substitutions yields

$$\int \frac{1}{\sqrt{x(1-x)}}\,dx = \theta + C = \arcsin\left(\frac{u}{a}\right) + C = \arcsin\left(\frac{x - 1/2}{1/2}\right) + C = \arcsin(2x - 1) + C. \blacksquare$$

Example 2.9 Evaluate $\displaystyle\int_0^3 \frac{1}{(4x^2 - 12x + 13)^{3/2}}\,dx$.

Here we have

$$4x^2 - 12x + 13 = 4(x^2 - 3x) + 13 = 4(x^2 - 3x + 9/4) - 9 + 13 = (2x - 3)^2 + 4$$

and so the denominator has the form $(a^2 + u^2)^{3/2}$ for $a = 2$ and $u = 2x - 3$. We make this substitution followed by the tangent substitution $u = a\tan\theta$ illustrated in Figure 2.8(a). Evaluating the indefinite integral first yields

$$\int \frac{1}{(4x^2 - 12x + 13)^{3/2}}\,dx = \int \frac{(1/2)du}{(a^2 + u^2)^{3/2}} = \frac{1}{2}\int \frac{a\sec^2\theta\,d\theta}{(a\sec\theta)^3}$$

$$= \frac{1}{2a^2}\int \cos\theta\,d\theta = \frac{1}{2a^2}\sin\theta + C.$$

Undoing both substitutions yields

$$\int \frac{1}{(4x^2 - 12x + 13)^{3/2}}\,dx = \frac{1}{8}\sin\theta + C = \frac{u}{8\sqrt{a^2 + u^2}} + C = \frac{2x - 3}{8\sqrt{4x^2 - 12x + 13}} + C,$$

and thus

$$\int_0^3 \frac{1}{(4x^2 - 12x + 13)^{3/2}}\,dx = \left[\frac{2x - 3}{8\sqrt{4x^2 - 12x + 13}}\right]_{x=0}^3 = \frac{3}{4\sqrt{13}} \approx 0.208. \quad\blacksquare$$

It will take some practice for you to become proficient with trigonometric substitutions and in evaluating the trigonometric integrals that result from them.

2.2.1 Exercises

In Exercises 1 through 12, evaluate the indefinite integral. Be sure to check your work by differentiation.

1. $\displaystyle\int \frac{1}{\sqrt{1+x^2}}\,dx$ 2. $\displaystyle\int \frac{x}{(4-x^2)^{3/2}}\,dx$ 3. $\displaystyle\int \frac{dt}{\sqrt{t^2-4}}$

4. $\displaystyle\int \frac{\sqrt{1-x^2}}{x}\,dx$ 5. $\displaystyle\int \frac{\sqrt{y^2-9}}{y^3}\,dy$ 6. $\displaystyle\int \sqrt{u^2+9}\,du$

7. $\displaystyle\int \frac{du}{(1-u^2)^2}$ 8. $\displaystyle\int \frac{\sqrt{y^2+4}}{y^2}\,dy$ 9. $\displaystyle\int \frac{dx}{x\sqrt{x^4-1}}$

10. $\displaystyle\int \frac{dx}{x^2+2x+5}\,dx$ 11. $\displaystyle\int \sqrt{4\theta-\theta^2}\,d\theta$ 12. $\displaystyle\int \frac{dy}{\sqrt{y^2-2y}}$

In Exercises 13 through 18, evaluate each definite integral.

13. $\displaystyle\int_1^2 \frac{d\theta}{(5-\theta^2)^{3/2}}$ 14. $\displaystyle\int_1^4 \frac{\sqrt{x^2+1}}{x}\,dx$ 15. $\displaystyle\int_2^3 \frac{dt}{t^2\sqrt{t^2-1}}$

16. $\displaystyle\int_0^2 \frac{dz}{z^2-2z+2}$ 17. $\displaystyle\int_1^3 \frac{dx}{\sqrt{4x-x^2}}$ 18. $\displaystyle\int_{-2}^2 x^3(x^2+1)^{5/2}\,dx$

19. Use appropriate trigonometric substitutions to evaluate the integrals

     (a) $\displaystyle\int_0^1 \frac{dx}{4-x^2}$ (b) $\displaystyle\int_3^4 \frac{dx}{4-x^2}$

     (Hint: you will need different substitutions in the two integrals. Why?)

20. Find the arc length of the portion of the graph of $y = \ln x$ between $(1, 0)$ and $(e, 1)$.

2.3 Applications of trigonometric substitutions

In this section we consider some applications of integration for which trigonometric substitutions are useful.

Example 2.10 Find the area of the *hyperbolic sector* bounded by the right-hand branch of the hyperbola $x^2 - y^2 = 1$, the positive x-axis, and the line segment joining the origin to the point (c, s) on the hyperbola. See Figure 2.10 for the case where $s > 0$, where the sector is shaded dark gray.

The area A of the sector equals the area to the left of the hyperbola for y between 0 and s minus the area of the light gray right triangle with vertices $(0, 0)$, $(0, s)$, and (c, s):

$$A = \int_0^s \sqrt{y^2+1}\,dy - \frac{1}{2}cs.$$

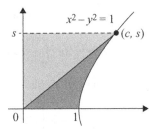

Figure 2.10. Finding the area of a hyperbolic sector

The form of the integrand calls for the tangent substitution $y = \tan\theta$, $dy = \sec^2\theta\,d\theta$, and $\sqrt{y^2 + 1} = \sec\theta$, hence

$$A = \int_0^{\tan^{-1} s} \sec^3\theta\,d\theta - \frac{1}{2}cs.$$

Using (1.9) from Example 1.9 yields

$$A = \frac{1}{2}\left[\sec\theta\tan\theta + \ln|\sec\theta + \tan\theta|\right]_{\theta=0}^{\tan^{-1} s} - \frac{1}{2}cs.$$

But $\tan\left(\tan^{-1} s\right) = s$ and $\sec(\tan^{-1} s) = \sqrt{s^2 + 1} = c$, and thus

$$A = \frac{1}{2}\left[cs + \ln|c + s|\right] - \frac{1}{2}cs = \frac{1}{2}\ln(c + s).$$

If $s < 0$ then the point (c, s) lies on the portion of the right-hand branch of the hyperbola in the fourth quadrant, and the formula $A = \frac{1}{2}\ln(c + s)$ yields the negative of the area of the sector, that is, $A = \frac{1}{2}\ln(c + s)$ is the *signed area* of the sector, and the actual area is $-A$. The result of this example will be useful when we study the hyperbolic functions in Chapter 6. ∎

Example 2.11 Find the volume of the solid obtained by rotating the region bounded by the graphs of the curve $y^2 = 1 - (1/x^2)$ and the line $x = \sqrt{2}$ about the y-axis. See Figure 2.11.

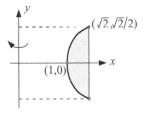

Figure 2.11. Generating a solid of revolution

The portion of the curve in the first quadrant is given by $y = \sqrt{x^2 - 1}/x$, so that using the shell method to compute the volume V we have

$$V = \int_1^{\sqrt{2}} 2\pi x \frac{\sqrt{x^2 - 1}}{x}\,dx = 2\pi \int_1^{\sqrt{2}} \sqrt{x^2 - 1}\,dx.$$

The trigonometric substitution $x = \sec\theta$ with $dx = \sec\theta\tan\theta\,d\theta$ and $\sqrt{x^2 - 1} = \tan\theta$ yields

$$V = 2\pi \int_0^{\pi/4} \tan^2\theta\sec\theta d\theta = 2\pi \int_0^{\pi/4} (\sec^3\theta - \sec\theta)d\theta.$$

Using (1.9) and (1.10) to evaluate this integral yields

$$V = \pi\left[\sec\theta\tan\theta - \ln|\sec\theta + \tan\theta|\right]_{\theta=0}^{\pi/4} = \pi\left[\sqrt{2} - \ln(\sqrt{2} + 1)\right] \approx 1.6740. \quad \blacksquare$$

Example 2.12 Find the arc length of the curve $y = e^{-x}$ from $(0, 1)$ to $(1, 1/e)$. We'll let you graph this one on your calculator. When you do, you'll see that the arc length L is a little more than $\sqrt{1 + (1 - 1/e)^2} \approx 1.183$, the straight line distance from $(0, 1)$ to $(1, 1/e)$. Using (2.4) L is given by

$$L = \int_0^1 \sqrt{1 + e^{-2x}}\,dx = \int_0^1 \frac{\sqrt{1 + e^{2x}}}{e^x}\,dx.$$

The substitution $u = e^x$ yields

$$L = \int_1^e \frac{\sqrt{1 + u^2}}{u^2}\,du.$$

We now use a trigonometric substitution $u = \tan\theta$ with $du = \sec^2\theta\,d\theta$ and $\sqrt{1 + u^2} = \sec\theta$, hence

$$L = \int_{\pi/4}^{\arctan e} \frac{\sec\theta}{\tan^2\theta}\sec^2\theta\,d\theta.$$

To simplify an integrand such as this one, it often helps to express it in terms of sines and cosines:

$$\frac{\sec^3\theta}{\tan^2\theta} = \frac{1}{\cos\theta\sin^2\theta},$$

and if there are sines and cosines in the denominator, multiply the numerator by $\sin^2\theta + \cos^2\theta = 1$ in order to write the integrand without trigonometric functions in the denominator:

$$\frac{\sec^3\theta}{\tan^2\theta} = \frac{1}{\cos\theta\sin^2\theta} = \frac{\sin^2\theta + \cos^2\theta}{\cos\theta\sin^2\theta} = \sec\theta + \csc\theta\cot\theta.$$

Thus

$$L = \int_{\pi/4}^{\arctan e} [\sec\theta + \csc\theta\cot\theta]\,d\theta$$

$$= \left[\ln(|\sec\theta + \tan\theta|) - \csc\theta\right]_{\theta=\pi/4}^{\arctan e} = \ln\frac{e + \sqrt{1 + e^2}}{1 + \sqrt{2}} - \frac{\sqrt{1 + e^2}}{e} + \sqrt{2} \approx 1.1927. \quad \blacksquare$$

☺ All six trigonometric functions appeared in the solution. Success with trigonometric substitution requires facility with the trigonometric functions!

2.3.1 Exercises

1. The equation $a^2 y^2 + b^2 x^2 = a^2 b^2$ for positive a and b describes an ellipse with x-intercepts a and $-a$ and y-intercepts b and $-b$, as shown in Figure 2.12(a).

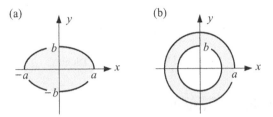

Figure 2.12. The ellipse and annulus in Exercise 1

(a) Find the area of the ellipse.
(b) In Figure 2.12(b) we see an *annulus* (the region between two concentric circles) with outer radius a and inner radius b (the same a and b as in the ellipse). Find the ratio a/b if the area of the annulus and the area of the ellipse are the same. (Hint: such an ellipse is called a *golden* ellipse.)

2. If the portion of the ellipse in Exercise 1 to the right of the y-axis is revolved about the y-axis, the resulting solid is called a *spheroid* (if $a > b$ it's called an *oblate spheroid*; if $b > a$ it's called a *prolate spheroid*). Find the volume of the solid.

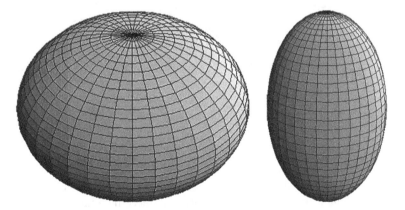

Figure 2.13. Oblate and prolate spheroids

3. Find the area of the region enclosed by one branch of the hyperbola $y^2 - x^2 = 1$ and the line $y = 2$ (Such a region is sometimes called a *right hyperbolic segment*).

4. In the plane let O be the origin, C the circle of radius a centered at $(0, a)$, and T the line tangent to C at $(0, 2a)$. Draw a ray from O intersecting C at A and T at B. The *cissoid of Diocles* is the set of points P such that the distance between O and P is equal to the distance between A and B. See Figure 2.14.
An equation of the cissoid is $x^2 = y^3/(2a - y)$. Find the volume of the solid formed by rotating the cissoid about the line T.

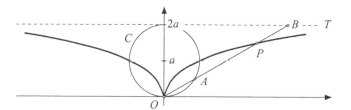

Figure 2.14. The cissoid of Diocles

2.4 Area of a surface of revolution

When the graph of a function in the plane is revolved about a line in the plane (that doesn't intersect the graph), the result is a *surface of revolution*. Examples of surfaces of revolution (and the generating graph) include a sphere (generated by a semicircle), cones and cylinders (generated by a line segment), and a torus (generated by a circle).

Example 2.13 If we rotate the graph of $y = x^2$ between $(0, 0)$ and $(2, 4)$ about the y-axis, the curve traces out a surface in space, as shown in Figure 2.15. The surface is a portion of an object known as a *paraboloid of revolution*. What is its surface area? We shall return to this example later. ▨

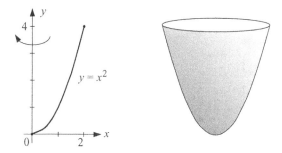

Figure 2.15. A paraboloid of revolution

To find the area of a surface of revolution, we proceed as we did in finding arc length. We partition the curve into n pieces, and approximate each piece by a line segment of length L_i joining its endpoints (x_{i-1}, y_{i-1}) and (x_i, y_i). When the line segment is revolved about the y-axis, it traces out a *frustum* of a cone (frustum is Latin for "a piece"), the portion of a cone lying between the two planes parallel to the base of the cone, as shown in Figure 2.16(a). Hence our first task is to find the area of a frustum, given the radii r_1 and r_2 and the slant height s $(= L_i)$.

If we cut the frustum along a line on its surface and lay it flat, then we have Figure 2.16(b), and the region shaded gray is the difference of two circular sectors, the smaller of radius x (currently unknown, but we will evaluate x shortly), and the larger of radius $x + s$. The smaller sector has area

$$\frac{2\pi r_1}{2\pi x} \cdot \pi x^2 = \pi x r_1,$$

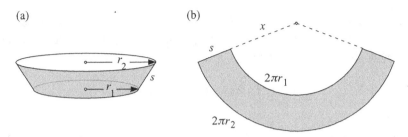

Figure 2.16. The area of a frustum of a cone

and similarly the larger sector has area $\pi(x+s)r_2$. Thus the area of the frustum is

$$\pi[(x+s)r_2 - xr_1] = \pi[x(r_2-r_1)+sr_2].$$

But the two sectors are similar, so that $x/r_1 = (x+s)/r_2$, hence $x(r_2-r_1) = sr_1$ and so the area of the frustum is

$$\pi[sr_1 + sr_2] = \pi(r_1+r_2)s = 2\pi \bar{r} s,$$

where $\bar{r} = (r_1+r_2)/2$, the average of the two radii. Thus when we rotate the line segment joining the two points (x_{i-1}, y_{i-1}) and (x_i, y_i) about the y-axis, the resulting frustum has radii $r_1 = x_{i-1}$ and $r_2 = x_i$ and slant height $s = L_i$. But recall from Section 2.1 that $L_i = \sqrt{1+[f'(x_i^*)]^2}\,\Delta x$ for some point x_i^* between x_{i-1} and x_i, so that the area A_i of the ith frustum is

$$A_i = 2\pi \bar{x}_i L_i = 2\pi \bar{x}_i \sqrt{1+[f'(x_i^*)]^2}\,\Delta x,$$

where $\bar{x}_i$ is the midpoint of the interval $[x_{i-1}, x_i]$. To find the area A of the surface we now add the areas of the n frusta (the plural of frustum) and take the limit as $n \to \infty$, yielding

$$A = \lim_{n\to\infty} \sum_{i=1}^{n} 2\pi \bar{x}_i \sqrt{1+[f'(x_i^*)]^2}\,\Delta x. \tag{2.6}$$

The sum in (2.6) is *not* a Riemann sum, since it is evaluated at two points $\bar{x}_i$ and x_i^* in each subinterval $[x_{i-1}, x_i]$. But it seems reasonable (and can be proved in an advanced calculus course) that $\bar{x}_i$ and x_i^* approach the same value in the limit since both are trapped in an interval whose length Δx approaches 0 as $n \to \infty$. Hence we have

Theorem 2.3 *If a function f has a continuous derivative on the interval $[a, b]$, $0 \le a < b$, then the area A of the surface obtained when the graph of $y = f(x)$ from $(a, f(a))$ to $(b, f(b))$ is rotated about the y-axis is given by*

$$A = \int_a^b 2\pi x \sqrt{1+[f'(x)]]^2}\,dx. \tag{2.7}$$

Switching the roles of x and y in Theorem 2.3 yields

Corollary 2.4 *If a function g has a continuous derivative on the interval $[c, d]$, $0 \leq c < d$, then the area A of the surface obtained when the graph of $x = g(y)$ from $(g(c), c)$ to $(g(d), d)$ is rotated* <u>about the x-axis</u> *is given by*

$$A = \int_c^d 2\pi y \sqrt{1 + [g'(y)]]^2} \, dy. \tag{2.8}$$

Back to Example 2.13 Since the derivative of $f(x) = x^2$ is continuous on $[0, 2]$, the area of the paraboloid of revolution is

$$A = \int_0^2 2\pi x \sqrt{1 + 4x^2} \, dx = \cdots = \frac{\pi}{6}[17\sqrt{17} - 1] \approx 36.1769. \quad \blacksquare$$

☺ You should fill in the gap indicated by the ellipses ($\cdots$)!

Example 2.14 If we rotate the graph of the hyperbola $y = 1/x$ between $(1/2, 2)$ and $(1, 1)$ about the y-axis, we trace a portion of an object that resembles a lampshade, as shown in Figure 2.17. What is the area A of the resulting surface of revolution?

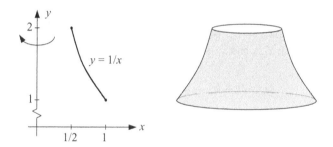

Figure 2.17. The surface of revolution in Example 2.13

Using Theorem 2.3, the area A is given by

$$A = \int_1^2 2\pi x \sqrt{1 + \frac{1}{x^4}} \, dx = \pi \int_1^4 \frac{\sqrt{1 + u^2}}{u^2} \, du,$$

the same integral (but with different limits) that we saw in Example 2.12. Hence

$$A = \pi \int_{\pi/4}^{\arctan 4} [\sec \theta + \csc \theta \cot \theta] \, d\theta$$

$$= \pi \left[\ln|\sec \theta + \tan \theta| - \csc \theta \right]_{\theta = \pi/4}^{\arctan 4} = \pi \left[\ln \frac{4 + \sqrt{17}}{1 + \sqrt{2}} - \frac{\sqrt{17}}{4} + \sqrt{2} \right] \approx 5.0164. \quad \blacksquare$$

In Theorem 2.3, we emphasized the phrase *about the y-axis*. The next example will illustrate why.

Example 2.15 Find the area of the surface of revolution obtained by rotating the curve $y = x^3$ between $(0, 0)$ and $(2, 8)$ about the x-axis. We could use Corollary 2.4 for this problem, but we will use this example to motivate the derivation of a new formula. We will return to this example later. ∎

If we rotate the curve in Figure 2.15 about the x-axis rather than the y-axis, then the radii of the frusta are $r_1 = f(x_{i-1})$ and $r_2 = f(x_i)$. But the number $[f(x_{i-1}) + f(x_i)]/2$ lies between the numbers $f(x_{i-1})$ and $f(x_i)$, and hence by the intermediate value theorem there exists a number $\hat{x}_i$ in (x_{i-1}, x_i) such that $f(\hat{x}_i) = [f(x_{i-1}) + f(x_i)]/2$. Thus the area A_i of the ith frustum in this case is given by

$$A_i = 2\pi f(\hat{x}_i)L_i = 2\pi f(\hat{x}_i)\sqrt{1 + \left[f'(x_i^*)\right]^2}\,\Delta x.$$

To find the area A of the surface we now add the areas of the n frusta and take the limit as $n \to \infty$, yielding

$$A = \lim_{n\to\infty} \sum_{i=1}^{n} 2\pi f(\hat{x}_i)\sqrt{1 + \left[f'(x_i^*)\right]^2}\,\Delta x. \tag{2.9}$$

The sum in (2.9) is again not a Riemann sum, but as before both $\hat{x}_i$ and x_i^* approach the same number in the limit, and we have

Theorem 2.5 *If a positive function f has a continuous derivative on the interval $[a, b]$, then the area A of the surface is obtained when the graph of $y = f(x)$ from $(a, f(a))$ to $(b, f(b))$ is rotated about the x-axis is given by*

$$A = \int_a^b 2\pi f(x)\sqrt{1 + [f'(x)]^2}\,dx. \tag{2.10}$$

Back to Example 2.15 With $f(x) = x^3$ and $[a, b] = [0, 2]$ we have

$$A = 2\pi \int_0^2 x^3\sqrt{1 + 9x^4}\,dx = \frac{\pi}{18}\int_1^{145} u^{1/2}\,du = \frac{\pi}{27}\left(145\sqrt{145} - 1\right) \approx 203.04. \quad \blacksquare$$

Switching the roles of x and y in Theorem 2.5 yields

Corollary 2.6 *If a positive function g has a continuous derivative on the interval $[c, d]$, then the area A of the surface obtained when the graph of $x = g(y)$ from $(g(c), c)$ to $(g(d), d)$ is rotated about the y-axis is given by*

$$A = \int_c^d 2\pi g(y)\sqrt{1 + [g'(y)]^2}\,dy. \tag{2.11}$$

Example 2.16 A *spherical zone* is the portion of a sphere bounded by two parallel planes intersecting the sphere, as shown in gray in Figure 2.18(a). The distance h between the planes is called the *altitude* of the zone.

(a)   (b)

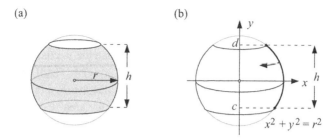

Figure 2.18. A spherical zone

To find the surface area of a spherical zone, we use (2.11) with $g(y) = \sqrt{r^2 - y^2}$ and $h = d - c$. Since $\sqrt{1 + [g'(y)]^2} = r/g(y)$, we have

$$A = \int_c^d 2\pi g(y) \cdot \frac{r}{g(y)}\, dy = 2\pi r \int_c^d dy = 2\pi r(d - c) = 2\pi r h.$$

As a consequence, the surface area of the entire sphere is $A = 2\pi r(2r) = 4\pi r^2$.

When one of the planes bounding a spherical zone is tangent to the sphere, the resulting portion of the sphere is called a *spherical cap*. See Exercise 15. ∎

2.4.1 Exercises

In Exercises 1 through 4, set up, but do not evaluate, integrals for the area of the surface obtained when the graph of the given function is revolved about the indicated axis.

1. $y = \sin x$, $\quad 0 \le x \le \pi$, $\quad x$-axis $\qquad$ 2. $y = \cos x$, $\quad 0 \le x \le \pi/2$, $\quad y$-axis

3. $y = \ln x$, $\quad 1 \le x \le e$, $\quad x$-axis $\qquad$ 4. $y = \sqrt{x^2 + 1}$, $\quad 0 \le x \le \sqrt{3}$, $\quad y$-axis

In Exercises 5 through 10, find the area of the surface obtained when the graph of the given function is revolved about the indicated axis.

5. $y = x^3$, $\quad 0 \le x \le 2$, $\quad x$-axis $\qquad\qquad$ 6. $y = \cos x$, $\quad -\frac{\pi}{2} \le x \le \frac{\pi}{2}$, $\quad x$-axis

7. $y = \sqrt{x}\left(x - \frac{1}{3}\right)$, $\quad 1 \le x \le 4$, $\quad y$-axis $\qquad$ 8. $y = \frac{x^3}{6} + \frac{1}{2x}$, $\quad 1 \le x \le 2$, $\quad x$-axis

9. $y = \frac{1}{8}x^2 - \ln x$, $\quad 1 \le x \le e$, $\quad y$-axis $\qquad$ 10. $y = e^x$, $\quad 0 \le x \le 1$, $\quad x$-axis

11. Derive a formula for the area of the curved portion of a cone with base radius r and height h by revolving a right triangle about one of its legs.

12. Find the surface area of the spheroid in Exercise 2 in Section 2.3.1.

13. Find the surface area of the torus in Exercise 25 in Section 1.1.1

14. In Section 2.1.1, Exercise 19, you found the arc length of the astroid $x^{2/3} + y^{2/3} = a^{2/3}$. Find the volume and surface area of the solid formed when the astroid is revolved about either the x- or y-axis.

15. At Abigail's *bat mitzvah* (the "coming of age" ceremony for Jewish girls at age 12), a *kippah* (a traditional head covering for Jewish men) was worn by each male guest, as shown in Figure 2.19(a).

(a)   (b)

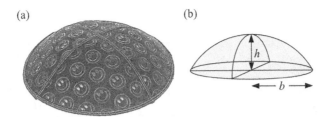

Figure 2.19. A kippah and a spherical cap

If the base radius of the kippah is b cm and the altitude h cm, find the surface area of the kippah in terms of b and h. (Hint: see Example 2.16 since a spherical cap is a special case of a spherical zone.)

16. On the Earth, the tropics are generally considered to be the zone lying between the Tropic of Cancer (latitude 23.4378° N) and the Tropic of Capricorn (latitude 23.4378° S), the gray band on the world map in Figure 2.20. The map seems to indicate that perhaps less than one-third of the Earth's surface is tropical. Maps can be deceiving. Assuming the Earth's surface is a sphere, what percent of the Earth's surface really does lie in the tropics?

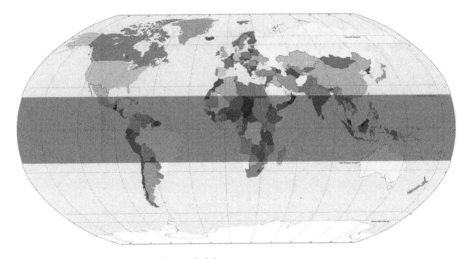

Figure 2.20. A map of the world

2.5 Explorations

2.5.1 The number π and the circle.

The number π is often defined as the ratio of the circumference C of a circle to its diameter. So if r denotes the radius of the circle, then $\pi = C/2r$, or $C = 2\pi r$. And "we all know" that the area A of the circle is given by $A = \pi r^2$. But we ask: *Why should the same constant π appear in both formulas?*

   An ancient and intuitive answer is given in Figure 2.21. If we slice a circular disk of radius r into a large number of congruent pie-shaped wedges, then we can arrange them in a figure that resembles a parallelogram with base πr (one half the circumference) and height r, thus A should be πr^2.

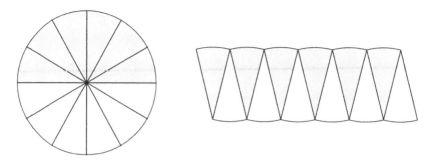

Figure 2.21. An intuitive way to find the area of a circle

Of course, that is not a proof, but it can be made rigorous with the use of limits. But if we are going to use limits, we might as well use calculus, which we now do in this exploration with integrals.

Exercise 1. Since the circumference C and area A of a circle with radius r are each four times the arc length and area of the quarter circle in the first quadrant, show that

$$C = 4 \int_0^r \frac{r}{\sqrt{r^2 - x^2}}\, dx \quad \text{and} \quad A = 4 \int_0^r \sqrt{r^2 - x^2}\, dx.$$

Exercise 2. Do you notice anything strange about the integral for C?

We hope you noticed that the integrand is undefined at $x = r$, and recalled from your PCE that the existence of the definite integral $\int_a^b f(x)\, dx$ requires that the integrand be defined on $[a, b]$. The integral for C is an example of an *improper integral*, which we will study in Chapter 9.

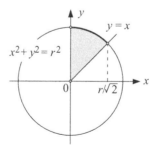

Figure 2.22. Using calculus to find the area of a circle

Exercise 3. Repeat Exercise 1 using only the one-eighth of the circle shown in Figure 2.22 to show that

$$C = 8 \int_0^{r/\sqrt{2}} \frac{r}{\sqrt{r^2 - x^2}}\, dx \quad \text{and} \quad A = 8 \int_0^{r/\sqrt{2}} \left(\sqrt{r^2 - x^2} - x \right) dx.$$

Exercise 4. Use the change of variables $x = rt$ in both integrals to obtain

$$C = 2\left(4\int_0^{1/\sqrt{2}} \frac{1}{\sqrt{1-t^2}}\, dt\right) r \quad \text{and} \quad A = \left(8\int_0^{1/\sqrt{2}} \sqrt{1-t^2}\, dt - 2\right) r^2.$$

Exercise 5. Since π is *defined* as the ratio of a circle's circumference C to its diameter $2r$, the expression in the first set of parentheses is equal to π. To conclude that the expression in the second set of parentheses is also π, use the identity $1 = (1 - t^2) + t^2$ followed by integration by parts to show

$$4\int_0^{1/\sqrt{2}} \frac{1}{\sqrt{1-t^2}}\, dt = 8\int_0^{1/\sqrt{2}} \sqrt{1-t^2}\, dt - 2.$$

Thus the same constant π appears in both the area and the circumference formulas for the circle. You were able to prove this by using integration techniques, but without actually evaluating any integrals!

2.5.2 Derivatives of area and volume

You may have noticed that the derivative of the area A of a circle with respect to its radius r equals its circumference C, and that the derivative of the volume V of a sphere with respect to its radius equals its surface area S:

$$\frac{dA}{dr} = \frac{d}{dr}\pi r^2 = 2\pi r = C \quad \text{and} \quad \frac{dV}{dr} = \frac{d}{dr}\frac{4}{3}\pi r^3 = 4\pi r^2 = S.$$

Are there other regions in the plane and solids with these properties?

Exercise 1. (a) An annulus (see Exercise 1 in Section 2.1) is the region between two concentric circles. If the outer radius is twice the inner radius r, as shown in Figure 2.23(a), show that $dA/dr = P$ where P is the sum of the circumferences of the two circles.

(b) In Figure 2.23(b) we see a circular cylinder whose height h equals the diameter of the base. Show that $dV/dr = S$ where the surface area S is the sum of the areas of the top, bottom, and lateral (side) portions.

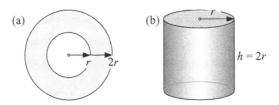

Figure 2.23. The annulus and cylinder in Exercise 1

Are the circle, sphere, annulus and cylinder special in this regard, or do these relationships hold for other plane figures and solids? In this Exploration you will learn the answer to this question, and that it depends on how you choose to measure the linear dimension of the object.

Exercise 2. Consider a square and a cube each with side length s. Does the derivative of the area A of the square with respect to s equal the perimeter P? Is the derivative of the volume V of the cube with respect to s equal to the surface area S?

You discovered the answer to both questions is "no!" since $dA/ds = 2s \neq 4s = P$ and $dV/ds = 3s^2 \neq 6s^2 = S$. So maybe the circle, sphere, annulus and cylinder are special. But let's repeat the calculations using a different linear measurement.

Exercise 3. Repeat Exercise 2, but take the derivatives with respect to half the side length, i.e., set $s = 2x$ and find dA/dx and dV/dx.

Aha! Now $A = (2x)^2 = 4x^2$ and $P = 4(2x) = 8x$ so that $dA/dx = P$; and $V = (2x)^3 = 8x^3$ and $S = 6(2x)^2 = 24x^2$ so that $dV/dx = S$. So perhaps the area-perimeter and volume-surface area relationships hold for objects other than the circle, sphere, annulus, and cylinder!

Consider a region in the plane whose perimeter P and area A are defined and finite, and expressible in terms of some linear dimension s of the region (such as side length, diameter, or radius) as $P = ks$ and $A = cs^2$ for constants k and c. In the next exercise you will find a linear dimension x (as a function of s) such that $dA/dx = P$.

Exercise 4. Show that (a) if $dA/dx = P$, then $x = (2c/k)s = 2A/P$; and (b) if $x = 2A/P$, then $dA/dx = P$. (Hint: in (a) let $x = ts$ and solve for the constant t.)

Exercise 5. Suppose the region in the plane is an equilateral triangle with side length s. Since its perimeter $P = 3s$ and area $A = (\sqrt{3}/4)s^2$, show that if $dA/dx = P$ then (a) $x = h/3$ where $h = s\sqrt{3}/2$ is the altitude of the triangle (geometrically, x is the radius of the inscribed circle), and (b) express P and A in terms of x and verify that $dA/dx = P$. See Figure 2.24(a).

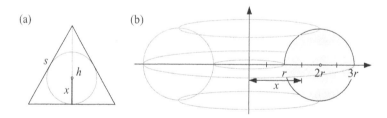

Figure 2.24. The linear dimension x for (a) the equilateral triangle and (b) the torus

Exercise 6. Now consider a region in space whose surface area S and volume V are defined, finite, and expressible in terms of a linear dimension s as $S = ks^2$ and $v = cs^3$ for constants k and c. Show that in this case $dV/dx = S$ if and only if $x = 3V/S$.

Let's apply this result to the torus, which we studied in Exercises 25 in Section 1.1 and 13 in Section 2.4.

Exercise 7. If you haven't already done those exercises, do them now to find the volume and surface area of the torus.

Exercise 8. Consider the special case of the torus where $R = 2r$. Show that in terms of the radius r, $dV/dr \neq S$. Then show that the linear dimension x for which $dV/dx = S$ is $x = 3r/2$, half the distance from the center of the torus to the outermost edge. See Figure 2.24(b).

2.6 Acknowledgments

1. Exercise 20 in Section 2.1.1 is from C. Moen, Arc length and Pythagorean triples, *College Mathematics Journal*, **38** (2007), pp. 222–223.
2. Exercise 1b in Section 2.3.1 is from A. D. Rawlins, A note on the golden ratio, *Mathematical Gazette*, **79** (1995), p. 104.
3. The motivation for Exploration 2.5.1 is from E. F. Assmus Jr., Pi *American Mathematical Monthly*, **92** (1985), pp. 213–214.
4. Exploration 2.5.2 is adapted from J. Tong, Area and perimeter, volume and surface area, *College Mathematics Journal*, **28** (1997), p. 57.

3

Differential Equations: Graphical, Numerical, and Symbolic Solutions

Scientific laws, when we have reason to think them accurate, are different in form from the common-sense rules which have exceptions: they are always, at least in physics, either differential equations, or statistical averages.

Bertrand Russell

In your PCE you studied the motion of objects near the surface of the Earth in free fall. In your studies, acceleration due to gravity was the only force acting on the object and it was approximately constant. That is, at sea level, the acceleration due to gravity a_{grav} is approximately equal to the constant g given by $g = 32$ ft/sec^2 or $g = 9.8$ m/sec^2.

But if we consider two objects, a bowling ball and a feather, observing their velocities as they fall to Earth, experience tells us the bowling ball will achieve a much greater velocity and strike the ground first. Why is this? Earth's atmosphere gets in the way! In a vacuum, two objects dropped from the same height would have the same velocities at all times since the only force acting on them is gravity. But in reality when they are dropped near the Earth's surface, each experiences an additional force, air resistance, which leads to the bowling ball hitting the ground before the feather.

In general, air resistance is a force that induces acceleration in the direction opposite the motion of the object. One reasonable model for air resistance is that the magnitude of the induced acceleration is proportional to the velocity of the object, that is $a_{\text{air}} = -kv$. The constant k here depends on the object (think of the bowling ball versus the feather).

Combining the effects of gravity and air resistance, we have that net acceleration a is given by

$$a = a_{\text{grav}} + a_{\text{air}} = g - kv. \qquad (3.1)$$

Because acceleration is the derivative of velocity over time, we can rewrite (3.1) as

$$v' = g - kv \ \text{ or } \ v'(t) = g - kv(t) \ \text{ or } \ \frac{dv}{dt} = g - kv, \qquad (3.2)$$

Figure 3.1. Which falls faster?

an example of a *differential equation*, an equation that relates an unknown function (in this case $v(t)$) with one or more of its derivatives (here $v'(t)$). We might think an equation such as (3.2) would be called a *derivative equation*. We refer to it as a differential equation because one solution technique involves rewriting the equation in differential form (see Section 3.4).

Differential equations are powerful tools for mathematically describing, analyzing, and understanding many physical, biological, economic, and social phenomena. In this chapter, we introduce methods for analyzing and solving differential equations graphically, numerically, and symbolically.

3.1 Differential equations: definitions and examples

In your PCE you were introduced to differential equations, and we review some definitions and examples.

Definition 3.1 A *differential equation*, or *DE*, is an equation relating an unknown function and one or more of its derivatives. A function is a *solution* to a differential equation if, upon substitution, the function makes the equation true. The highest order derivative that appears in the differential equation is called the *order* of the differential equation.

Example 3.1 In (3.2) is an example of a *first-order differential equation* since it involves only up to the first derivative of the unknown function. One solution is given by

$$v = \frac{g}{k} + e^{-kt}.$$

We can (and should) check that this function indeed is a solution by substituting into (3.2). On the one hand, we have

$$\frac{dv}{dt} = -ke^{-kt}.$$

On the other hand,

$$g - kv = g - k\left(\frac{g}{k} + e^{-kt}\right) = -ke^{-kt},$$

and therefore, $v = \frac{g}{k} + e^{-kt}$ is a solution to $\frac{dv}{dt} = g - kv$. ∎

Example 3.2 Consider the DE

$$\frac{dy}{dx} = x(y - 1). \tag{3.3}$$

Here, our unknown function is y with independent variable x. (3.3) can be written as

$$y'(x) = x(y(x) - 1) \quad \text{or} \quad y' = x(y - 1).$$

You can check that $y(x) = 1 + e^{x^2/2}$ is a solution (see Exercise 2). We will learn several methods for analyzing and solving this equation. ∎

Example 3.3 The functions $y(x) = \sin x + x$ and $y(x) = 3\cos x + x$ are both solutions to the second order DE

$$y'' + y = x. \tag{3.4}$$

We use substitution to check. For $y = \sin x + x$ we have

$$y'(x) = \cos x + 1, \quad y''(x) = -\sin x \quad \Rightarrow \quad y'' + y = -\sin x + (\sin x + x) = x.$$

For $y = 3\cos x + x$ we have

$$y'(x) = -3\sin x + 1, \quad y''(x) = -3\cos x \Rightarrow y'' + y = -3\cos x + (3\cos x + x) = x. \quad ∎$$

⊘ (3.4) has other solutions as well. For example, $y(x) = 3\cos x - \sqrt{2}\sin x + x$ is a solution (you should check this).

Example 3.4 (Newton's law of cooling) On a cold day, a child gets a cup of hot cocoa and takes it outside to play in the snow. The cocoa is too hot to drink at first, but after a few minutes, the cocoa is drinkable. What happens after a few more minutes? Of course, if it's truly cold outside, the cocoa cools to the point where it's more like chocolate milk than hot cocoa. Eventually, it may even freeze.

One model for the temperature of an object in a constant temperature environment is Newton's law of cooling.

Newton's law of cooling: *The rate of change in temperature of an object in a constant temperature environment is proportional to the difference between the environment's temperature and the object's temperature.*

Let A be the constant temperature of the environment, and $F(t)$ the temperature of the object at time t. The phrase "is proportional to" translates mathematically to "equals a constant times." Therefore, Newton's law of cooling can be written as

$$F'(t) = k(A - F(t)) \quad \text{or} \quad \frac{dF}{dt} = k(A - F), \tag{3.5}$$

where k is a positive constant. This equation is strikingly similar to (3.2). Observe that when $F(t) > A$, then $\frac{dF}{dt} < 0$ and we have cooling, while when $F(t) < A$, then $\frac{dF}{dt} > 0$ and we have warming. ∎

Example 3.5 (The snowplow problem) One day it started snowing at a heavy and steady rate. A snowplow started out at noon, going 2 miles the first hour and 1 mile the second hour. What time did it start snowing?

Figure 3.2. A snow blower

Assume that there was no snow on the road at the time it started snowing. Let t denote time in hours, with $t = 0$ at noon, and let t_0 denote the time it started snowing. Note that $t_0 < 0$. Let $y(t)$ be the distance traveled by the snowplow, in miles, so that $y(1) - y(0) = 2$ and $y(2) - y(1) = 1$. Let $h(t)$ be the height of the snow in feet at time t. Then $h(t_0) = 0$. Let s be the (constant) rate of snowfall measured in feet per hour, so that $h(t) = s(t - t_0)$. Finally let w denote the width of the snowplow in feet and k the amount of snow that the plow can remove per hour, measured in cubic feet per hour. Then

$$w \cdot h(t) \cdot \frac{dy}{dt} \cdot 5280 = k \qquad \text{so that} \qquad \frac{dy}{dt} = \frac{c}{t - t_0}$$

where $c = 5280\frac{k}{sw}$. This differential equation can be solved by integrating with respect to t to get $y(t) = c \ln(t - t_0) + a$ where a is an arbitrary constant.

Now we make use of the initial conditions $y(1) - y(0) = 2$ and $y(2) - y(1) = 1$ to see

$$2 = c \ln(1 - t_0) + a - c \ln(-t_0) - a = c \ln\left(\frac{1 - t_0}{-t_0}\right) = c \ln\left(\frac{t_0 - 1}{t_0}\right)$$

and

$$1 = c \ln(2 - t_0) + a - c \ln(1 - t_0) - a = c \ln\left(\frac{2 - t_0}{1 - t_0}\right) = c \ln\left(\frac{t_0 - 2}{t_0 - 1}\right).$$

Thus

$$\ln\left(\frac{t_0 - 1}{t_0}\right) = 2 \ln\left(\frac{t_0 - 2}{t_0 - 1}\right) = \ln\left[\left(\frac{t_0 - 2}{t_0 - 1}\right)^2\right]$$

which yields

$$\frac{t_0 - 1}{t_0} = \left(\frac{t_0 - 2}{t_0 - 1}\right)^2.$$

This last equation simplifies to $t_0^2 - t_0 - 1 = 0$. Since $t_0 < 0$, we see that $t_0 = \frac{1-\sqrt{5}}{2}$ or $t_0 \approx -.618$. Hence it began snowing at about 11:23 am. ■

3.1.1 Exercises

In Exercises 1–5, you are given a DE and several functions. Check each function to determine if it is a solution to the DE.

1. $\dfrac{dy}{dx} = 2y$ $y = e^{2x}, y = e^{4x}, y = -5e^{2x}, y = e^{2x} + 5$

2. $y'(x) = x \cdot (y(x) - 1)$ $y(x) = 1 + e^{x^2/2}, y = 7e^{x^2/2} + 1, y = 1 - 2^{-x^2/2}$

3. $t^2 y'' - 3ty' + 4y = 0$ $y = t^2, y = t^2 \ln t, y = t^2 + \ln t, y = \ln t$

4. $\dfrac{d^2\theta}{dt^2} + \omega^2\theta = 0$ $\theta = \cos(\omega t), \theta = \cos(\omega t + \pi/4), \theta = \cos(\omega t) + \pi/4$

5. $y' + y \tan x = 0$ $y = \cos x, y = \cos 2x, y = 2\cos x, y = \cos x + 2$

In Exercises 6–20, find a function that is a solution to the given DE. To do so, try to find a function whose derivative is the specified right-hand side. You can try educated guessing or integration methods. In each exercise, $y' = \frac{dy}{dx}$. Check your answer by differentiating and substituting.

6. $y' = 2$

7. $y' = x^3 - 2x^2 + 3$

8. $y' = e^x$

9. $y' = \sin x$

10. $y' = \dfrac{1}{1+x^2}$

11. $y' = \dfrac{x}{2x^2 - 1}$

12. $y' = \dfrac{1}{\sqrt{1-x^2}}$

13. $y' = \tan x$

14. $y' = \dfrac{x}{\left(1+x^2\right)^2}$

15. $y' = \sqrt{36 - x^2}$

16. $y' = xe^x$

17. $y' = \dfrac{\ln x}{x^3}$

18. $y' = y$

19. $y' = -y$

20. $y'' = -y$

21. In Example 3.4, suppose that the DE governing the temperature $F(t)$ of the cocoa is given by $\frac{dF}{dt} = (30 - F)\ln(4/3)$ where t is measured in minutes and F in degrees Fahrenheit.

   (a) Use the DE to determine $\frac{d^2 F}{dt^2}$ in terms of F. What conclusions can you draw about any solution to the DE given your result?

   (b) Show that $F(t) = 30 + 120\,(3/4)^t$ satisfies the DE and therefore is a feasible model for the temperature of the cocoa.

   (c) With the solution in part (b), time $t = 0$ is the time when the child steps outside. What is the temperature of the cocoa at this time? If the cocoa is drinkable when it reaches a temperature of 100°, how long does it take for the child to be able to drink the cocoa?

22. The motion of an object attached to the end of a vertical spring can be modeled by the differential equation $my'' = -ky$ with initial conditions $y(0) = 1$ and $y'(0) = 0$ where y is

the position of the object at time t. The constant k depends upon the spring, and m is the mass of the object.

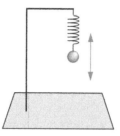

Figure 3.3. Motion of an object attached to a vertical spring

(a) Show that $y = \cos\left(\sqrt{\frac{k}{m}}t\right)$ is a solution to this differential equation.
(b) What does this model say happens to the motion of the object over time in the long run?

23. When a drug is given to a patient, it enters the bloodstream, and then is absorbed by the body. Studies suggest that the rate that a drug is absorbed by the body is proportional to the amount of drug present in the bloodstream. Let $A(t)$ be the amount of drug present in the bloodstream at time t. Write a DE involving A that illustrates this model. Your DE will have an unknown constant of proportionality. Is it positive or negative? Explain your answer.

24. Consider a simple electric circuit, with a capacitor, resistor, and voltage source (for example a battery) as illustrated in Figure 3.4. The capacitor has a charge Q that varies over time t and a capacitance C measured in farads.

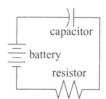

Figure 3.4. A simple circuit

The resistor has a constant resistance R measured in ohms, and the voltage supplied by the source is measured in volts. Kirchhoff's laws can be used to conclude that

$$RQ' + \frac{1}{C}Q = V.$$

Assume C and V are constant.
(a) Show that $Q(t) = CV + ke^{-t/CR}$ where k is a positive constant is a solution to this equation.
(b) What is the limiting charge, assuming the voltage remains constant. That is, what is $\lim_{t\to\infty} Q(t)$?

3.2 Slope fields: graphical solutions

When a differential equation is written in the form $\frac{dy}{dx} = f(x, y)$, we are able to compute the value of $\frac{dy}{dx}$ whenever we have values for x and y. Since $\frac{dy}{dx}$ represents the slope of a curve, when we compute its value we gain information about the solution curve $y(x)$. This information allows us to visualize solutions to the differential equation by using a *slope field*.

Back to Example 3.2 (3.3) has the form $\frac{dy}{dx} = f(x, y)$ where $f(x, y) = x(y - 1)$. Given specific values of x and y we can compute the value of $\frac{dy}{dx} = f(x, y)$. Furthermore, if $y = y(x)$ is a solution to the equation, then $\frac{dy}{dx}$ gives the slope of the tangent line to the graph of the solution. This observation allows us to use $f(x, y)$ to sketch small sections of the graphs of solutions to the DE. To do so, we compute slopes of solutions at several points and draw short segments with appropriate slopes at each of the points. The final graph we obtain is known as a slope field for the DE.

Table 3.1. Some values of slope for $\frac{dy}{dx} = x(y - 1)$

(x, y)	$(-1, -1)$	$(-1, 0)$	$(-1, 1)$	$(-1, 2)$	$(0, -1)$	$(0, 0)$
Slope	$-1(-1-1) = 2$	1	0	-1	0	0
(x, y)	$(0, 1)$	$(0, 2)$	$(1, -1)$	$(1, 0)$	$(1, 1)$	$(1, 2)$
Slope	0	0	-2	-1	0	1

We usually use technology to create slope fields, but let's practice building a slope field by hand. Table 3.1 gives values of the slopes of tangent lines to solutions to $\frac{dy}{dx} = x(y - 1)$ at various points. For example, at the point $(-1, 0)$, $\frac{dy}{dx}\big|_{(-1,0)} = -1(0 - 1) = 1$.

Any solution passing through one of these points must have a graph that heads in the direction of the slope line. If we know a point through which a solution passes, we can trace the solution by following the slope lines. Using a computer algebra system, we are able to produce much more extensive slope fields. Figure 3.5 shows both the slope field from our hand calculations, and a more detailed computer-generated slope field together with the graph of the solution to the equation that satisfies the initial condition $y(-1) = 0$. In other words, the solution curve traced in the graph passes through the point $(-1, 0)$. The differential equation together with the initial condition is known as an initial value problem.

Definition 3.2 A first-order *initial value problem*, or *IVP*, is a differential equation together with a point through which the solution must pass (the *initial condition*). A first-order IVP has the form

$$\frac{dy}{dx} = f(x, y), \quad y(x_0) = y_0$$

Note on the number of solutions

A differential equation that has a solution has many solutions, but a first-order initial value problem typically has a unique solution. Conditions that guarantee the existence and uniqueness of solutions to IVPs can be found in more advanced texts on differential equations. The IVPs we investigate have unique solutions.

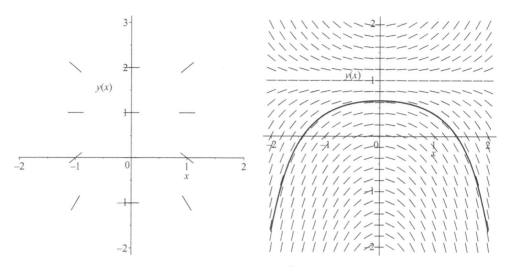

Figure 3.5. Slope fields for $\frac{dy}{dx} = x(y-1)$

Example 3.6 A skydiver jumps from a plane at an altitude of 13,000 feet. Initially, her vertical velocity is 0 ft/sec. But, she immediately experiences an acceleration due to gravity of 32 ft/sec^2. During the free fall part of her jump, the skydiver experiences acceleration from air resistance as well. As in (3.2), we assume the acceleration due to air resistance is proportional to the skydiver's velocity. Experiments suggest that $k = 0.15$ is a good approximation for the constant of proportionality, so that our IVP is

$$\frac{dv}{dt} = 32 - 0.15v, \qquad v(0) = 0. \tag{3.6}$$

Figure 3.6. A skydiver

Soon, we will find a symbolic expression for the unique solution to this IVP, but for now, let's examine a slope field for the DE together with the traced solution for the IVP.

A careful inspection of the slope field indicates that the skydiver's velocity is bounded and approaches a limit over time, about 210 feet per second or about 143 miles per hour. The

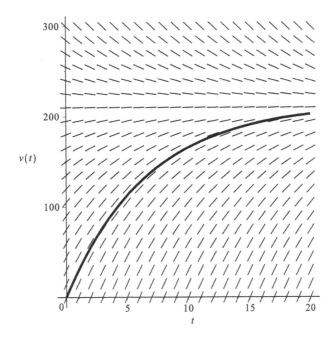

Figure 3.7. Skydiver's velocity

limiting value is known as *terminal velocity* and we will investigate more when we see methods for solving (3.6) symbolically. ∎

Example 3.7 Consider the differential equation

$$\frac{dy}{dx} = x + y. \tag{3.7}$$

The form of this equation guarantees a unique solution for any initial condition $y(x_0) = y_0$. So, whenever we sketch solutions in the slope field, they either never intersect or they coincide exactly. Figure 3.8 illustrates solutions to (3.7) with different initial values. One of the solutions sketched is $y = -x - 1$ (you should check that this satisfies the differential equation). Our observation about distinct solutions tells us that any other solution has a graph that either lies above the line $y = -x - 1$ or below it. Furthermore, our slope field suggests that solutions whose graphs lie below the line have graphs that are concave down while those above the line are concave up. We can check this by considering the second derivative:

$$\frac{d^2y}{dx^2} = \frac{d}{dx}\left(\frac{dy}{dx}\right) = \frac{d}{dx}(x+y) = 1 + \frac{dy}{dx} = 1 + x + y.$$

A solution $y(x)$ to the differential equation whose graph is below the line $1 + x + y = 0$ satisfies $y(x) < -x - 1$. For this solution, $\frac{d^2y}{dx^2} < 0$, and the graph is concave down. Similarly, for a solution whose graph is above the line, $\frac{d^2y}{dx^2} > 0$, making the graph concave up. ∎

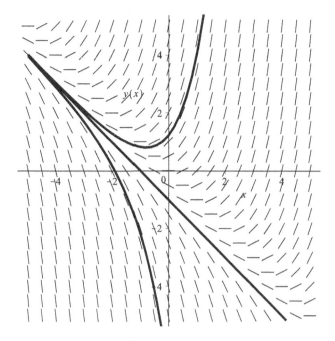

Figure 3.8. Solutions to $\frac{dy}{dx} = x + y$ with $y(-2) = 0.2$, 1.0, and 1.3

3.2.1 Exercises

1. Perhaps the simplest differential equations are $\frac{dy}{dx} = x$ and $\frac{dy}{dx} = y$. Sketch slope fields for each in the square $[-2, 2] \times [-2, 2]$. Solve each differential equation. (Hint: you know functions of x whose derivatives are x, and you also know functions equal to their own derivatives.)

2. For each DE, match the slope field with the DE. Give an explanation for each of your choices.

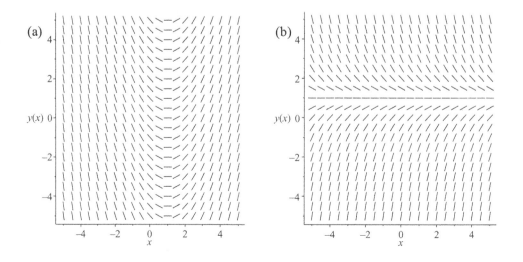

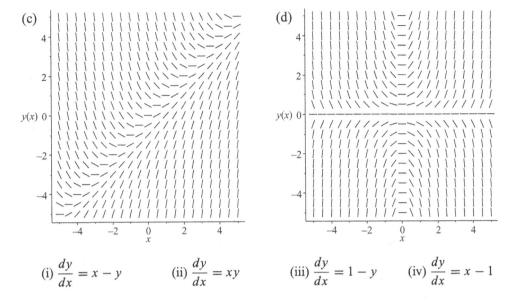

(i) $\dfrac{dy}{dx} = x - y$ (ii) $\dfrac{dy}{dx} = xy$ (iii) $\dfrac{dy}{dx} = 1 - y$ (iv) $\dfrac{dy}{dx} = x - 1$

For Exercises 3–6, use technology to sketch slope fields in the square $[-2, 2] \times [-2, 2]$. For each slope field try to guess the form of the solution, and verify your guess by differentiation. You will solve these equations exactly in Section 3.4.

3. $\dfrac{dy}{dx} = \dfrac{y}{x}$ 4. $\dfrac{dy}{dx} = -\dfrac{x}{y}$ 5. $\dfrac{dy}{dx} = \dfrac{x}{y}$ 6. $\dfrac{dy}{dx} = -\dfrac{y}{x}$

For Exercises 7–10, find $\dfrac{d^2y}{dx^2}$ for the indicated DE. Does the sign of the second derivative in various portions of $[-2, 2] \times [-2, 2]$ agree with the concavity you observe in the slope field?

7. The DE in Exercise 3.

8. The DE in Exercise 4.

9. The DE in Exercise 5.

10. The DE in Exercise 6.

11. Use technology to sketch a slope field for the DE $\dfrac{dy}{dx} = x^2/(1 - y^2)$ in the window $[-4, 4] \times [-4, 4]$. What does the slope field show along the y-axis? How is this consistent with the ODE? What does the slope field show along both $y = 1$ and $y = -1$? Explain this behavior.

12. Use technology to sketch a slope field in various portions of $[-2, 2] \times [-2, 2]$ for the differential equation $\dfrac{dy}{dx} = -xy$, and sketch several solutions, including the one that passes through the point $y(0) = 1/\sqrt{2\pi} \approx 0.4$. Use the second derivative to verify the concavity of the solutions and to locate inflection points for each solution. Do you see why the curves are called "bell-shaped"? The solution passing through $y(0) = 1/\sqrt{2\pi}$ is known as the *standard normal probability density function*, which we will study further in Chapter 9. You will solve this differential equation in Exercise 8 in Section 3.4.

3.3 Euler's method: numerical solutions

Slope fields motivate a method for numerically approximating solutions to initial value problems. Starting at the point determined by the initial condition, move along a short line segment in the direction indicated by the slope line at the point. In doing this, we usually do not stay on the solution curve. But if we move only a short distance, we do stay close to the solution curve. After moving a short distance, we repeat the steps until reaching the input at which we wish to approximate the solution's value. This process is known as *Euler's method* and is illustrated in the following example.

Back to example 3.7 Consider the IVP

$$\frac{dy}{dx} = x + y, \quad y(0) = 0.5. \tag{3.8}$$

We estimate $y(1)$ by taking four steps, each with the same change in x-coordinate, working our way from $x = 0$ to $x = 1$. Each step is $\Delta x = 0.25$ in the x-direction. Δx is known as the *stepsize*. Each step is along a line segment whose slope is determined by the differential equation at the starting point of the step.

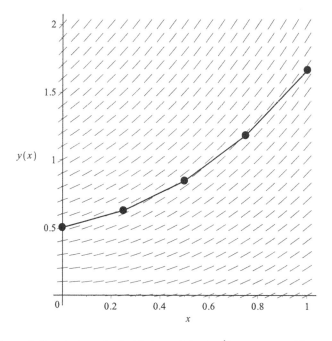

Figure 3.9. Euler's method with $\Delta x = 0.25$ for $\frac{dy}{dx} = x + y$, $y(0) = 0.5$

Figure 3.9 illustrates how we proceed. At our starting point $(x_0, y_0) = (0, 0.5)$, the slope of our first segment is

$$m_0 = \left.\frac{dy}{dx}\right|_{(x_0, y_0)} = x_0 + y_0 = 0 + 0.5 = 0.5 .$$

It ends at the point $(x_1, y_1) = (0.25, y_1)$. We compute y_1 by observing

$$y_1 = y_0 + m_0 \Delta x = 0.5 + 0.5 \cdot 0.25 = 0.625.$$

Now repeat the process starting at the point $(x_1, y_1) = (0.25, 0.625)$. We move to the point $(0.5, y_2)$ along a segment of slope $0.25 + 0.625 = 0.875$. We find that $y_2 = 0.625 + 0.875 \cdot 0.25 = 0.84375$. Continuing in this fashion, our estimate is $y(1) \approx 1.662109375$. We can organize this information in a table such as Table 3.2.

Table 3.2. Euler's method with $\Delta x = 0.25$ for $\frac{dy}{dx} = x + y$, $y(0) = 0.5$

step, i	x_i	$y_i = y_{i-1} + m_i \Delta x$	slope ($m_i = x_i + y_i$)
0	0	0.5	$0 + 0.5 = 0.5$
1	0.25	$0.5 + 0.5 \cdot 0.25 = 0.625$	$0.25 + 0.625 = 0.875$
2	0.5	$0.625 + 0.875 \cdot 0.25 = 0.84375$	1.34375
3	0.75	1.1796875	1.9296875
4	1	1.662109375	

In general, Euler's method gives us a straightforward algorithm for estimating the value of a solution to an IVP at particular points.

Euler's method

Consider the initial value problem

$$\frac{dy}{dx} = F(x, y), \quad y(a) = y_0.$$

To estimate $y(b)$ using n equal sized steps, set $\Delta x = (b - a)/n$ and $x_0 = a$, and for each i from 0 to $n - 1$, compute

$$m_i = \frac{dy}{dx}\bigg|_{(x_i, y_i)} = F(x_i, y_i),$$

$$x_{i+1} = x_i + \Delta x,$$

$$y_{i+1} = y_i + m_i \Delta x.$$

Then y_n is our Euler's method estimate of $y(b)$.

3.3.1 Exercises

In Exercises 1–4 use Euler's method by hand (no technology) with stepsize $\Delta x = 0.5$ to estimate $y(3)$.

1. $\dfrac{dy}{dx} = \dfrac{y}{x}$, $\quad y(1) = 2$

2. $\dfrac{dy}{dx} = -\dfrac{y}{x}$, $\quad y(1) = 2$

3. $\dfrac{dy}{dx} = \dfrac{x}{y}$, $\quad y(1) = 2$

4. $\dfrac{dy}{dx} = -\dfrac{x}{y}$, $\quad y(1) = 2$

5. Consider $\frac{dy}{dx} = x(y - 1)$, $\quad y(0) = -3$.
   (a) Use Euler's method to estimate $y(1)$ with stepsize $\Delta x = 0.5$.
   (b) Use Euler's method to estimate $y(1)$ with stepsize $\Delta x = 0.25$.
   (c) Use Euler's method to estimate $y(1)$ with stepsize $\Delta x = 0.05$.

6. Consider $\frac{dy}{dx} = x^2/(1 - y^2)$, $\quad y(1) = 2$.
   (a) Use Euler's method to estimate $y(2)$ with stepsize $\Delta x = 0.5$.
   (b) Use Euler's method to estimate $y(2)$ with stepsize $\Delta x = 0.2$.
   (c) Use Euler's method to estimate $y(2)$ with stepsize $\Delta x = 0.05$.

7. Euler's method can produce erroneous results. Consider the DE $\frac{dy}{dx} = x^2/(1 - y^2)$.
   (a) Explain why no solution to the DE can touch either of the horizontal lines $y = 1$ and $y = -1$.
   (b) Suppose that $y(1) = 1.1$. Why is it inappropriate to use Euler's method with $\Delta x = 0.5$ to approximate $y(2)$?

8. Euler's method with small stepsize does a good job of estimating solutions to most first-order ODEs. But, using large stepsize can yield some very interesting results. Consider the basic scaled logistic IVP $y' = y(1 - y)$, $\quad y(0) = 1.3$. For each value of Δx below, use 20 iterations of Euler's method and plot the values computed on an ny-coordinate system, where n is the number of the iteration.

   (a) $\Delta x = 1.65$ (b) $\Delta x = 2.1$ (c) $\Delta x = 2.5$ (d) $\Delta x = 2.65$

   What do you observe about the behavior of the Euler iterates?

3.4 Separation of variables: symbolic solutions

In your PCE, you studied separable differential equations. With y an unknown function of the variable x, a differential equation is *separable* if it can be written in the form

$$\frac{dy}{dx} = (\text{expression in } x) \cdot (\text{expression in } y).$$

We can solve such an equation by integrating both sides and solving for y. Often, we write separable equations in *separated differential form*

$$(\text{expression in } y)\, dy = (\text{expression in } x)\, dx.$$

We illustrate several separable differential equations in Table 3.3.

The last two equations involve some factoring to write them in separated differential form.

Most differential equations are not separable. For example, $\frac{dy}{dx} = x + y$ is not separable. This is because the right-hand side cannot be written as a *product* of an *expression in y* and an *expression in x*.

Once we have separated the variables in a separable differential equation, we integrate both sides of the equation and attempt to solve for the appropriate variable (y in the above examples).

Example 3.8 The DE

$$\frac{dy}{dx} = x + xy - y - 1$$

Table 3.3. Some separable differential equations

Equation	Separated differential form
$\dfrac{dy}{dx} = x^2 y$	$\dfrac{1}{y} dy = x^2 \, dx$
$\dfrac{dy}{dx} = \dfrac{\ln x}{\sin y}$	$\sin y \, dy = \ln x \, dx$
$\dfrac{dy}{dx} = x + xy$	$\dfrac{dy}{1+y} = x \, dx$
$\dfrac{dy}{dx} = x + xy - y - 1$	$\dfrac{dy}{1+y} = (x-1) \, dx$

is separable and appears in the last row of Table 3.3. After factoring the right-hand side, we see that

$$\frac{dy}{dx} = x(1+y) - (1+y) = (x-1)(1+y).$$

Note that $y(x) = -1$ is a solution to this differential equation and no other solution can pass through $y = -1$. So, with $y \neq -1$,

$$\frac{dy}{1+y} = (x-1) \, dx. \tag{3.9}$$

We can integrate both sides of (3.9) to obtain

$$\int \frac{dy}{1+y} = \int (x-1) \, dx \quad \Rightarrow \quad \ln|1+y| = \frac{1}{2}x^2 - x + C.$$

At this stage, we have just one constant of integration. We only need one because the constant from the left-hand side of the equation is combined with the constant from the right. Exponentiating both sides yields

$$|1+y| = e^C e^{\frac{1}{2}x^2 - x}.$$

Note that e^C is an abitrary positive constant (do you see why it is positive?). Removing the absolute values from the left side, we have

$$1 + y = ke^{\frac{1}{2}x^2 - x},$$

where k is an arbitrary nonzero constant (though in this case $k = 0$ will also give a solution to the DE). The general solution to our equation is thus

$$y = -1 + ke^{\frac{1}{2}x^2 - x}. \tag{3.10}$$

We call this the *general solution* of the differential equation since we have a family of functions with a parameter k, and all solutions are of this form. (You will prove this in your differential equations course.) ∎

Example 3.9 Consider again the DE

$$\frac{dy}{dx} = x + xy - y - 1.$$

If this DE is part of an IVP then we can find a particular solution by substituting the initial condition into our solution (3.10) to determine the constant k.

For example, if $y(2) = 5$, we substitute $x = 2$ and $y = 5$ into (3.10) and get $5 = -1 + k$ so that $k = 6$. Therefore, the solution to the IVP

$$\frac{dy}{dx} = x + xy - y - 1, \quad y(2) = 5$$

is

$$y = -1 + 6e^{\frac{1}{2}x^2 - x}. \qquad \blacksquare$$

Back to Example 3.6 We can use separation of variables to find the velocity for the skydiver in Example 3.6. To solve the IVP

$$\frac{dv}{dt} = 32 - 0.15v, \qquad v(0) = 0, \tag{3.11}$$

we separate variables and integrate to obtain

$$\int \frac{dv}{32 - 0.15v} = \int dt \qquad \Rightarrow \qquad -\frac{1}{0.15} \ln|32 - 0.15v| = t + C.$$

We can either substitute the initial condition to find C right now, or we can solve the equation first then substitute the initial condition. We will solve first to get

$$v = \frac{32}{0.15} - ke^{-0.15t}.$$

Substituting $v(0) = 0$, that is $t = 0$ and $v = 0$, we find that $k = 32/0.15$ so that the skydiver's velocity is given by

$$v(t) = \frac{32}{0.15} \left(1 - e^{-0.15t}\right).$$

Since $\lim_{t \to \infty} v(t) = 32/0.15$, this model predicts the skydiver's terminal velocity as approximately 213 feet per second or 145 miles per hour. $\blacksquare$

Example 3.10 In Figure 3.10 we see police investigating the death of a crime victim. One indicator of the time of death is body temperature. Assuming the victim was healthy with a body temperature of 98.6 degrees Fahrenheit, we may be able to use Newton's law of cooling (see Example 3.4) to estimate the time of death.

Suppose that the room temperature is a constant $75°$ Fahrenheit, and a body with temperature $80°$ is discovered at midnight. Also suppose that 1 hour later the temperature of the body is $77°$. Let $F(t)$ be the temperature of the body t hours after midnight. We seek the time t_d at which the body temperature was last $98.6°$.

Time	t	Body Temperature
time of death	t_d	98.6° F
midnight	0	80° F
1 am	1	77° F

Figure 3.10. From a 1905 issue of the French newspaper *Le Petit Journal*

Using Newton's law of cooling, we have the IVP

$$\frac{dF}{dt} = k(75 - F) \quad F(0) = 80, F(1) = 77. \tag{3.12}$$

This IVP has two initial conditions, but they are needed since we must find the values of two constants: the unknown constant k, and the constant of integration, C.

The DE is separable and in Exercise 18 you will show that its general solution is

$$F(t) = 75 - Ce^{kt}.$$

To determine C and k, we substitute the initial conditions to arrive at the system of equations

$$80 = 75 - C$$
$$77 = 75 - Ce^{k}.$$

The first equation tells us $C = -5$, so the second equation becomes $77 = 75 + 5e^{k}$. Thus, $k = \ln(0.4)$. The solution to (3.12) is $F(t) = 75 + 5e^{\ln(0.4)t} = 75 + 5(0.4)^{t}$.

Using a calculator to solve $75 + 5(0.4)^{t_d} = 98.6$, we find $t_d \approx -1.7$. That is, the time of death was approximately 1 hour 42 minutes before midnight, or 10:18 pm ∎

Example 3.11 Consider the initial value problem

$$\frac{dy}{dt} = \frac{ty}{\sqrt{1 - t^2}}, \quad y(0) = e. \tag{3.13}$$

As long as $-1 < t < 1$ the derivative $\frac{dy}{dx}$ exists. The equation is separable, and we can write

$$\int \frac{1}{y}\, dy = \int \frac{t}{\sqrt{1-t^2}}\, dt. \tag{3.14}$$

The integrand on the left side of (3.14) is the derivative of $\ln|y|$. The integral on the right requires a substitution.

Let $u = 1 - t^2$ so that $du = -2t\, dt$ or $-\dfrac{1}{2}\, du = t\, dt$. Then

$$\int \frac{t}{\sqrt{1-t^2}}\, dt = -\frac{1}{2}\int \frac{1}{\sqrt{u}}\, du = -\sqrt{u} + C = -\sqrt{1-t^2} + C.$$

Therefore,

$$\ln|y| = -\sqrt{1-t^2} + C. \tag{3.15}$$

Since we are not interested in the general solution to the differential equation, but rather the solution to an initial value problem, we can determine the constant C by using the initial condition before solving for y. Substituting $t = 0$ and $y = e$, we see

$$1 = \ln e = -\sqrt{1} + C,$$

so $C = 2$. Using this and (3.15), we get $y(t) = e^2 e^{-\sqrt{1-t^2}}$ or $y = e^{2-\sqrt{1-t^2}}$. ∎

3.4.1 Exercises

In Exercises 1–12, use separation of variables to find the general solution.

1. $\dfrac{dy}{dx} = \dfrac{y}{x}$
        2. $\dfrac{dy}{dx} = -\dfrac{x}{y}$
        3. $\dfrac{dy}{dx} = \dfrac{x}{y}$

4. $\dfrac{dy}{dx} = -\dfrac{y}{x}$
        5. $\dfrac{dy}{dx} = x^2 y$
        6. $\dfrac{dy}{dx} = \dfrac{\ln x}{\sin y}$

7. $\dfrac{dy}{dx} = x + xy$
        8. $\dfrac{dy}{dx} = -xy$
        9. $\dfrac{dy}{dx} = \dfrac{x^2}{\cos y}$

10. $\dfrac{dy}{dx} = \dfrac{xy}{\sqrt{1+x^2}}$
        11. $\dfrac{dy}{dx} = \dfrac{1}{y\sqrt{1-x^2}}$
        12. $\dfrac{dy}{dx} = y\ln(x)$

In Exercises 13–17, use separation of variables to solve each IVP.

13. $\dfrac{dy}{dx} = 2y$, $y(1) = 3$ 14. $y' + y\tan x = 0$, $y(0) = 1$ 15. $Q' + \frac{1}{2}Q = 100$, $Q(0) = 0$

16. $y'(x) = xy^2 - y^2 + xy - y$, $y(2) = 1$ 17. $x(1-x)y'(x) = 1 - x - y + xy$, $y(2) = 5$

18. Use separation of variables to show that the DE from Example 3.10, $F' = k(75 - F)$, has solution $F = 75 - Ce^{kt}$.

3.5 First-order linear equations and integrating factors

Many differential equations are not separable. Example 3.7, $y' = x + y$, is not separable, but we can solve it with a clever use of the product rule. First, we rewrite the equation in the form

$$y' - y = x. \tag{3.16}$$

Now we multiply both sides of (3.16) by e^{-x} and obtain

$$e^{-x} y' - e^{-x} y = x e^{-x}.$$

While it may seem we've made the equation more complicated, the left side is now the derivative of a product:

$$e^{-x} y' - e^{-x} y = \frac{d}{dx} \left(e^{-x} y \right).$$

So, we rewrite (3.16) as

$$\frac{d}{dx} \left(e^{-x} y \right) = x e^{-x}.$$

We integrate both sides with respect to x using integration by parts on the right to obtain

$$e^{-x} y = \int x e^{-x} \, dx = -x e^{-x} - e^{-x} + C.$$

So, the general solution to (3.16) is $y = -x - 1 + Ce^x$. (You should check that such functions satisfy the DE.) In the following section, we develop a general procedure based on what we've just done.

3.5.1 Integrating factors

Consider a differential equation of the form

$$y'(x) + r(x)y(x) = s(x) \tag{3.17}$$

where $r(x)$ and $s(x)$ are known functions, and the unknown function is $y(x)$. For example, in (3.16) we have $r(x) = -1$ and $s(x) = x$. Equations of this form are known as *linear first-order differential equations*. While some linear equations are separable, others are not. However, we can still find the general solution to the equation by using our knowledge of the product rule and the exponential function.

Recall that the product rule is given by

$$\frac{d}{dx} (f(x)g(x)) = f'(x)g(x) + f(x)g'(x).$$

The left-hand side of (3.17) is very close in form to the right-hand side of the product rule. We can put it into this form exactly by multiplying both sides of the equation by an appropriate factor, $h(x)$, called the *integrating factor*. We select $h(x)$ so that

$$\frac{d}{dx} (h(x)y(x)) = h(x)y'(x) + h'(x)y(x) = h(x)y'(x) + h(x)r(x)y(x). \tag{3.18}$$

The key is to select h so that $h'(x) = h(x)r(x)$. It looks very much like h might be an exponential function. Indeed, if $h(x) = e^{R(x)}$ where $R'(x) = r(x)$, then using the chain rule, we have

$$h'(x) = \frac{d}{dx}e^{R(x)} = e^{R(x)}R'(x) = h(x)r(x).$$

Multiplying both sides of (3.17) by $h(x) = e^{R(x)}$, we see

$$h(x)y'(x) + h(x)r(x)y(x) = h(x)s(x)$$
$$\frac{d}{dx}(h(x)y(x)) = h(x)s(x).$$

At this stage, we antidifferentiate both sides and solve for $y(x)$.

Integrating Factor Method
Given a linear first-order differential equation in the form

$$y'(x) + r(x)y(x) = s(x)$$

multiply both sides by the *integrating factor* $h(x) = e^{R(x)}$ where $R'(x) = r(x)$ to obtain

$$\frac{d}{dx}\left(e^{R(x)}y(x)\right) = e^{R(x)}s(x).$$

Integrate both sides of this equation and solve for $y(x)$ to solve the differential equation.

Example 3.12 Consider the DE $y' + \frac{1}{x}y = \cos x$ with $x > 0$. This is a linear DE in y, with $r(x) = \frac{1}{x}$. Setting $R(x) = \ln x$, we find that the integrating factor is $e^{\ln x} = x$. Multiplying both sides by this integrating factor, we have $xy' + y = x\cos x$, so that

$$\frac{d}{dx}(xy) = x\cos x \quad \text{or} \quad xy = \int x\cos x\, dx.$$

We apply integration by parts to the integral to obtain

$$xy = x\sin x + \cos x + C,$$

and so the general solution to the DE is given by $y = \sin x + \frac{1}{x}\cos x + \frac{C}{x}$. As always, you should check that this is indeed the general solution to the DE. ∎

Newton's law of cooling with non-constant environmental temperature
Recall that the DE $\frac{dF}{dt} = k(A - F)$ models the temperature $F(t)$ of an object in an environment of constant temperature A (see Example 3.4). This equation is separable and can be solved just as in Example 3.6. But what if the environmental temperature is changing over time? That is, the temperature is given by a function $A(t)$ that is not necessarily constant. The equation then becomes

$$F'(t) = k(A(t) - F(t)) \quad \text{or} \quad F'(t) + kF(t) = kA(t), \tag{3.19}$$

a linear equation solvable by the integrating factor method.

Back to Example 3.10 We return to considering the time of death, t_d, of a body discovered at midnight with a body temperature of $80°$ F. Suppose that the air temperature in which the body was discovered was not a constant 75° F, but rather decreased at a constant rate from 85° at 7 pm to 75° at midnight. Our information is summarized in the following table.

Time	t	Body Temperature	Air Temperature
7 pm	-5		85° F
time of death	t_d	$98.6°\,F$	
midnight	0	80°	75°
1 am	1	77°	

Since we assumed the air temperature decreased at a constant rate, we can model it by a linear function $A(t)$ whose graph has slope $-10/5 = -2$ and that passes through the point $(75, 0)$. That is, $A(t) = 75 - 2t$. Thus, we model the temperature of the body with the IVP

$$F' + kF = k(75 - 2t), \qquad F(0) = 80, F(1) = 77. \tag{3.20}$$

To solve (3.20), we use the integrating factor method. Here we have $r(t) = k$ so that $R(t) = kt$. Thus, the integrating factor is e^{kt}. Multiplying through by e^{kt} transforms the equation into

$$\frac{d}{dt}\left(e^{kt}F(t)\right) = e^{kt}k(75 - 2t) = 75ke^{kt} - 2kte^{kt}.$$

Therefore,

$$e^{kt}F(t) = \int \left(75ke^{kt} - 2kte^{kt}\right) dt.$$

Using integration by parts, we obtain

$$e^{kt}F(t) = 75e^{kt} - 2te^{kt} + \frac{2}{k}e^{kt} + C$$

$$F(t) = 75 - 2t + \frac{2}{k} + Ce^{-kt}.$$

To determine the values of the constants k and C, we use the initial conditions $F(0) = 80$ and $F(1) = 77$:

$$F(0) = 80 = 75 + \frac{2}{k} + C, \qquad\qquad F(1) = 77 = 73 + \frac{2}{k} + Ce^{-k}.$$

Solving for C in the first equation and substituting into the second gives us

$$4 = \frac{2}{k} + \left(5 - \frac{2}{k}\right)e^{-k}.$$

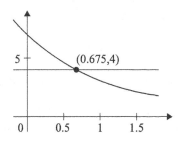

Figure 3.11. Solving $4 = \frac{2}{k} + \left(5 - \frac{2}{k}\right)e^{-k}$

Using the graphs of $y = 4$ and $y = (2/x) + (5 - (2/x))e^{-x}$ (see Figure 3.11), we estimate the solution of this equation to be $k \approx 0.675$ and so $C \approx 2.037$. Thus, the temperature of the body from the time of death to midnight, with $t = 0$ corresponding to midnight, is given by

$$F(t) \approx 77.963 - 2t - 2.037e^{-0.675t}.$$

To determine the time of death, we solve the equation $F(t) = 98.6$ and obtain $t \approx -2.934$. We estimate the time of death at approximately 9:04 pm. ■

3.5.2 Exercises

For Exercises 1–6, use the method of integrating factors to find the general solutions of the DE.

1. $y' + \dfrac{y}{x} = \cos x$ 2. $y' - 2xy = x$ 3. $xy' + y = \dfrac{x}{1 + x^2}$

4. $\dfrac{dy}{dx} - 2xy = e^{x^2}$ 5. $\dfrac{dy}{dx} + \dfrac{y}{x} = e^{x^2}$ 6. $xy' + \dfrac{y}{\ln x} = 1$

For Exercises 7–11, use the method of integrating factors to solve the IVP.

7. $y' + y = 1$, $y(0) = -3$ 8. $y' + xy = e^{-x^2/2}$, $y(0) = 1$ 9. $xy' - y = 2x^3$, $y(1) = 4$

10. $y' \sec x = y + \sin x \cos x$, $y(1) = 0$ 11. $y'\sqrt{1 - x^2} + y = 1$, $y(0) = 3$

12. Suppose that in Example 3.10, air temperature is more accurately given by the function $A(t) = 5 \sin\left(\frac{\pi}{6}(t - 4)\right) + 80$. Determine the time of death of the body discovered at midnight.

3.6 Explorations

3.6.1 A differential equation modeling weight loss

Our weight depends upon many factors including genetic disposition, exercise, and diet. In this exploration, you will investigate a model for weight loss involving diet and exercise.

To maintain current weight, an average adult male must consume approximately 15 calories per pound per day. Letting w be weight in pounds, we have consumption of $15w$ calories per day needed to maintain weight w. Consuming less than this will result in weight loss.

It is estimated that an individual must consume a total of 3500 calories fewer than normal over a period of time to lose one pound. With this estimate and under the assumption that the rate of change in weight is proportional to the difference between the number of calories required to maintain weight and the actual calories consumed, we have that

$$\frac{dw}{dt} = 3500(C - 15w) \tag{3.21}$$

where C is the calories consumed per day and and t is time in days.

Exercise 1. Using (3.21), without solving for w, explain how the differential equation implies that an individual will lose weight consuming fewer than 15 calories per pound per day.

Exercise 2. Use separation of variables to solve for w.

Exericse 3. Suppose an individual weighs 170 pounds and cuts caloric intake to 2200 calories per day. According to your solution, how long will it take for the individual to lose 10 pounds? What would the calorie intake per day need to be for this individual to lose 10 pounds in 60 days?

An adult male who weighs 180 pounds decides to cut his calorie intake to 2200 calories per day. At the same time, he begins jogging, starting at 1 mile per day and working up to 10 miles per day by the end of 6 months (180 days).

Exercise 4. Suppose the adult male begins, at $t = 0$ days, running 1 mile and increases his distance linearly over the 180 days to reach 10 miles per day. Write an expression for the distance he runs each day in terms of t.

Exercise 5. A fairly accurate estimate of the number of calories burned jogging is $0.73 \cdot w \cdot$ (miles run). Using your expression from Exercise 4, write an expression that gives the number of calories burned per day at time t for this adult male.

Exercise 6. With this additional number of calories burned, (3.21) becomes

$$\frac{dw}{dt} = 3500 \left(0.73w \left(1 + \frac{t}{20} \right) - 15w \right).$$

This equation is also separable. Use separation of variables to solve it, remembering that $w(0) = 180$ pounds.

3.6.2 A differential equation modeling a mixture

Differential equations can be used to model the amount or concentration of a substance in a mixture when there is material entering or exiting the mixture.

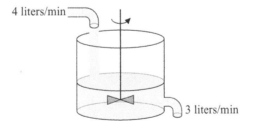

Figure 3.12. A tank of brine

Let's consider a specific example. Suppose that a brine containing 100 grams of salt per liter flows into a tank initially filled with 500 liters of water containing 12,000 grams of salt. The brine enters the tank at a rate of 4 liters minute, and the mixture is flowing out of the tank at the rate of 3 liters per minute. To determine the amount (mass) of salt in the tank at any given time, we construct a differential equations model.

Let $A(t)$ be the mass in grams of salt in the tank at time t minutes.

Exercise 1. Given the concentration of salt in the brine and the rate at which brine enters the tank, what is the rate at which salt is entering the tank?

Exercise 2. As the brine enters the tank, we assume that the mixture in the tank mixes instantly. Use the information to determine the concentration (in g/L) at time f of salt in the tank in terms of $A(t)$.

Exericse 3. The mixture in the tank flows out at 3 liters per minute. Given this information and the answers to Exercises 1 and 2, determine $A'(t)$, the rate of change of salt in the tank in g/min at time t minutes.

You should have arrived at the equation

$$A'(t) = 400 - \frac{A(t)}{500 + t} \cdot 3$$

which we rewrite as

$$A'(t) + \frac{3}{500 + t} A(t) = 400. \tag{3.22}$$

Exercise 4. Use the method of integrating factors to find the general solution of (3.22), and use the initial condition $A(0) = 12,000$ to find $A(t)$.

3.6.3 A more accurate numerical method for IVPs.

In Section 3.3 we studied Euler's method for approximating solutions to a IVP of the form

$$y'(x) = f(x, y), \quad y_0 = y(x_0). \tag{3.23}$$

Using Euler's method to estimate values of the solution, $y(x)$, we begin at the initial condition point (x_0, y_0) and move along a line segment of slope $f(x_0, y_0)$ to a new point $f(x_0 + \Delta x, y_0 + \Delta x \cdot f(x_0, y_0))$. This is repeated using the new point as the new initial condition point. The result is a collection of points that are approximations for the values of the solution at specific x values.

In this exploration, we introduce another method, related to Euler's, that generally produces better results. The method is known as the *fourth-order Runge-Kutta* method. We will abbreviate it $RK4$.

Let's proceed with a specific example:

$$\frac{dy}{dx} = y'(x) = y - 2x + 3, \quad y(1) = 2. \tag{3.24}$$

We will use the stepsize $\Delta x = 0.2$ and approximate $y(1.2)$.

Exercise 1. Use one step of Euler's method to approximate $y(1.2)$.

Exercise 2. Explain why $y(1.2) = 2 + \int_1^{1.2} y'(x)\, dx$. *Do not attempt to determine the integral.*

To determine the value of the integral in Exercise 2, we would need an expression for $y'(x)$ in terms of just x. But our expression for $y'(x)$ involves both x and y. So we will proceed by approximating the integral.

Exercise 3. Using one left endpoint rectangle to approximate the integral, approximate the value of $y(1.2)$. How does your answer compare with the estimate you obtained using Euler's method? Explain why.

In Chapter 6, we will learn a more robust method for approximating integrals, Simpson's rule. Using this rule in its simplest form, we can approximate the integral with

$$\int_1^{1.2} y'(x)\,dx \approx \frac{0.2}{6}\left(y'(1) + 4y'(1.1) + y'(1.2)\right). \tag{3.25}$$

We can determine $y'(1) = f(1,2) = 3$. Alas, since our expression for y' involves both x and y values, we have no way to exactly compute $y'(1.1)$ and $y'(1.2)$. So, we must approximate them. In doing so, we will average two approximations of $y'(1.1)$. That is, we find two approximations of y(1.1) and then compute $y'(1.1)$ at each of these points and average these values.

Exercise 4. Using $y' = y - 2x + 3$, $\quad y(1) = 2$ approximate $y(1.1)$ with one step of Euler's method ($\Delta x = 0.1$). Use your answer to approximate $y'(1.1)$. Label your answer A_2.

Exercise 5. Now, redo your estimate of $y(1.1)$ by using A_2 as the slope of the line segment in Euler's method (rather than $y'(1)$). Use this new estimate of $y(1.1)$ to approximate $y'(1.1)$ at this new y value. Label your answer A_3 and determine the average of A_2 and A_3.

Exercise 6. Finally, we approximate $y'(1.2)$ using the point obtained by one step of Euler's method (to the end of the interval using $\Delta x = 2$), but using A_3 as our slope rather than $y'(1)$. Approximate $y'(1.2)$ using this method and label your answer A_4.

We now compute the estimate of $y(1.2)$ using (3.25).

Exercise 7. Approximate $\int_1^{1.2} y'(x)\,dx$ using the values obtained in Exercises 5 and 6. Use your approximation to estimate $y(1.2)$ (see Exercise 2).

Exercise 8. (3.24) is a first-order linear differential equation. Using the method of integrating factors, solve this IVP and compute $y(1.2)$ exactly. How many digits of accuracy did Euler's method (from Exercise 1) have? How many digits of accuracy did the new estimate from Exercise 7 have?

The general fourth-order Runge Kutta method
The general method contained in this example is called the fourth-order Runge Kutta method, abbreviated RK4. Given an IVP of the form $y' = f(x, y)$, $y(x_0) = y_0$, we fix Δx, and determine $y_1 = y(x_0 + \Delta x)$ by the following algorithm:

$$A_1 = f(x_0, y_0)$$
$$A_2 = f\left(x_0 + \frac{\Delta x}{2}, y_0 + \frac{\Delta x}{2}A_1\right)$$
$$A_3 = f\left(x_0 + \frac{\Delta x}{2}, y_0 + \frac{\Delta x}{2}A_2\right)$$
$$A_4 = f\left(x_0 + \Delta x, y_0 + \Delta x \cdot A_3\right)$$
$$y_1 \approx y_0 + \frac{\Delta x}{6}\left(A_1 + 4\frac{A_2 + A_3}{2} + A_4\right).$$

Exercise 9. Consider the IVP from Exercise 5 in Section 3.3.1:

$$\frac{dy}{dx} = x \cdot (y - 1), \ y(0) = -3.$$

Use the fourth-order Runge Kutta method with $\Delta x = 0.25$ to approximate $y(1)$. Compare your results with Exercise 5.

3.7 Acknowledgments

1. Example 3.5 is adapted from X. P. Xu, The snowplow problem revisited, *College Mathematics Journal*, **22** (1991), p. 139.
2. Exploration 3.6.1 is adapted from A. C. Segal, A linear diet model, *College Mathematics Journal*, **18** (1987), pp. 44–45.

4

The Logistic Model, Partial Fractions, and Least Squares

Anyone who believes exponential growth can go on forever in a finite world is either a madman or an economist.

Kenneth Boulding

One model for the growth of a population—of bacteria, fish, rabbits, or even humans—is the *exponential growth model*, which you encountered in your PCE: If $P(t)$ denotes the size of a population at time t, then

$$P(t) = P_0 e^{kt},$$

where $P_0 = P(0)$ is the initial population and $k > 0$ is the growth constant. This model was derived from the differential equation

$$\frac{dP}{dt} = kP, \tag{4.1}$$

i.e., *the rate of growth of the population is proportional to the population size*. When k is negative, (4.1) yields an exponential decay model.

Is an exponential growth model appropriate for all populations? In this chapter we will learn that the answer is definitely *no*, and so we will develop the *logistic model*, a more sophisticated model for describing certain populations. In the process we develop an integration technique called *partial fractions*, and learn a method for fitting a straight line to a data set called *least squares*.

4.1 The logistic model for the population of the United States

Article 1, Section 2 (as modified by the Fourteenth Amendment) in the Constitution of the United States reads

> Representatives and direct Taxes shall be apportioned among the several States which may be included within this Union, according to their respective Numbers. The actual Enumeration shall be made within three Years after the first Meeting of the Congress

of the United States, and within every subsequent Term of ten Years, in such Manner as they shall by Law direct.

This enumeration is known as the *Decennial Census*, and the table below gives the population of the United States every ten years since 1900:

Table 4.1. U.S. Population (in millions), 1900–2010

1900	76.303	1940	131.669	1980	226.542
1910	91.972	1950	151.326	1990	248.718
1920	105.711	1960	179.323	2000	281.422
1930	122.775	1970	203.302	2010	308.746

The census is used for more than the apportionment of the House of Representatives. Census data are used to make decisions about Federal spending on education, hospitals, housing, highways, and many more programs. So it is important to forecast population trends accurately to enable the government to plan for future revenue and spending.

Before applying the exponential model (or any model, for that matter), we must check to see if the assumptions from which the model was derived are reasonable. Dividing both sides of (4.1) by P yields

$$\frac{1}{P}\frac{dP}{dt} = k. \tag{4.2}$$

The term to the left of the equals sign in (4.2) is called the *relative growth rate*, its units are "people per year per person," and is often expressed as a percent (i.e., people per year per 100 persons). The exponential growth model assumes a constant relative growth rate k. Are the census data in Table 4.1 consistent with the assumption that the relative growth rate is constant?

To answer, we need to estimate dP/dt from the data in Table 4.1. Geometrically, $dP/dt = P'(t)$ is the slope of the tangent line to the graph of $y = P(t)$ at each year, and we will approximate it by the slope of the secant line between two points, one ten years earlier and one ten years later, e.g., $P'(1910) \approx [P(1920) - P(1900)]/20$ (for $P'(1900)$ we use the 1890 population 62.980; while for $P'(2010)$ we use the slope of the secant line over the interval $[2000, 2010]$.) Dividing by P yields the data in Table 4.2

Table 4.2. Estimated annual relative growth rates, 1900–2010

1900	0.018998	1940	0.010842	1980	0.010024
1910	0.015987	1950	0.015745	1990	0.011033
1920	0.014569	1960	0.014492	2000	0.010665
1930	0.010571	1970	0.011613	2010	0.008850

Obviously the relative growth rate is not constant, but is it approximately constant for this time period? In Figure 4.1 we plot the relative growth rate as a function of the population size P.

Several events in the history of the United States during the 20th century are reflected in Figure 4.1: The Great Depression (the 1930s), the post-war births of the baby boomers (1946–1964), and even the echo boom (children of the boomers, born 1982–1995). But clearly the

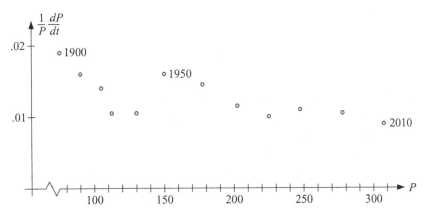

Figure 4.1. Relative growth rate as a function of population size

relative growth rate is not even approximately constant, so we will not use the exponential growth model.

Observe, however, that the relative growth rate from 1950 to the present is approximately a linear function of population size with a negative slope. Perhaps a better model might be one where the relative growth rate is given by

$$\frac{1}{P} \cdot \frac{dP}{dt} = (\text{slope})P + (\text{intercept}) = \left(-\frac{k}{L}\right)P + k = k\left(1 - \frac{P}{L}\right) \tag{4.3}$$

for appropriate positive constants k and L, that is, a linear function of P with intercept k and slope $-k/L$. A model based on the differential equation in (4.3) is called a *logistic model*. While this form of a linear equation may appear a bit strange, in the next section we will see that the constants k and L have a natural interpretation in terms of population growth.

Why the name *logistic*?
In the 1830s and 1840s the Belgian mathematician Pierre François Verhulst (1804–1849) published several papers on population growth models. He called one of the models—the one we are considering here—"la courbe logistique" (the logistic curve), for reasons that are not clear. The word "logistic" comes from the Greek λογος, meaning "speech," "reasoning," "discourse," but also "reckoning," "ratio," "proportion." Although the word is not very descriptive of the model, it is the name (along with the Verhulst model) most commonly used for this model.

Before solving (4.3) for P as a function of t, we make a few qualitative observations. First, we rewrite the equation in the form

$$\frac{dP}{dt} = kP\left(\frac{L - P}{L}\right) \tag{4.4}$$

and note that dP/dt is positive if and only if P lies between 0 and L. So P (which must be positive to be interesting) increases when $P < L$ (and decreases if $P > L$); consequently L is sometimes called the *limiting value* of the population (or the *carrying capacity* of the population). We will assume that $0 < P < L$, and so $dP/dt > 0$.

To solve (4.4) and in the process better understand the dynamics of the logistic model, we rewrite (4.4) in terms of the variable

$$R = \frac{L - P}{P}.$$

The quantity R, which we call the *logistic ratio*, is the ratio of the potential additional population $L - P$ to the current population P. We now show that in the logistic model the logistic ratio R decays exponentially, which enables us to solve (4.4).

Beginning with the chain rule, we have

$$\frac{dR}{dt} = \frac{dR}{dP}\frac{dP}{dt} = \frac{d}{dP}\left(\frac{L - P}{P}\right)\frac{dP}{dt} = -\frac{L}{P^2}\frac{dP}{dt},$$

and hence

$$\frac{dR}{dt} = -\frac{L}{P^2}\frac{dP}{dt} = -k\frac{L}{P}\left(\frac{L - P}{L}\right) = -k\left(\frac{L - P}{P}\right) = -kR.$$

The PCE solution to the exponential decay equation $dR/dt = -kR$ is $R = Ae^{-kt}$ for some positive constant A. Replacing R by $(L - P)/P$ and solving for P yields

$$P = P(t) = \frac{L}{1 + Ae^{-kt}} \tag{4.5}$$

as the general solution to (4.4). All that remains is to find values of the constants A, L, and k to fit the model to the population data.

Our method of solving (4.4) is not the one commonly found in other calculus texts, as the substitution $R = (L - P)/P$ is certainly not an obvious one. You may have been tempted to apply the separation of variables technique to (4.4), and that will work as well. Separating the variables in (4.4) yields

$$\frac{L}{P(L - P)}dP = k\,dt. \tag{4.6}$$

The integral of the right-hand side is easy, and yields $\int k\,dt = kt + C_1$. If the integrand in the left-hand side were $1/P$ or $1/(L - P)$, integration would yield a natural logarithm. But the denominator $P(L - P)$ in the fraction is the common denominator you would use when adding $1/P$ and $1/(L - P)$, and in fact we have

$$\frac{L}{P(L - P)} = \frac{1}{P} + \frac{1}{L - P}. \tag{4.7}$$

Hence

$$\int \frac{L}{P(L - P)}dP = \int \frac{1}{P} + \frac{1}{L - P}\,dP = \ln|P| - \ln|L - P| + C_2 = \ln\left|\frac{P}{L - P}\right| + C_2.$$

Since $0 < P < L$ we can remove the absolute value signs, then move the constant of integration C_2 to the right-hand side and set $C = C_1 - C_2$ to obtain

$$\ln\frac{P}{L - P} = kt + C.$$

Exponentiation now yields

$$\frac{P}{L-P} = e^{kt}e^{C}, \quad \text{or} \quad \frac{L-P}{P} = e^{-kt}e^{-C} = Ae^{-kt}$$

where $A = e^{-C}$ is a positive constant. Solving for P now yields (4.5). This technique—replacing a complex rational function by a sum of simpler ones, as we did in (4.7)—is called *partial fraction decomposition*, or simply *partial fractions*. See Section 4.2.

The constants L and k come from the assumption that the relative growth rate $(1/P)(dP/dt)$ is a linear function of the population size P as expressed in (4.3). So we only need to fit a line (in slope-intercept form) to the data in Tables 4.1 and 4.2 , and set the intercept equal to k and the slope equal to $-k/L$.

There are several ways to fit a line $y = mx + b$ to a set of n points $\{(x_1, y_1),$ $(x_2, y_2), \ldots, (x_n, y_n)\}$. When x is considered the independent variable and y the dependent variable, a common procedure, called *least squares*, is to find values for m and b to minimize the sum of the squared distances $[y_i - (mx_i + b)]^2$ between the y-values in the data set and the corresponding y-values on the line. This is easily done (see Section 4.4) and the result is

$$m = \frac{\sum_{i=1}^{n}(x_i - \bar{x})(y_i - \bar{y})}{\sum_{i=1}^{n}(x_i - \bar{x})^2} = \frac{\sum_{i=1}^{n}x_i y_i - n\bar{x}\bar{y}}{\sum_{i=1}^{n}x_i^2 - n\bar{x}^2} \quad \text{and} \quad b = \bar{y} - m\bar{x}$$

where $\bar{x}$ and $\bar{y}$ denote the *arithmetic means* of the x- and y-values in the data set, i.e., $\bar{x} = (1/n)\sum_{i=1}^{n}x_i$ and $\bar{y} = (1/n)\sum_{i=1}^{n}y_i$. Today m and b are usually computed on a calculator or with computer software. Doing so with the population data in Tables 4.1 and 4.2 for the years 1950–2010 yields

$$m = -0.00003970383 \quad \text{and} \quad b = 0.02084621$$

and the resulting line $y = mx + b$ is plotted in Figure 4.2, along with the seven data points.

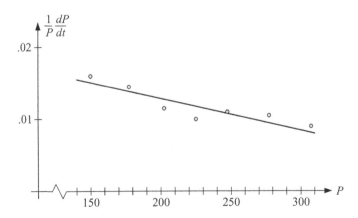

Figure 4.2. The least squares line for the population data, 1950–2010

Hence the constants k and L for the logistic model (4.5) are

$$k = b = 0.02084621 \quad \text{and} \quad L = -\frac{b}{m} = 525.0428.$$

Table 4.3. True versus logistic model U.S. population (in millions), 1950–2020

year	t	P(true)	P(logistic model)
1950	0	151.326	151.326
1960	10	179.323	174.728
1970	20	203.302	199.814
1980	30	226.542	226.177
1990	40	248.718	253.308
2000	50	281.422	280.638
2010	60	308.746	307.579
2020	70	?	333.576

To evaluate the remaining constant A in (4.5), we now let t denote time in years after 1950, and P_0 the initial population, i.e., $P_0 = P(0) = 151.326$, the population in 1950 (in millions). Then $P_0 = L/(1 + A)$, so that $A + (L - P_0)/P_0 \approx 2.46961$. Hence the logistic growth model for U.S. population for 1950–2010 is

$$P(t) = \frac{525.0428}{1 + 2.46961e^{-0.020846t}}. \tag{4.8}$$

How well does the model fit the data? In Table 4.3 and Figure 4.3 we compare the true population to the population predicted by (4.8).

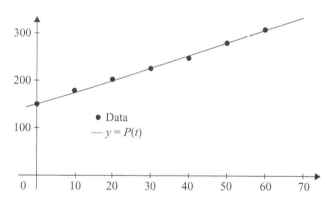

Figure 4.3. True versus logistic model of U.S. population

The model predicts a U.S. population of approximately 333.6 million in the 2020 decennial census. Such a prediction may be risky, as the model is not a perfect fit to the data. The relative growth rate in the U.S. population is not precisely linear, as in the basic assumption for the logistic model.

In Figure 4.4 we see the graph of a typical logistic growth function $y = P(t)$ where $P(t)$ is given by (4.5). In the model the population grows at an increasing rate for a while, followed by growth at a decreasing rate, with $y = L$ as a horizontal asymptote. The graph has an inflection point, which you will locate in Exercise 1.

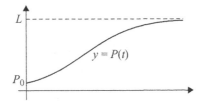

Figure 4.4. A graph of $y = P(t) = L/(1 + Ae^{-kt})$

Example 4.1 (Harvested logistic) A population of fish in a lake can be modeled using the so-called *harvested logistic equation*.

Figure 4.5. A population of fish (courtesy of Gary Halvorson, Oregon State Archives)

For a particular lake, this equation is given in (4.9). Here P is measured in thousands of fish and time t is measured in years. The first term on the right is identical to the standard form for logistic growth (see (4.4)) where the carrying capacity is 90,000 fish. The second term, -11.25, represents the harvesting of 11,250 fish per year.

$$\frac{dP}{dt} = 0.9P \left(\frac{90 - P}{90} \right) - 11.25. \tag{4.9}$$

The slope field in Figure 4.6 shows that there are two constant solutions to our DE. We can find them exactly by setting $\frac{dP}{dt} = 0$ and solving the resulting quadratic equation $P^2 - 90P + 1125 = 0$ to get $P(t) = 75$ and $P(t) = 15$.

Any solution to our DE that has initial value $P(0)$ above 15 (starting with more than 15,000 fish) will approach $P = 75$ over time. This says that our fish population will stabilize at 75,000 fish. However, if we start with a fish population of fewer than 15,000 fish, the slope field indicates that our harvesting 11,250 per year will destroy the population. Both solutions $P = 75$ and $P = 15$ are known as *equilibrium* solutions since when P is one of these values, P' is always 0 and hence P is constant, that is, the population is in equilibrium. The solutions differ in that solutions that at some time are near 75 will be drawn to 75 over time, while solutions that are near 15 at some time move away from 15. For this reason $P = 75$ is known as a *stable equilibrium solution* and $P = 15$ is called an *unstable equilibrium solution*. ■

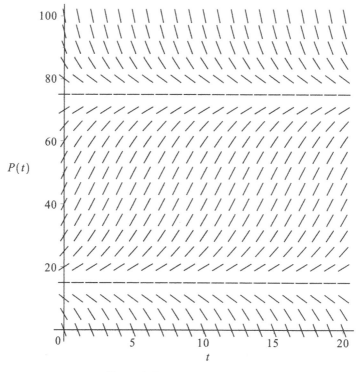

Figure 4.6. Slope field for (4.9)

4.1.1 Exercises

1. Suppose a population P is growing according to the logistic differential equation (4.3).
   (a) Show that the population is growing fastest at a time $t_{1/2}$ that satisfies $P(t_{1/2}) = L/2$, i.e., the time at which the population size is one-half its limiting value. (Hint: to maximize dP/dt, first find d^2P/dt^2.)
   (b) Express $t_{1/2}$ in terms of A and k.
   (c) Show that the graph of $y = P(t)$ has an inflection point at $(t_{1/2}, L/2)$.
   (d) Show that the logistic function (4.5) can be written in terms of $t_{1/2}$ as

$$P(t) = \frac{L}{1 + e^{-k(t-t_{1/2})}}.$$

2. *Rumors on campus.* Rumors on the ESU (The Enormous State University) campus spread quickly. In fact, they spread at a rate jointly proportional to the fraction x of the students who have heard the rumor and the fraction $1 - x$ of the students who have not yet heard the rumor.
   (a) Let t denote time in hours. Write a differential equation expressing dx/dt as a function of x, and solve it for x as a function of t.
   (b) At noon 5% of the students have heard the rumor, and by 1 pm 20% have heard the rumor. At what time will 95% have heard the rumor?

3. *Epidemics.* One model for the spread of a contagious disease such as the flu is that it spreads at a rate jointly proportional to the fraction x of the population who are infected and the fraction $1 - x$ who are not infected.

   (a) Let t denote time in weeks. Write a differential equation expressing dx/dt as a function of x, and solve it for x as a function of t.

   (b) In Springfield (population 50,000) 500 people have the flu at the beginning of the epidemic, and two weeks later 4000 are infected. How long after the outbreak will 10,000 be infected? When will 90% of the population be infected?

4. *Bacterial growth.* In a limited environment where 1 million bacteria is the maximum number the environment can support, the rate of bacteria growth is jointly proportional to the number present and the difference of 1 million and the number present.

   (a) If 500 bacteria are initially present and the rate of growth is 60 bacteria per minute when 1000 bacteria are present, express the number P of bacteria present as a function of the number t of minutes the bacteria have been growing. (Hint: let P be the number of bacteria in 1000s.)

   (b) How many bacteria are present after 30 minutes? 1 hour? 2 hours? 3 hours?

   (c) When is the rate of growth of bacteria the greatest?

   (d) Show that the maximum number of bacteria the environment can support is present within 7 hours.

5. The U.S. Census Bureau computes an annual growth rate $r(t) = \ln[P(t + 1)/P(t)]$. What assumptions about population growth lead to this formula? With the same assumptions, how would you calculate a ten-year growth rate?

6. Consider the general form of the harvested logistic equation from Example 4.1

$$\frac{dP}{dt} = kP\left(\frac{C - P}{C}\right) - H.$$

   (a) Show that the equilibrium solutions, if they exist, are given by

$$P(t) = \frac{C}{2}\left(1 \pm \sqrt{1 - \frac{4H}{Ck}}\right).$$

   (b) In this model, the *maximal sustainable harvest* is the largest value of H such that there is an equilibrium solution. Find the maximal sustainable harvest under this model. What happens to the population over time if H is larger than this value?

4.2 Integration by partial fractions

Consider the rational function $f(x)/g(x)$ where f and g are polynomials with no non-constant common factors and for which the degree of f is less than the degree of g. *Partial fractions* is a technique for decomposing $f(x)/g(x)$ into a sum of rational functions each of whose denominators is one of the factors of g. (These rational functions are the "partial fractions.").

   The standard method for expressing a rational function as a sum of partial fractions—called the method of undetermined coefficients—is algebra intensive and rather tedious. Hence we will replace some of that algebra with a method based on a technique from calculus, one based on finding limits.

We first consider the simple but common case where g is a product of distinct linear factors of the form $(x - a)$. For example, we write the rational function

$$\frac{4x}{2x^2 + 3x - 2} = \frac{4x}{(x + 2)(2x - 1)} \quad \text{as} \quad \frac{2x}{(x + 2)(x - 1/2)}.$$

In this case each partial fraction is a constant divided by one of the linear factors. Suppose $g(x) = (x - a_1)(x - a_2) \cdots (x - a_n)$, where $f(a_i) \neq 0$ for all i. Our task is to find real numbers $A_1, A_2, \ldots, A_n$ so that

$$\frac{f(x)}{g(x)} = \frac{A_1}{x - a_1} + \frac{A_2}{x - a_2} + \cdots + \frac{A_n}{x - a_n}. \tag{4.10}$$

The expression on the right in (4.10) is called the *partial fraction decomposition* of $f(x)/g(x)$. We will find A_1; the pattern for the other A_is is similar. Let $g_1(x) = g(x)/(x - a_1)$ so that $g(x) = (x - a_1)g_1(x)$, i.e., $g_1(x)$ is the product of all the factors of g except $(x - a_1)$. Hence $g_1(a_1) \neq 0$, and we can write $f(x)/g(x)$ as

$$\frac{f(x)}{g(x)} = \frac{A_1}{x - a_1} + \frac{p(x)}{g_1(x)}$$

for some polynomial p. We now multiply both sides by $x - a_1$ and take the limit as $x \to a_1$. On the left side we have

$$\lim_{x \to a_1} \frac{(x - a_1)f(x)}{g(x)} = \lim_{x \to a_1} \frac{f(x)}{g_1(x)} = \frac{f(a_1)}{g_1(a_1)}$$

while on the right side we have

$$\lim_{x \to a_1}(x - a_1)\left[\frac{A_1}{x - a_1} + \frac{p(x)}{g_1(x)}\right] = \lim_{x \to a_1}\left[A_1 + \frac{(x - a_1)p(x)}{g_1(x)}\right] = A_1 + 0 = A_1,$$

and hence $A_1 = f(a_1)/g_1(a_1)$. In words, to find A_1 drop the factor $(x - a_1)$ from g and evaluate the remaining rational function $f(x)/g_1(x)$ at a_1. Here is a simple example:

Example 4.2 Find the partial fraction decomposition of $\dfrac{2x}{(x - 1)(x - 2)(x - 3)}$.
Since $A_1 = \frac{2(1)}{(1-2)(1-3)} = 1$, $A_2 = \frac{2(2)}{(2-1)(2-3)} = -4$, and $A_3 = \frac{2(3)}{(3-1)(3-2)} = 3$, we have

$$\frac{2x}{(x - 1)(x - 2)(x - 3)} = \frac{1}{x - 1} - \frac{4}{x - 2} + \frac{3}{x - 3}. \blacksquare$$

☺ You can always check your work in a partial fraction decomposition by adding the partial fractions; you should obtain the original rational function.

We can now easily integrate the rational function in Example 4.2:

Example 4.3 Evaluate $\displaystyle\int \frac{2x}{(x-1)(x-2)(x-3)}\,dx$.

$$\int \frac{2x}{(x-1)(x-2)(x-3)}\,dx = \int \left(\frac{1}{x-1} - \frac{4}{x-2} + \frac{3}{x-3}\right)dx$$

$$= \ln|x-1| - 4\ln|x-2| + 3\ln|x-3| + C$$

$$= \ln\left|\frac{(x-1)(x-3)^3}{(x-2)^4}\right| + C. \ \blacksquare$$

Example 4.4 Find the partial fraction decomposition of $\dfrac{13}{6x^2 - 5x - 6}$.

Since $6x^2 - 5x - 6 = (2x-3)(3x+2)$ we have

$$\frac{13}{6x^2 - 5x - 6} = \frac{13}{(2x-3)(3x+2)} = \frac{13/6}{(x-3/2)(x+2/3)} = \frac{A_1}{x-3/2} + \frac{A_2}{x+2/3}.$$

Thus $A_1 = \frac{13/6}{3/2+2/3} = 1$, $A_2 = \frac{13/6}{-2/3-3/2} = -1$, and

$$\frac{13}{6x^2 - 5x - 6} = \frac{1}{x-3/2} - \frac{1}{x+2/3} = \frac{2}{2x-3} - \frac{3}{3x+2}. \ \blacksquare$$

Example 4.5 Evaluate $\displaystyle\int \frac{13}{6x^2 - 5x - 6}\,dx$.

From Example 4.4, we have

$$\int \frac{13}{6x^2 - 5x - 6}\,dx = \int \frac{2}{2x-3} - \frac{3}{3x+2}\,dx$$

$$= \ln|2x-3| - \ln|3x+2| + C = \ln\left|\frac{2x-3}{3x+2}\right| + C. \ \blacksquare$$

> ☺ Don't forget to check both the partial fraction decomposition (by algebra) *and* the integration (by differentiation)!

As a second case, suppose that g has a single irreducible quadratic factor in addition to the distinct linear factors (a quadratic is irreducible if it has no real roots). The partial fraction corresponding to the quadratic factor has a linear numerator. We illustrate with an example:

Example 4.6 Evaluate $\displaystyle\int \frac{4x^2}{x^4 - 1}\,dx$.

First we find the partial fraction decomposition of $4x^2/(x^4 - 1)$. Since the denominator $x^4 - 1$ factors into $(x-1)(x+1)(x^2+1)$, we write

$$\frac{4x^2}{x^4 - 1} = \frac{A_1}{x-1} + \frac{A_2}{x+1} + \frac{B_1 x + B_2}{x^2 + 1}.$$

We find A_1 and A_2 as before: $A_1 = \frac{4(1)^2}{(1+1)(1+1)} = 1$ and $A_2 = \frac{4(-1)^2}{(-1-1)(1+1)} = -1$, so that

$$\frac{4x^2}{x^4 - 1} = \frac{1}{x - 1} - \frac{1}{x + 1} + \frac{B_1 x + B_2}{x^2 + 1}. \tag{4.11}$$

To find B_1, multiply both sides of (4.11) by x and take the limit as $x \to \infty$: on the left we have $\lim\limits_{x \to \infty} \frac{4x^2}{x^4 - 1} = 0$, while on the right we have

$$\lim_{x \to \infty} \left[\frac{x}{x - 1} - \frac{x}{x + 1} + \frac{B_1 x^2 + B_2 x}{x^2 + 1} \right] = 1 - 1 + B_1 = B_1,$$

hence $B_1 = 0$. To find B_2, set $x = 0$ in (4.11): $0 = -1 - 1 + B_2$, and hence $B_2 = 2$ (if one of the roots of the denominator is 0, use any real number other than the roots of the distinct linear factors). Thus

$$\frac{4x^2}{x^4 - 1} = \frac{1}{x - 1} - \frac{1}{x + 1} + \frac{2}{x^2 + 1}$$

and so

$$\int \frac{4x^2}{x^4 - 1} \, dx = \int \frac{1}{x - 1} - \frac{1}{x + 1} + \frac{2}{x^2 + 1} \, dx = \ln \left| \frac{x - 1}{x + 1} \right| + 2 \arctan x + C. \blacksquare$$

☺ Don't forget about simpler techniques for integration! If the preceding example were $\int \frac{4x^3}{x^4 - 1} \, dx$, the substitution $u = x^4 - 1$ suffices and is *much easier* than using the partial fraction decomposition!

Example 4.7 Evaluate $\int \frac{7x + 3}{x^3 + x^2 - 2} \, dx$.

The denominator factors into $x - 1$ times the irreducible quadratic $x^2 + 2x + 2$, and we have

$$\frac{7x + 3}{x^3 + x^2 - 2} = \frac{A}{x - 1} + \frac{B_1 x + B_2}{x^2 + 2x + 2}$$

so $A = (7 + 3)/(1 + 2 + 2) = 2$, and thus

$$\frac{7x + 3}{x^3 + x^2 - 2} = \frac{2}{x - 1} + \frac{B_1 x + B_2}{x^2 + 2x + 2}. \tag{4.12}$$

Multiplying (4.12) by x and taking the limit of both sides as $x \to \infty$ yields $0 = 2 + B_1$, so that $B_1 = -2$. Setting $x = 0$ in (4.12) yields $-3/2 = -2 + B_2/2$ so that $B_2 = 1$. Hence

$$\frac{7x + 3}{x^3 + x^2 - 2} = \frac{2}{x - 1} + \frac{-2x + 1}{x^2 + 2x + 2} = \frac{2}{x - 1} - \frac{2x + 2}{x^2 + 2x + 2} + \frac{3}{(x + 1)^2 + 1}.$$

In the last step we decomposed the quadratic partial fraction into two partial fractions, one whose numerator is the derivative of the denominator (thinking ahead to the integration), and a second with a constant numerator in whose denominator we have completed the square (so that the substitution $u = x + 1$ will yield an arctangent integral). Hence

$$\int \frac{7x+3}{x^3+x^2-2}\,dx = \int \frac{2}{x-1} - \frac{2x+2}{x^2+2x+2} + \frac{3}{(x+1)^2+1}\,dx$$
$$= 2\ln|x-1| - \ln(x^2+2x+2) + 3\arctan(x+1) + C. \;\blacksquare$$

Example 4.8 Evaluate $\displaystyle\int_1^3 \frac{x^4+1}{x^3+x}\,dx$.

First we find the partial fraction decomposition of the integrand. Since the degree of the numerator is greater than the degree of the denominator, we first use long division to write

$$\frac{x^4+1}{x^3+1} = x - \frac{x^2-1}{x(x^2+1)} \quad\text{with}\quad \frac{x^2-1}{x(x^2+1)} = \frac{A}{x} + \frac{B_1 x + B_2}{x^2+1}.$$

As before, $A = \frac{0-1}{0+1} = -1$, so that

$$\frac{x^2-1}{x(x^2+1)} = \frac{-1}{x} + \frac{B_1 x + B_2}{x^2+1}. \tag{4.13}$$

Multiplying both sides by x and taking the limit as $x \to \infty$ yields $1 = -1 + B_1$ so that $B_1 = 2$. Setting $x = 1$ (rather than $x = 0$, since 0 is a root of the denominator) in (4.13) yields $0 = -1 + (B_1 + B_2)/2$ so that $B_2 = 0$ and we have

$$\frac{x^4+1}{x^3+x} = x + \frac{1}{x} - \frac{2x}{x^2+1}. \tag{4.14}$$

Thus

$$\int_1^3 \frac{x^4+1}{x^3+x}\,dx = \int_1^3 x + \frac{1}{x} - \frac{2x}{x^2+1}\,dx = \left[\frac{1}{2}x^2 + \ln\frac{x}{x^2+1}\right]_{x=1}^3 = 4 + \ln\frac{3}{5} \approx 3.489. \;\blacksquare$$

Example 4.9 *The Law of Mass Action.* In chemistry, the law of mass action is a mathematical model for elementary reactions describing the rate at which certain substances combine to form other substances. In its simplest form the model states that the rate at which substance P reacts with substance Q to form substance R is jointly proportional to the amounts of P and Q remaining at time t. Suppose p grams of P react with q grams of Q to form $p + q$ grams of R, and that initially (i.e., at $t = 0$) P_0 grams of P, Q_0 grams of Q, and 0 grams of R are present. If x denotes the amount (in grams) of R present at time t, then the amounts of P and Q consumed by time t are $px/(p+q)$ and $qx/(p+q)$ grams, respectively. Hence the rate dx/dt at which R is produced is given by

$$\frac{dx}{dt} = k\left(P_0 - \frac{px}{p+q}\right)\left(Q_0 - \frac{qx}{p+q}\right) = \frac{kpq}{(p+q)^2}\left(\frac{p+q}{p}P_0 - x\right)\left(\frac{p+q}{q}Q_0 - x\right).$$

Now replace the constant $kpq/(p+q)^2$ by simply k, and set $a = (p+q)P_0/p$ and $b = (p+q)Q_0/q$; then the differential equation is

$$\frac{dx}{dt} = k(a-x)(b-x), \quad\text{or}\quad \frac{1}{(a-x)(b-x)}\,dx = k\,dt.$$

We can solve the equation using integration with an ordinary substitution (when $a = b$) or partial fractions (when $a \neq b$). See Exercise 21 for a continuation of this example. $\blacksquare$

4.2.1 Partial fraction differentiation

You have seen a number of instances where a differentiation technique is useful in integration—the chain rule for differentiation leads to the substitution technique for integration, and the product rule for differentiation leads to integration by parts. Here is an example of the reverse, where an integration technique (partial fractions) is useful in differentiation.

Example 4.10 Find y''' if $y = \dfrac{2x}{x^3 - 6x^2 + 11x - 6}$.

Of course we can compute y''' using the quotient rule three times, but the algebra required for simplification after each step is rather messy (to say the least). But we have seen this function before. It is the function in Example 4.2 since $x^3 - 6x^2 + 11x - 6 = (x - 1)(x - 2)(x - 3)$. Using algebra to express y as a sum of partial fractions *before* differentiating will simplify the algebra required *after* each differentiation:

$$y = \frac{2x}{x^3 - 6x^2 + 11x - 6} = \frac{2x}{(x - 1)(x - 2)(x - 3)} = \frac{1}{x - 1} - \frac{4}{x - 2} + \frac{3}{x - 3}.$$

The partial fractions are now easily differentiated three times to yield

$$y' = \frac{-1}{(x - 1)^2} + \frac{4}{(x - 2)^2} - \frac{3}{(x - 3)^2},$$

$$y'' = \frac{2}{(x - 1)^3} - \frac{8}{(x - 2)^3} + \frac{6}{(x - 3)^3},$$

$$y''' = \frac{-6}{(x - 1)^4} + \frac{24}{(x - 2)^4} - \frac{18}{(x - 3)^4}. \ \blacksquare$$

To conclude this section, we note that it is possible to express *every* rational function as a sum of a polynomial and partial fractions each with a denominator in one of four forms: a linear function, a positive integer power of a linear function, an irreducible quadratic function, or a positive integer power of an irreducible quadratic function. Consequently, it is possible to integrate every rational function, and the antiderivative will be a sum or difference of at most four types of functions: polynomials, rational functions, logarithms, and arctangents. However, the algebra to find the partial fraction decomposition can be tedious, and we have not considered here all the possible forms of partial fractions. However, nearly every computer algebra system has been programmed to find the required decomposition and the resulting antiderivative.

4.2.2 Exercises

In Exercises 1 through 9, evaluate the indefinite integral. Be sure to check your work.

1. $\displaystyle\int \frac{dx}{x^2 - 4}$

2. $\displaystyle\int \frac{dx}{x^3 - 4x}$

3. $\displaystyle\int \frac{x^2 + 1}{x^3 - 1} \, dx$

4. $\displaystyle\int \frac{x^2 - 1}{x^3 + 1} \, dx$

5. $\displaystyle\int \frac{3x^2 - 2x - 1}{x^2 - 2x - 3} \, dx$

6. $\displaystyle\int \frac{dx}{x^3 - x^2 + x - 1}$

7. $\displaystyle\int \frac{3 \, dx}{2x^2 + 5x + 2}$

8. $\displaystyle\int \frac{6x^2}{x^4 - 5x^2 + 4} \, dx$

9. $\displaystyle\int \frac{3 \, dx}{x^3 - 1}$

10. Find the volume of the solid obtained by revolving about the y-axis the region bounded by the x-axis, the y-axis, $y = 4/(3 + 2x - x^2)$, and $x = 2$.

11. Derive the following useful formulas for finding the partial fraction decomposition of certain rational expressions: If u is a function of x and $a \neq b$, then

$$\frac{1}{(u-a)(u-b)} = \frac{1}{a-b}\left(\frac{1}{u-a} - \frac{1}{u-b}\right) \text{ and } \frac{u}{(u-a)(u-b)} = \frac{1}{a-b}\left(\frac{a}{u-a} - \frac{b}{u-b}\right).$$

Use Exercise 11 to evaluate the integrals in Exercises 12 through 14.

12. $\displaystyle\int \frac{dx}{(x^2+1)(x^2+4)}$

13. $\displaystyle\int \frac{x^2\,dx}{(x^2+1)(x^2+4)}$

14. $\displaystyle\int \frac{(x^2+x+1)\,dx}{(x^2+1)(x^2+4)}$

Evaluate the integrals in Exercises 15 through 18 using algebra, a u-substitution, or both to simplify the integrand before considering the use of partial fractions.

15. $\displaystyle\int \frac{x^3+2x^2+3x+4}{(x-1)^4}\,dx$

16. $\displaystyle\int \frac{x^2\,dx}{(x^2+4)^2}$

17. $\displaystyle\int \frac{dx}{x(x^3+1)}$

18. $\displaystyle\int \frac{dx}{(x^2+x)^2}$

19. *The many faces of $\int \sec x\,dx$.* In Example 1.9 in Section 1.2 we used integration by parts to show that $\int \sec x\,dx = \ln|\sec x + \tan x| + C$. Now use partial fractions to show that

$$\int \sec x\,dx = \frac{1}{2}\ln\left(\frac{1+\sin x}{1-\sin x}\right) + C$$

(Hint: substitute $u = \sin x$ to obtain $\int du/(1-u^2)$. Justify that absolute value bars are not required in the formula.) Are the two formulas for $\int \sec x\,dx$ equivalent?

20. Prove that $\pi < 22/7$. (Hint: consider $\displaystyle\int_0^1 \frac{x^4(1-x)^4}{1+x^2}\,dx$.)

21. Solve the differential equation in Example 4.9 for the $a \neq b$ case.

Use technology to evaluate the integrals in Exercises 22 through 24 and verify the results by differentiation.

22. $\displaystyle\int \frac{x^2}{(x^2+1)^3}\,dx$

23. $\displaystyle\int \frac{4x^2}{x^4+4}\,dx$

24. $\displaystyle\int \frac{x^2-2}{(x^3+x)^3}\,dx$

In Exercises 25 and 26 find y''' when

25. $y = \dfrac{3x^2+3x-12}{6x^3+5x^2-6x}$

26. $y = \dfrac{2x^2-11x-9}{x^3-2x^2-3x}$

27. Find a formula for the nth derivative of $y = (x^2-x-1)^{-1}$. (Hint: use the first identity in Exercise 11 above letting a and b represent the two roots of x^2-x-1, one of which is the golden ratio φ from Example 2.2.)

28. In Example 3.6, we used a model for velocity of a skydiver that assumed air resistance is proportional to velocity. Another model for air resistance is that the acceleration due to air resistance is proportional to the square of velocity. Thus, the DE governing the velocity of the skydiver becomes $\frac{dv}{dt} = 32 - cv^2$, where velocity v is in feet per second, time t is in

seconds, and the constant c has units 1/feet. Experiments show that a skydiver in a standard position reaches a terminal velocity of approximately 143 miles per hour (or approximately 210 feet per second).

(a) Without solving the DE, explain how we know $c \approx 7.256 \times 10^{-4}$.

(b) If $v(0) = 0$, use separation of variables technology to solve for $v(t)$. Plot your solution on the same axis as the solution to Example 3.6. Describe the differences in the solutions.

4.3 Substitutions to rationalize the integrand

The partial fractions technique can be extended to many other integrals by using a substitution to rationalize the integrand, that is, to express the integrand as a quotient of two polynomials. We illustrate the idea with some examples.

The first example concerns integrands with fractional powers of the variable x. In this case we use a substitution of the form $x = u^n$, where n is the least common multiple of the denominators of the exponents on x. In this example we will use partial fractions to integrate after the substitution.

Example 4.11 Evaluate $\displaystyle\int \frac{dx}{\sqrt{x}(1 - \sqrt[3]{x})}$.

The powers of x appearing in the integrand are $1/2$ and $1/3$, the least common multiple of 2 and 3 is 6, so we let $x = u^6$. Then $\sqrt{x} = u^3$, $\sqrt[3]{x} = u^2$, and $dx = 6u^5\, du$ so that

$$\int \frac{dx}{\sqrt{x}(1 - \sqrt[3]{x})} = \int \frac{6u^5\, du}{u^3(1 - u^2)} = \int \frac{6u^2}{1 - u^2} du = 6 \int \left(\frac{1}{1 - u^2} - 1\right) du$$

$$= 3 \int \left(\frac{1}{1 + u} + \frac{1}{1 - u} - 2\right) du = 3 \ln\left|\frac{1 + u}{1 - u}\right| - 6u + C$$

$$= 3 \ln\left|\frac{1 + x^{1/6}}{1 - x^{1/6}}\right| - 6x^{1/6} + C. \; \blacksquare$$

In our next example we substitute for a square root in the integrand and hope for the best.

Example 4.12 Evaluate $\displaystyle\int \frac{dx}{(x - 2)\sqrt{x + 2}}$.

If we set $u = \sqrt{x + 2}$ then $x - 2 = u^2 - 4$ and $dx = 2u\, du$ so that

$$\int \frac{dx}{(x - 2)\sqrt{x + 2}} = \int \frac{2u\, du}{(u^2 - 4)u} = \int \frac{2\, du}{(u - 2)(u + 2)} = \frac{1}{2} \int \left(\frac{1}{u - 2} - \frac{1}{u + 2}\right) du$$

$$= \frac{1}{2} \ln\left|\frac{u - 2}{u + 2}\right| + C = \frac{1}{2} \ln\left|\frac{\sqrt{x + 2} - 2}{\sqrt{x + 2} + 2}\right| + C.$$

Our final example is an area problem. The substitution to rationalize the integrand leads to an integral requiring a trigonometric substitution.

Example 4.13 Find the area of the region bounded by the graphs of $y = \sqrt{x/(1 - x)}$, $x = 1/2$, and the x-axis. See Figure 4.7.

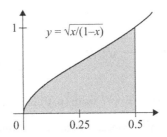

Figure 4.7. The area under the graph of $y = f(x) = \sqrt{x/(1-x)}$

From Figure 4.7, the area appears to be slightly more than $1/4$. If we let A denote the area, then

$$A = \int_0^{1/2} \sqrt{\frac{x}{1-x}} \, dx.$$

If we set $u = \sqrt{x/(1-x)}$, then $x = u^2/(1+u^2)$, $dx = (2u \, du)/(1+u^2)^2$, and u goes from 0 to 1 as x goes from 0 to $1/2$. Hence

$$A = \int_0^1 u \frac{2u}{(1+u^2)^2} \, du = 2 \int_0^1 \frac{u^2}{(1+u^2)^2} \, du.$$

This integral calls for the tangent substitution $\theta = \arctan u$, so that $u^2 = \tan^2 \theta$, $(1+u^2)^2 = \sec^4 \theta$, $du = \sec^2 \theta d\theta$, and θ goes from 0 to $\pi/4$ as u goes from 0 to 1. Thus

$$A = 2 \int_0^{\pi/4} \frac{\tan^2 \theta}{\sec^4 \theta} \sec^2 \theta \, d\theta = 2 \int_0^{\pi/4} \sin^2 \theta d\theta.$$

In Example 2.3 we learned that $\int \sin^2 \theta \, d\theta = (1/2)[\theta - \sin\theta \cos\theta] + C$, hence

$$A = \left[\theta - \sin\theta \cos\theta\right]_{\theta=0}^{\pi/4} = \frac{\pi}{4} - \frac{\sqrt{2}}{2} \cdot \frac{\sqrt{2}}{2} = \frac{\pi-2}{4} \approx 0.2854. \blacksquare$$

While in general there is no rule for finding a substitution that will simplify an integrand, in one instance there is: when the integrand is a rational expression in $\sin x$ and $\cos x$. See Exploration 4.5.2.

4.3.1 Exercises

In Exercises 1 through 6 evaluate the indefinite integrals.

1. $\displaystyle\int \frac{dx}{x - x^{1/3}}$ 2. $\displaystyle\int \frac{dx}{\sqrt{x} - \sqrt[3]{x}}$ 3. $\displaystyle\int \frac{\sqrt{x}\,dx}{1 + \sqrt[3]{x}}$

4. $\displaystyle\int \frac{dx}{x\sqrt{x} - 1}$ 5. $\displaystyle\int \frac{1 - \sqrt{x}}{x - x^{3/4}} \, dx$ 6. $\displaystyle\int \frac{x\,dx}{(1+x)^{2/3}}$

In Exercises 7 through 9, evaluate the definite integral.

7. $\displaystyle\int_{\sqrt{2}/2}^{1} \frac{x^3}{\sqrt{2-x^2}}\,dx$ 8. $\displaystyle\int_{0}^{9} \sqrt{3-\sqrt{x}}\,dx$ 9. $\displaystyle\int_{0}^{3} \frac{dx}{(x+2)\sqrt{x+1}}$

10. Find the arc length of the curve $y = x^{2/3}$ from $(0, 0)$ to $(1, 1)$.

11. Find the area of the surface of revolution obtained by rotating the semicubical parbola $y = x^{3/2}$ (see Example 2.1) between $(0, 0)$ and $(1, 1)$ about the y-axis.

4.4 The least squares equations

Given a set $\{(x_1, y_2), (x_2, y_2), \ldots, (x_n, y_n)\}$ of ordered pairs, we will find the line $y = mx + b$, commonly called the *least squares line*, that minimizes the sum

$$S(m, b) = \sum_{i=1}^{n} [y_i - (mx_i + b)]^2$$

of the squared distances between the y-values in the data set and the corresponding y-values on the line. To simplify the algebra, it will be convenient to let $u_i = x_i - \bar{x}$ and $v_i = y_i - \bar{y}$, where $\bar{x}$ and $\bar{y}$ denote the *arithmetic means* of the x- and y-values in the data set, i.e., $\bar{x} = (1/n)\sum_{i=1}^{n} x_i$ and $\bar{y} = (1/n)\sum_{i=1}^{n} y_i$. Observe that $\sum_{i=1}^{n} u_i = \sum_{i=1}^{n}(x_i - \bar{x}) = n\bar{x} - n\bar{x} = 0$, and similarly $\sum_{i=1}^{n} v_i = 0$. Thus we have

$$S(m, b) = \sum_{i=1}^{n} [(v_i + \bar{y}) - (mu_i + m\bar{x} + b)]^2 = \sum_{i=1}^{n} [(v_i - mu_i) + (\bar{y} - m\bar{x} - b)]^2$$

$$= \sum_{i=1}^{n} [(v_i - mu_i)^2 + 2(\bar{y} - m\bar{x} - b)(v_i - mu_i) + (\bar{y} - m\bar{x} - b)^2]$$

$$= \sum_{i=1}^{n} (v_i - mu_i)^2 + 2(\bar{y} - m\bar{x} - b)\sum_{i=1}^{n}(v_i - mu_i) + n(\bar{y} - m\bar{x} - b)^2$$

$$= \sum_{i=1}^{n} (v_i - mu_i)^2 + n(\bar{y} - m\bar{x} - b)^2.$$

To minimize $S(m, b)$ we now find the value of m for which the first term $s(m) = \sum_{i=1}^{n}(v_i - mu_i)^2$ in $S(m, b)$ is an absolute minimum, and then set $b = \bar{y} - m\bar{x}$ so that the second term $n(\bar{y} - m\bar{x} - b)^2 = 0$. Now

$$s(m) = \sum_{i=1}^{n}(v_i - mu_i)^2 = \sum_{i=1}^{n} v_i^2 - 2m\sum_{i=1}^{n} v_i u_i + m^2 \sum_{i=1}^{n} u_i^2$$

$$= Am^2 + Bm + C,$$

where $A = \sum_{i=1}^{n} u_i^2$, $B = -2\sum_{i=1}^{n} u_i v_i$, and $C = \sum_{i=1}^{n} v_i^2$. Since the graph of s (as a function of m) is a parabola opening upwards, its absolute minimum value occurs at $m = -B/(2A)$, and hence the slope and intercept of the least squares line are

$$m = \frac{-B}{2A} = \frac{\sum_{i=1}^{n} u_i v_i}{\sum_{i=1}^{n} u_i^2} = \frac{\sum_{i=1}^{n}(x_i - \bar{x})(y_i - \bar{y})}{\sum_{i=1}^{n}(x_i - \bar{x})^2} = \frac{\sum_{i=1}^{n} x_i y_i - n\bar{x}\bar{y}}{\sum_{i=1}^{n} x_i^2 - n\bar{x}^2} \quad \text{and} \quad b = \bar{y} - m\bar{x},$$

as claimed in Section 4.1. If you prefer to use differential calculus rather than analytic ge-ometry for the final step, notice that $s(m)$ has a single critical point at $m = -B/(2A)$ and $s''(m) = 2A > 0$ for all m, so that the absolute minimum value of s occurs at the critical point.

4.4.1 Exercises

1. (a) Show that the point $(\bar{x}, \bar{y})$ always lies on the least squares line.
   (b) When is the least squares line vertical?

2. For the data set $\{(x_1, y_1), (x_2, y_2), \ldots, (x_n, y_n)\}$, show that

$$\left[\sum_{i=1}^{n}(x_i - \bar{x})(y_i - \bar{y}) \right]^2 \le \sum_{i=1}^{n}(x - \bar{x})^2 \sum_{i=1}^{n}(y_i - \bar{y})^2.$$

(Hint: in the notation of this section, first show that $s(-B/2A) \ge 0$ implies $B^2 \le 4AC$.) This inequality is used in statistics to show that the sample correlation coefficient r_{xy} satisfies $-1 \le r_{xy} \le 1$.

3. Prove the *Cauchy-Schwarz inequality* for finite sums: If $\{(a_1, b_1), (a_2, b_2), \ldots, (a_n, b_n)\}$ is a set of ordered pairs of real numbers, then

$$\left(\sum_{i=1}^{n} a_i b_i \right)^2 \le \left(\sum_{i=1}^{n} a_i^2 \right) \left(\sum_{i=1}^{n} b_i^2 \right).$$

(Hint: apply the result of Exercise 2 to the set of $2n$ ordered pairs:

$$\{(x_1, y_1), (x_2, y_2), \ldots, (x_{2n}, y_{2n})\} =$$
$$\{(a_1, b_1), (a_2, b_2), \ldots, (a_n, b_n), (-a_1, -b_1), (-a_2, -b_2), \ldots, (-a_n, -b_n)\},$$

noting that $\bar{x} = \bar{y} = 0$.) You will see this inequality again in courses in multivariable calculus and linear algebra.

4. In some modeling situations we require y to be proportional to x. In such models we fit a line $y = mx$ to the given data set, that is, we require the line to pass through the origin. Find the least squares estimate of the slope m based on a data set $\{(x_1, y_1), (x_2, y_2), \ldots, (x_n, y_n)\}$.

5. Is there a relationship between straight-line distances and road distances between two points in a city? If we assume that the city is flat, the points are chosen randomly, and that roads are laid out in a rectangular grid, then the road distance should be about $4/\pi \approx 1.273$ times the straight-line distance (this is the average value of the function $f(\theta) = \sin\theta + \cos\theta$ on the interval $[0, \pi/2]$).

Table 4.4 gives the distance by road and the straight-line distance between twenty pairs of points in Sheffield, England. Do you think that the road distance can be predicted (approximately) from the straight-line (linear) distance?

Table 4.4. Distances in miles between twenty pairs of points in Sheffield, England

Linear	Road	Linear	Road	Linear	Road	Linear	Road
9.5	10.7	11.8	19.7	9.8	11.7	26.5	31.2
5.0	6.5	12.1	16.6	19.0	25.6	4.8	6.5
23.0	29.4	22.0	29.0	14.6	16.3	21.7	25.7
15.2	17.2	28.2	40.5	8.3	9.5	18.0	26.5
11.4	18.4	12.1	14.2	21.6	28.8	28.0	33.1

(a) Plot the data in Table 4.4 using the straight-line distance as the x-coordinate and the road distance as the y-coordinate. Does it appear that the road distance is roughly proportional to the linear distance?

(b) Use the result in Exercise 4 to estimate the constant of proportionality. Is it close to the value $4/\pi \approx 1.273$ in the model?

6. Find the least squares estimates for the parameters k and A in the exponential growth or decay model $y = Ae^{kt}$ based on the data set $\{(t_1, y_1), (t_2, y_2), \ldots, (t_n, y_n)\}$. (Hint: consider logarithms.)

7. Table 4.5 presents annual world oil crude oil production in millions of barrels (mbbl) from 1880 to 1984.

Table 4.5. Annual world crude oil production, 1880–1984

Year	Mbbl	Year	Mbbl	Year	Mbbl	Year	Mbbl
1880	30	1925	1069	1960	7674	1974	20389
1890	77	1930	1412	1962	8882	1976	20188
1900	149	1935	1655	1964	10310	1978	21922
1905	215	1940	2150	1966	12016	1980	21732
1910	328	1945	2595	1968	14104	1982	19403
1915	432	1950	3803	1970	16690	1984	19608
1920	689	1955	5626	1972	18584		

(a) Does it appear that the increase in annual world crude oil production follows an exponential pattern? Do the logarithms of annual world crude oil production follow a linear pattern? (Hint: plot the data.)

(b) Political turmoil in the oil-producing regions of the Middle East affected the pattern of oil production after 1973. Use the results of Exercise 6 above to fit an exponential growth model to the data from 1880 to 1972. In a statistics course you will learn how to judge how well the exponential model fits these data.

4.5 Explorations

4.5.1 World population growth

In Figure 4.8 we see a graph produced by the U.S. Census Bureau showing actual and projected world population for the 100-year period 1950–2050.

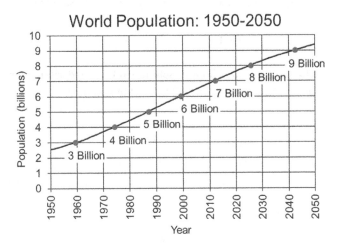

Figure 4.8. World population 1950–2050

Exercise 1. Does it appear that the Census Bureau might be using a logistic model?

Comparing Figures 4.4 and 4.8 suggests the answer might be yes. However, the Census Bureau actually uses a more sophisticated model, taking into account the age distribution of the population, not solely the total population size.

Exercise 2. Using the methodology in Section 4.1, compute the estimated annual relative growth rates for the years 1960–2010, and graph them as a function of population size. Does it appear that the relative growth rate is approximately a linear function of population size for the years 1960–2010?

Your graph should show that the six data points do indeed appear to be close to linear, and thus the logistic model might be appropriate for the data. Here are data for the world population for the years 1950–2010:

Table 4.6. World population (in billions), 1950–2010

1950	2.5565	1980	4.4529	2010	6.8525
1960	3.0424	1990	5.2890		
1970	3.7130	2000	6.0896		

Exercise 3. Find the least squares line that fits your data, and estimate the constants k and L in (4.5).

Depending on the number of decimals and the software used, your line should be approximately $y = -0.0022x + 0.02669$, so that $k \approx 0.02669$ and $L \approx 12.132$. Since L represents the carrying capacity, a logistic model based on the data predicts that an upper bound on world population is just over 12 billion people, less than twice the current population.

Exercise 4. Evaluate the constant A in (4.5), and use your model to predict the world population in 2020. When will the world population reach 9 billion?

    With t equal to the number of years since 1960, your model should be something like

$$P(t) = \frac{12.132}{1 + 2.998e^{-0.0267t}}.$$

So in 2020, $t = 60$ and $P(60) \approx 7.573$ billion. Setting $P(t) = 9$ and solving for t yields $t \approx 81$, so the model predicts the world population will reach 9 billion in about 2041, close to the year in the graph in Figure 4.8.

4.5.2 The method of last resort

How would you evaluate $\int dx/(3 \sin x - 4 \cos x)$? After realizing that trigonometric identities don't help much, you might try a computer algebra system or an online integration program.

Exercise 1. Use a computer algebra system or online integration site to evaluate

$$\int \frac{dx}{3 \sin x - 4 \cos x}.$$

Depending on the technology, you will obtain an answer something like

$$\frac{1}{5} \ln \left| \frac{2 \tan(x/2) - 1}{\tan(x/2) + 2} \right| + C \quad \text{or} \quad \frac{1}{5} \ln \left| \frac{2 \sin x - \cos x - 1}{\sin x + 2 \cos x + 2} \right| + C.$$

    In this Exploration you will discover a substitution technique used by the technology to evaluate the integral. It is designed to convert any rational expression in $\sin x$ and $\cos x$ into a rational function of another variable z, which you can then integrate using methods studied in this course. The appearance of $\tan(x/2)$ in one of the technology antiderivatives is the key to the substitution, which we state as a theorem for you to prove:

Theorem 4.1 *If an integrand is a rational expression in* $\sin x$ *and* $\cos x$, *then the substitution* $z = \tan(x/2)$ *yields an integrand that is a rational function of* z:

$$\sin x = \frac{2z}{1 + z^2}, \quad \cos x = \frac{1 - z^2}{1 + z^2}, \quad \text{and} \quad dx = \frac{2 \, dz}{1 + z^2}.$$

Before proving the theorem, here's an illustration for x in $(0, \pi/2)$:

Exercise 2. Do you see why $z = \tan(x/2)$ (in triangle OAB) implies $\sin x = 2z/(1 + z^2)$ and $\cos x = (1 - z^2)/(1 + z^2)$ (in triangle OCD)?

    While the picture in Figure 4.9 isn't a proof, it does suggest one: evaluate $\sin(x/2)$ and $\cos(x/2)$, and then use double-angle formulas to find $\sin x$ and $\cos x$. The proof consists of the next three exercises.

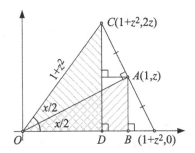

Figure 4.9. The $z = \tan(x/2)$ substitution

Exercise 3. Let $z = \tan(x/2)$ for x in $(-\pi, \pi)$. Show that

$$\cos(x/2) = \frac{1}{\sqrt{1+z^2}} \quad \text{and} \quad \sin(x/2) = \frac{z}{\sqrt{1+z^2}}.$$

Exercise 4. Show that

$$\sin x = \frac{2z}{1+z^2}, \quad \cos x = \frac{1-z^2}{z+z^2}, \quad \text{and} \quad \frac{\sin x}{1+\cos x} = z = \tan(x/2).$$

Exercise 5. Show that $dx = \frac{2\,dz}{1+z^2}$. (Hint: first show that $x = 2\arctan z$.)

Thus since $\sin x$ and $\cos x$ are rational functions of z, a rational expression in $\sin x$ and $\cos x$, times dx, will be a rational function of z times dz. We now return to the evaluation of $\int dx/(3\sin x - 4\cos x)$.

Exercise 6. Use the theorem and partial fractions to integrate and show that

$$\int \frac{dx}{3\sin x - 4\cos x} = \int \frac{dz}{(2z-1)(z+2)} = \frac{1}{5}\int \left(\frac{2}{2z-1} - \frac{1}{z+2} \right) = \frac{1}{5}\ln\left| \frac{2z-1}{z+2} \right| + C.$$

You can now undo the substitution to obtain the antiderivatives obtained using technology.

The reason we call this the method of last resort will become apparent in the next two exercises.

Exercise 7. Use the theorem and partial fractions to show that

$$\int \sec x \, dx = \ln\left| \frac{1 + \tan(x/2)}{1 - \tan(x/2)} \right| + C.$$

Is this result consistent with the forms for the antiderivative of the secant in (1.10) in Section 1.2 and Exercise 3 in Section 4.2?

Exercise 8. Evaluate $\displaystyle\int \frac{dx}{1 - \sin x}$.

We hope you did not use the substitution $z = \tan(x/2)$ in this exercise! Here trigonometric identities simplify the integrand:

$$\frac{1}{1 - \sin x} = \frac{1 + \sin x}{\cos^2 x} = \sec^2 x + \sec x \tan x,$$

and thus $\int dx/(1 - \sin x) = \tan x + \sec x + C.$

Here's one more for practice. (In its solution you will use a trigonometric substitution rather than partial fractions after the $z = \tan(x/2)$ substitution.)

Exercise 9. Evaluate $\displaystyle\int \frac{dx}{2 - \sin x + 2\cos x}$.

4.6 Acknowledgments

1. The Decennial Census data for 1790–2000 is from *Measuring America: The Decennial Censuses from 1790 to 2000*, U. S. Census Bureau, September 2002. The 2010 data is from 2010.census.gov/news/releases/operations/cb10-cn93.html.

2. The motivation for our discussion of the logistic ratio in Section 4.1 is from D. M. Bradley, Verhulst's logistic curve, *College Mathematics Journal*, **32** (2001), pp. 94–98.

3. The partial fractions technique using limits is adapted from P. T. Joshi, Efficient techniques for partial fractions, *Two-Year College Mathematics Journal*, **14** (1983), pp. 110–118.

4. The inspiration for using partial fractions to differentiate rational functions is from G. H. Hardy, *A Course of Pure Mathematics*, Tenth edition, Cambridge University Press, 1958.

5. The Sheffield data in Exercise 5 in Section 4.4.1 is from *A Handbook of Small Data Sets*, D. J. Hand, F. Daly, A. D. Lunn, K. J. McConway, and E. Ostrowski (editors), Chapman & Hall, London, 1994.

6. The oil production data in Exercise 7 in Section 4.4.1 is from *The Data and Story Library* website lib.stat.cmu.edu/DASL.

7. The proof of the Cauchy-Schwarz inequality in Exercise 3 in Section 4.4.1 is adapted from D. Rose, The Pearson and Cauchy-Schwarz inequalities, *College Mathematics Journal*, **39** (2008), p. 64.

5

Physical Applications of Integration

If one looks at the different problems of the integral calculus which arise naturally when one wishes to go deep into the different parts of physics, it is impossible not to be struck by the analogies existing.

Henri Poincaré

In the previous chapters, we saw several applications of integral calculus. In this chapter, we consider some questions that lend themselves to the technique of approximating with Riemann sums and then passing to the limit to arrive at a definite integral. While the applications covered in this chapter (work, centers of mass, fluid force) are important, we present them as models that can be transferred to many other areas of the physical and social sciences.

5.1 Work

When a force acts upon an object and that object moves in the direction of the force, the force is said to have done *work*. For example, when a spring is stretched from its natural length, the force that stretches the spring has done work.

The (amount of) work W done by a constant force F on an object that moves a distance of x units is defined to be

$$W = F \cdot d. \tag{5.1}$$

Example 5.1 Suppose that a weightlifter lifts barbells weighting 200 pounds from the floor to a height of 6 feet off the floor. In this process, the work done by the weightlifter on the barbells is $200 \cdot 6 = 1200$ ft-lbs. ■

In most cases, the force applied to an object is not constant, but varies with the position of the object.

Now, suppose that a force acting on an object, rather then being constant, is given by the continuous function $F(x)$ where x is the position of the object along a coordinate axis and that the object moves from $x = a$ to $x = b$. Let $P = \{x_0, x_1, x_2, \ldots, x_{n-1}, x_n\}$ be a regular partition of $[a, b]$ where $a = x_0$ and $b = x_n$. Set $\Delta x = \frac{b-a}{n}$. If Δx is sufficiently small, $F(x)$

Figure 5.1. An Israeli weightlifter at the 1960 Olympic Games in Rome

is approximately constant on the interval $[x_{i-1}, x_i]$ (note the important role the continuity of F plays here). We estimate the work done by the force in moving the object from $x = x_{i-1}$ to $x = x_i$ by

$$w_i = F(x_i^*)\Delta x$$

where x_i^* is some point in the interval $[x_{i-1}, x_i]$. The total work done by the force as the object moves from $x = a$ to $x = b$ is approximated by the sum of these w_i, so that

$$W \approx \sum_{i=1}^{n} w_i.$$

As n grows without bound (and thus $\Delta x \to 0$), we arrive at the total work

$$W = \lim_{n \to \infty} \sum_{i=1}^{n} F(x_i^*)\Delta x = \int_a^b F(x)\,dx. \tag{5.2}$$

Example 5.2 (Hooke's law for springs) As a spring is stretched or compressed from its natural length, the spring exerts a restoring force that varies with the amount of stretch or compression. The force it takes to stretch the spring has the same magnitude as this restoring force. A good model for this restoring force is given by *Hooke's Law*, named for the English physicist Robert Hooke (1635–1703). This physical principle states that for some springs, the restoring force exerted by the spring to return to its natural length is proportional to the displacement from its natural length. We will refer to a spring that behaves this way as a *Hooke's law spring*. Thus, if a Hooke's law spring is stretched by x units from its natural length, the restoring force of the spring is given by $f(x) = -kx$ where k is some positive constant (The negative sign accounts for the spring's restoring force being exerted in the opposite direction of the stretch). To actually stretch the spring, we must exert a force in the opposite direction of the restoring force. Thus, the force acting on the spring in order to stretch it is given by $F(x) = kx$.

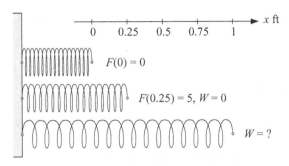

Figure 5.2. A Hooke's law spring

As a concrete example, suppose that when a Hooke's law spring is stretched by 3 inches (1/4 foot), the spring pulls with a restoring force of 5 pounds. Let's determine how much work is done in stretching this spring an additional 9 inches to a total extension of 1 foot. See Figure 5.2. We have

$$5 \text{ lbs} = k \cdot 0.25 \text{ ft}, \quad k = 20, \quad \text{so} \quad F(x) = 20x.$$

Thus, the work in stretching this spring an additional 9 inches is

$$W = \int_{1/4}^{1} 20x \, dx = 10x^2 \big|_{x=1/4}^{1} = 9.625 \text{ ft-lbs.}$$

Example 5.3 (Pumping water from a tank.) Consider a full water tank in the shape of a right circular cylinder, with base on the ground, no top, radius 4 feet, and height 10 feet. How much work is done is pumping the water out of the top of the tank?

We can imagine that the force is being applied to a piston at the bottom of the water and that as the piston rises, the amount of water above the piston is decreasing as water pours out the top of the tank. See Figure 5.3. Our task is to determine, at each height x above the bottom of the tank, the weight of the water above the piston, for this is the force needed to maintain the piston in its position. When the piston is x feet above the bottom of the tank, the height of water

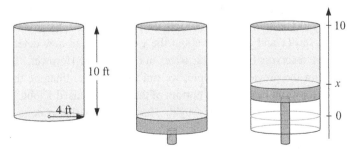

Figure 5.3. Water being pumped out of a cylindrical tank

above the piston is $10 - x$ feet and so the volume of water above the piston is

$$\pi 4^2 \cdot (10 - x) = 16\pi(10 - x) \text{ ft}^3.$$

Water has a weight-density of approximately $\rho_w = 62.4$ lbs/ft^3. Thus, the weight of the water above the piston is

$$F(x) = 16\rho_w \pi (10 - x) \text{ lbs.}$$

Thus, the work to move all of the water up and out of the tank is

$$W = \int_0^{10} 16\rho_w \pi (10 - x)\,dx = -8\rho_w \pi (10 - x)^2 \Big|_{x=0}^{10} = 800\rho_w \pi \approx 156{,}828 \text{ ft-lbs.} \ \blacksquare$$

While the method of integrating the force function applies to pumping water out of the cylindrical tank, with more complicated tanks, we usually revert to creating a Riemann sum to determine work.

Example 5.4 The city of Portland, Oregon has uncovered reservoirs for drinking water such as the one shown in Figure 5.4.

Figure 5.4. A reservoir in Portland, Oregon

Occasionally, someone does something they shouldn't do, and the city must drain the reservoir by pumping all the water out. When full a typical reservoir contains about 7 million gallons (about 1 million ft^3) of water. The shape of the reservoir can be modeled by a *paraboloid*, a shape formed by revolving a parabola about its axis (see Example 2.13). Since the reservoir in question is about 150 feet in radius and 30 feet deep at the center, we revolve the parabola $750y = x^2$, $0 \le x \le 150$ (x and y in feet) about the y-axis. We can now determine the work required to empty the reservoir by pumping water out at the top. However, if we attempt to apply the method from the previous example, we run into trouble. Imagine that the water is being pushed out the top from below. If the bottom of the water is raised 1 foot, the water in the reservoir rises less than 1 foot (do you see why?). Thus, the work that is done is smaller than the weight of the water (force) times 1 foot.

To attack this problem, we construct a Riemann sum. Let $P = \{y_0, y_1, y_2, \ldots, y_n\}$ be a regular partition of $[0, 30]$ where $y_0 = 0$ and $y_n = 30$. We estimate the volume of water in the

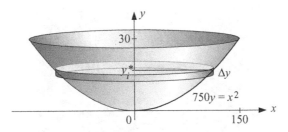

Figure 5.5. A paraboloidal model of a reservoir

reservoir that lies between height y_{i-1} and y_i with a cylinder of height Δy and radius determined at some point y_i^* between these heights. From Figure 5.5 we see that the radius of this cylinder is $\sqrt{750y_i^*}$. Thus, the weight w_i of water in this interval is

$$w_i = \rho_w \pi \left(\sqrt{750y_i^*}\right)^2 \Delta y = 750\rho_w \pi y_i^* \Delta y.$$

The water in this cylinder must be raised approximately $30 - y_i^*$ to reach the top and leave the reservoir. Thus, the work to pump this cylinder of water out of the reservoir is

$$w_i = 750\rho_w \pi y_i^*(30 - y_i^*)\Delta y.$$

Adding all these work elements gives us an estimate on the total work, and letting n grow without bound (and so $\Delta y \to 0$), we obtain

$$W = \lim_{n \to \infty} \sum_{i=1}^{n} w_i = \lim_{n \to \infty} \sum_{i=1}^{n} 750\rho_w \pi y_i^*(30 - y_i^*)\Delta y$$

$$= 750\rho_w \pi \int_0^{30} y(30 - y)\,dy$$

$$= 750\rho_w \pi \left(15y^2 - \frac{1}{3}y^3\right)\Big|_{y=0}^{30} = 3{,}375{,}000\rho_w \pi \approx 6.616 \times 10^8 \text{ ft-lbs.} \quad \blacksquare$$

Example 5.5 (Placing a satellite into geostationary orbit.) A satellite is said to be in geostationary orbit if its orbit is such that the satellite sits above the same spot on the earth. To place the satellite into orbit, we must work against the force of gravity and lift it to the geostationary altitude of 35,786 km from the earth's surface.

The magnitude of the gravitational force between two objects is given by $F = \frac{GmM}{r^2}$ where the objects have mass m and M respectively, r is the distance between the centers of mass of the objects, and G is the universal gravitational constant. The constant G has been estimated to be

$$G \approx 6.637 \times 10^{-11} \text{ m}^3/(\text{kg s}^2).$$

Assume the satellite has mass 500 kg, a medium sized satellite. The earth is approximately a sphere of radius 6370 km, and has mass 5.9742×10^{24} kg. The satellite must be lifted from the surface (6370 km from the center of the earth) to geostationary altitude. Let r be the distance, in meters, from the center of the earth to the satellite. The gravitational force on the satellite at

Figure 5.6. An artist's rendering of GOES (Geostationary Operational Environmental Satellite) of NOAA (the National Oceanic and Atmospheric Administration) observing earth

this distance is

$$F = F(r) = \frac{GmM}{r^2} = \frac{\left(6.673 \times 10^{-11}\right)\left(5 \times 10^2\right)\left(5.9742 \times 10^{24}\right)}{r^2} \approx \frac{1.983 \times 10^{17}}{r^2} \text{ newtons.}$$

Therefore, the work in lifting the satellite to this orbit is

$$W = \int_{6.370 \times 10^6}^{3.5786 \times 10^7} F(r)\,dr = 1.983 \times 10^{17} \int_{6.370 \times 10^6}^{3.5786 \times 10^7} \frac{1}{r^2}\,dr \approx 1.156 \times 10^{11} \text{ n-m.}$$

5.1.1 Exercises

In Exercises 1 through 6, determine the work accomplished for an object being acted on over the given interval by the given variable force function $F(x)$.

1. $F(x) = 16$ on $0 \leq x \leq 10$ 2. $F(x) = \cos x$ on $0 \leq x \leq \pi$

3. $F(x) = xe^{-x}$ on $1 \leq x \leq 3$ 4. $F(x) = \sec^2 x$ on $0 \leq x \leq \pi/4$

5. $F(x) = \frac{1}{4-x^2}$ on $0 \leq x \leq 1$ 6. $F(x) = \sec^3 x$ on $0 \leq x \leq \pi/4$

7. A large Hooke's law spring is stretched 6 inches when a force of 10 pounds is applied to one end. Determine the work done in stretching the spring from natural length to 2 feet beyond natural length. (Be careful with units.)

8. A Hooke's law spring has natural length 30 cm. When a force of 1 newton is applied to one end of the spring, the spring compresses to a length of 27 cm. Determine the work done in compressing the spring from a length of 27 cm to a length of 24 cm.

9. Assume Hooke's law applies to a spring of length l for displacements from natural length x where x is in the interval $[x_1, x_2]$. Determine the work done by a force that changes the displacement from natural length from $x = a$ to $x = b$.

10. Is it more work to raise a 70 pound bucket of water from a 50 foot well or a 50 pound bucket from a 70 foot well? (Assume the rope used to raise the buckets weighs 0.25 pounds per foot.)

Figure 5.7. Wells in Exercises 10 and 11

11. A bucket of water at the bottom of a 100 foot well was lifted to the top by a rope that weighs 0.5 pounds per foot. The bucket was leaking water (linearly) so that, although it started its journey at 40 pounds, it reached the top at only 25 pounds. How much work was done in lifting it to the top of the well?

12. A vertically oriented cylindrical water tank has radius r feet and height h feet. The water in the tank is initially a feet deep. Determine the work done in pumping all the water out the top of the tank.

13. A horizontally oriented cylindrical tank has radius 2 feet and length 6 feet. If the tank is half-full of water, how much work is required to pump the water out of a hole at the top of the tank?

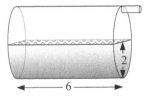

Figure 5.8. Tank in Exercise 13

14. A spherical tank of radius r feet is full of water. How much work is required to remove all the liquid out of a hole at the top of the tank?

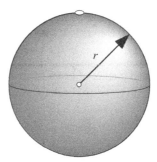

Figure 5.9. Tank in Exercise 14

15. In 1968, the crew of Apollo 8 were the first humans to reach a point where the gravitational pull of the moon and the gravitational pull of the earth are equal, a location sometimes called the *gravity neutral* point. This occurred when they were approximately 62,377 km above the surface of the moon. The distance between the center of the earth and the center of the moon is approximately 384,400 km. The moon has radius approximately 1737 km and the mass of the moon is approximately 7.3477×10^{22} kg.

Figure 5.10. A 1969 Hungarian stamp commemorating Apollo 8

(a) Considering only the gravitational pull of the earth, and assuming that the mass of the spacecraft was 28,817 kg, determine the work done in getting Apollo 8 to the gravity neutral point.

(b) As Apollo 8 left the surface of the earth, the earth's gravity exerted a force *pulling* the spacecraft back toward the earth. At the same time, the moon's gravity was pulling the spacecraft toward the moon. Assuming that the spacecraft moved in a straight line from the earth to the moon (an oversimplification, but sufficient for the current problem), the net gravitational force on the spacecraft is the difference of these two forces. Determine the work done in getting the Apollo 8 spacecraft to the gravity neutral point.

5.2 Mass and centers of mass

Mass-density, or just *density* for short, is a measurement of the concentration of mass within an object. We can measure density as mass per unit volume (e.g., g/cm^3, slugs/ft^3), or as mass per unit area (e.g., g/cm^2) or mass per unit length (e.g., g/cm). Sometimes, density is given in units of weight per unit volume (e.g., lbs/ft^3) in which case we are referring to weight-density.

An object is said to have *uniform density* if no matter which part of the object is considered, the density is the same. The mass of such an object is simply density times volume.

5.2.1 Mass

Suppose an object extends from $x = a$ to $x = b$ along the x-axis and that its density at position x is given by the continuous function $\rho(x)$. Let $P = \{x_0, x_1, x_2, \ldots, x_{n-1}, x_n\}$ be a regular partition of $[a, b]$ where $a = x_0$ and $b = x_n$. Set $\Delta x = \frac{b-a}{n}$. If Δx is sufficiently small, $\rho(x)$ is approximately constant on the interval $[x_{i-1}, x_i]$ (once again, the continuity of ρ is very important). We estimate the mass of the part of the object on the interval $[x_{i-1}, x_i]$ by

$$m_i = \rho(x_i^*)\Delta x$$

where x_i^* is some point in the interval $[x_{i-1}, x_i]$. Then the total mass of the object, m, is approximated by the sum of the m_i

$$m \approx \sum_{i=1}^{n} m_i.$$

As n grows, this approximation improves and we see total mass is given by

$$m = \int_a^b \rho(x)\,dx. \tag{5.3}$$

Example 5.6 Consider a thin tapered rod, 2 feet long, whose density at position x feet from the left end is given by the function $\rho(x) = \frac{1}{9-x^2}$ pounds per foot where $0 \le x \le 2$.

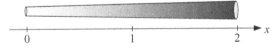

Figure 5.11. A tapered rod

The mass of the rod is computed using partial fractions:

$$\text{mass} = \int_0^2 \frac{1}{9-x^2}\,dx = \frac{1}{6}\int_0^2 \left(\frac{1}{3-x} + \frac{1}{3+x}\right) dx$$

$$= \frac{1}{6}\left(-\ln(3-x) + \ln(3+x)\right)\Big|_{x=0}^{2}$$

$$= \frac{1}{6}\ln\left(\frac{3+x}{3-x}\right)\Big|_{x=0}^{2}$$

$$= \frac{1}{6}\ln 5. \ \blacksquare$$

5.2.2 Center of mass

Suppose a father and child are playing on a see saw, with one person on either side of the fulcrum. To balance the see saw, the father must sit much closer to the fulcrum than the child. This comes about because of the torque, or rotational force, imparted at the fulcrum by each person. For each individual, this rotational force is $F = dw$ where w is the weight of the individual and d is the directed distance along the see saw to the fulcrum. To balance the see saw, these forces must be equal and opposite. In other words, they must sum to 0.

As a concrete example, suppose the father weighs 160 lbs, and the child weighs 40 lbs. Label the position of the fulcrum $\bar{x}$ (we'll see later why we use this notation). Let x_1 be the position of the father and x_2 the position of the child. Then to achieve balance, we have

$$160(x_1 - \bar{x}) + 40(x_2 - \bar{x}) = 0 \quad \text{or} \quad \bar{x} = \frac{160x_1 + 40x_2}{160 + 40}.$$

If another child who weighs 35 lbs is added to the see saw at position x_3, we would need the rotational forces imparted by the three individuals to add to 0 in order to achieve balance. Thus

$$210(x_1 - \bar{x}) + 40(x_2 - \bar{x}) + 35(x_3 - \bar{x}) = 0 \quad \text{or} \quad \bar{x} = \frac{160x_1 + 40x_2 + 35x_3}{160 + 40 + 35}.$$

The notation $\bar{x}$ was chosen because it appears that the position of the fulcrum or balance point is the weighted average or weighted mean of the positions of the three people. This observation can be generalized to find what is known as the *center of mass* of an object.

Suppose a bar has continuous mass density $\rho(x)$ where x in the interval $[a, b]$ measures linear position. Let $P = \{x_0, x_1, x_2, \ldots, x_n\}$ be a regular partition of $[a, b]$ with $\Delta x = \frac{b-a}{n}$. With small Δx, on each subinterval $[x_{i-1}, x_i]$ $\rho(x)$ is approximately constant and so the mass of this region is approximately $\rho(x_i^*)\Delta x$. Thus, the center of mass of the bar, $\bar{x}$, is approximated by

$$\bar{x} \approx \frac{\left(\rho(x_1^*)x_1 + \rho(x_2^*)x_2 + \cdots + \rho(x_n^*)x_n\right)\Delta x}{\left(\rho(x_1^*) + \rho(x_2^*) + \cdots + \rho(x_n^*)\right)\Delta x}.$$

Letting $\Delta x \to 0$, we obtain the center of mass of the bar

$$\bar{x} = \lim_{\Delta x \to 0} \frac{\left(\rho(x_1^*)x_1 + \rho(x_2^*)x_2 + \cdots + \rho(x_n^*)x_n\right)\Delta x}{\left(\rho(x_1^*) + \rho(x_2^*) + \cdots + \rho(x_n^*)\right)\Delta x} = \frac{\int_a^b x\rho(x)\,dx}{\int_a^b \rho(x)\,dx}. \tag{5.4}$$

The numerator, $\int_a^b x\rho(x)\,dx$, is known as the *first moment* of the function ρ. We summarize in the following theorem.

Theorem 5.1 (Center of mass) *Suppose an object lies along the x-axis on the interval $[a, b]$ and that the density of the object is the continuous function $\rho(x)$ on $[a, b]$. Then the center of mass of the object is*

$$\bar{x} = \frac{\int_a^b x\rho(x)\,dx}{\int_a^b \rho(x)\,dx}. \tag{5.5}$$

Back to Example 5.6 We determine the center of mass of the thin tapered rod in Example 5.6. The first moment of the function ρ is found using substitution. Letting $u = 9 - x^2$ so that $du = -2x\,dx$ we obtain

$$\int_0^2 x \frac{1}{9 - x^2}\,dx = -\frac{1}{2} \int_9^5 \frac{1}{u}\,du = \frac{1}{2} (\ln 9 - \ln 5).$$

Thus, the center of mass of the rod is

$$\bar{x} = \frac{\frac{1}{2}(\ln 9 - \ln 5)}{\frac{1}{6}\ln 5} = 3\frac{\ln 9}{\ln 5} - 3 \approx 1.0956. \quad \blacksquare$$

Centroid of a region between curves

Consider a two-dimensional region R bounded on the top by the continuous curve $y = f(x)$, on the bottom by the continuous curve $y = g(x)$, on the left by $x = a$, and on the right by $x = b$. If R has mass (so perhaps R is the base of a solid), it is often referred to as a *lamina*. We can compute the balance point of R. This balance point is also known as the *center of mass* of R and in the case of constant density, the *centroid* of R.

We can do a similar computation to that done in one dimension. But here, we assume that density (which is a function of *two variables x and y*) is constant. We will, in fact, assume this density is 1 unit of mass per unit area.

We begin with computing the y-coordinate of the centroid. Let $P = \{x_0, x_1, x_2, \ldots, x_n\}$ be a regular partition of $[a, b]$ with $\Delta x = \frac{b-a}{n}$. On the subinterval $[x_{i-1}, x_i]$ the vertical coordinate of the center of mass of the vertical strip between the graphs of f and g is approximately $\frac{1}{2}\left(f(x_i^*) + g(x_i^*)\right)$ where x_i^* is a point in the interval $[x_{i-1}, x_i]$.

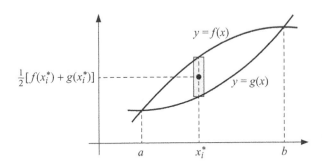

Figure 5.12. Determining the y-coordinate of the centroid of a region

Each vertical strip will contribute to the y-coordinate of the centroid according to its relative proportion of the total mass of the solid. Since our density is a constant 1 unit of mass per unit area, to determine the relative proportion of total mass we can simply compute the area of each vertical strip and divide by the total area of the region. On the interval $[x_{i-1}, x_i]$ the area of the vertical strip between the graphs of f and g is approximately $\left(f(x_i^*) - g(x_i^*)\right)\Delta x$. Thus, the proportion of overall area in this interval is approximately $\frac{(f(x_i^*) - g(x_i^*))}{A} \cdot \Delta x$ where A is the area of the region R.

Let $\bar{y}$ be the y-coordinate of the centroid. We approximate $\bar{y}$ by the weighted mean of the y-coordinates of the centroids of each vertical strip, so that

$$\bar{y} \approx \sum_{i=1}^{n} \left(\frac{(f(x_i^*) + g(x_i^*))}{2}\right) \cdot \left(\frac{(f(x_i^*) - g(x_i^*))}{A} \cdot \Delta x\right) = \frac{1}{A}\sum_{i=1}^{n} \frac{(f(x_1^*))^2 - (g(x_1^*))^2}{2} \Delta x.$$

This approximation improves as we shrink Δx (or increase n) and in the limit, as $\Delta x \to 0$, we obtain

$$\bar{y} = \frac{\int_a^b \frac{1}{2}\left([f(x)]^2 - [g(x)]^2\right)\,dx}{A} = \frac{\int_a^b \frac{1}{2}\left([f(x)]^2 - [g(x)]^2\right)\,dx}{\int_a^b (f(x) - g(x))\,dx}. \tag{5.6}$$

We now compute the x-coordinate of the centroid, $\bar{x}$. Once again, let $P = \{x_0, x_1, x_2, \ldots, x_n\}$ be a regular partition of $[a, b]$ with $\Delta x = \frac{b-a}{2}$. If we think of the mass of the part of the region R on the interval $[x_{i-1}, x_i]$ as being concentrated at a point x_i^* units from the y-axis (i.e., from $x = 0$), we can apply the one-dimensional method. Indeed, on the interval $[x_{i-1}, x_i]$ the mass of the vertical strip between the graphs of f and g is approximately $(f(x_i^*) - g(x_i^*))\,\Delta x$. With $(f(x_i^*) - g(x_i^*))$ playing the role of $\rho(x_i^*)$ in (5.4), we see that

$$\bar{x} = \frac{\int_a^b x\,(f(x) - g(x))\,dx}{\int_a^b (f(x) - g(x))\,dx}. \tag{5.7}$$

We summarize these results in the following theorem.

Theorem 5.2 (Centroid) *Let f and g be continuous on $[a, b]$ with $f(x) \geq g(x)$ for all x in $[a, b]$. The centroid, $(\bar{x}, \bar{y})$ of of the region bounded by $y = f(x)$ and $y = g(x)$ on $[a, b]$ is given by*

$$\bar{x} = \frac{\int_a^b x\,(f(x) - g(x))\,dx}{\int_a^b (f(x) - g(x))\,dx}, \qquad \bar{y} = \frac{\int_a^b \frac{1}{2}\left([f(x)]^2 - [g(x)]^2\right)\,dx}{\int_a^b (f(x) - g(x))\,dx}. \tag{5.8}$$

Example 5.7 (Example of centroid) Let's find the centroid of the region bounded below by the x-axis and above by the curve $y = \sin x$ where $0 \leq x \leq \pi$. By symmetry, it is clear that $\bar{x} = \pi/2$, but let's verify this with our formula. First, we compute the denominator

$$\int_0^\pi \sin(x)\,dx = -\cos x \Big|_{x=0}^\pi = 2.$$

Therefore

$$\bar{x} = \frac{1}{2}\int_0^\pi x \sin x\,dx.$$

The integral can be computed using integration by parts:

$$\int_0^\pi x \sin x = -x \cos x \Big|_{x=0}^\pi + \int_0^\pi \cos x\,dx = \pi.$$

Thus, we have verified that $x = \pi/2$. For $\bar{y}$ we compute

$$\bar{y} = \frac{1}{2} \int_0^\pi \frac{1}{2} \sin^2 x \, dx = \frac{1}{8} \int_0^\pi (1 - \cos 2x) \, dx = \frac{\pi}{8}.$$

Thus, the centroid of this region is $(\pi/2, \pi/8)$. ■

5.2.3 Exercises

1. Suppose an 80 pound child is sitting 3 feet from the fulcrum of a see saw and that the fulcrum is in the center of the see saw. How far from the fulcrum should a 50 pound child sit so that the see saw is balanced?

Figure 5.13. Balancing a see saw

In Exercises 2 through 7, a thin rod extending from $x = a$ to $x = b$ has the given density $\rho(x)$. Determine (a) the mass of the rod and (b) the center of mass of the rod.

2. $\rho(x) = 16$ on $0 \le x \le 10$ 3. $\rho(x) = \sin x$ on $0 \le x \le \pi$

4. $\rho(x) = e^{-x}$ on $1 \le x \le 3$ 5. $\rho(x) = x - x^3$ on $0 \le x \le 1$

6. $\rho(x) = xe^{-x}$ on $1 \le x \le 3$ 7. $\rho(x) = \ln(x)$ on $e \le x \le e^2$

In Exercises 8 through 11, determine the centroid of the region bounded above by $y = f(x)$, below by $y = g(x)$ and on the given interval.

8. $f(x) = x$, $g(x) = x^2$ on $0 \le x \le 1$.

9. $f(x) = \cos(x)$, $g(x) = \sin(x)$ on $0 \le x \le \pi/4$.

10. $f(x) = x$, $g(x) = xe^{x-1}$ on $0 \le x \le 1$.

11. $f(x) = \ln x$, $g(x) = 0$ on $1 \le x \le e$.

12. Consider a triangle with vertices $A = (a, 0)$, $B = (b, 0)$, and $C = (0, c)$
    (a) Determine the centroid of the triangle using (5.8).
    (b) Determine the midpoints M_1, M_2, and M_3 of the segments $\overline{AB}$, $\overline{BC}$, and $\overline{AC}$ respectively.
    (c) Show that the segments $\overline{CM_1}$, $\overline{AM_2}$, and $\overline{BM_3}$ intersect at the centroid of the triangle.

13. Determine the centroid of the region bounded above by the unit circle and bounded below by the x-axis on the interval $0 \le x \le a$ where $0 < a \le 1$. (The upper half of the unit circle is given by $y = \sqrt{1 - x^2}$.)

14. (a) Determine the centroid of the region bounded above by the graph of $y = e^{-x}$, bounded below by the x-axis and on the interval $0 \leq x \leq a$ where $a > 0$.

 (b) What happens to the coordinates of the centroid as $a \to \infty$? (Ideas such as this will be explored in Chapter 9.)

15. Figure 5.14 shows a circular disk of radius R from which we have removed a circle of radius r $(0 < r < R)$. The resulting region is called a *lune* or *crescent*, and the centroid of the lune is located at the edge of the removed disk (the black dot in the figure). What is the ratio R/r of the two radii?

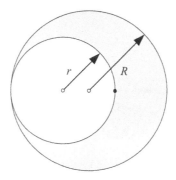

Figure 5.14. Lune for Exercise 15

5.3 Fluid force

Another important application of integration is computing the total force on an object that is submerged in a fluid.

Divers are well aware that as they descend deeper into a body of water, the pressure (force per unit area) on their bodies increases. This increase in pressure comes from the water that lies above them. It is the weight of water above the diver that creates this pressure. To determine the total force on an object submerged in a fluid, we must calculate the weight of the water that lies above the object.

In general, let g be acceleration due to gravity, and ρ the mass-density of the fluid (measured in mass per unit volume). Then the pressure P on an object at depth h is given by $P = \rho g h$. If ρ is the weight-density of the fluid, then $P = \rho h$.

Example 5.8 A rectangular pool is 10 feet wide and 20 feet long, and is filled with water to a constant depth of 6 feet. See Figure 5.15.

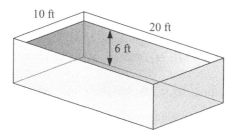

Figure 5.15. A swimming pool

The hydrostatic pressure at the bottom of the pool is $P = 62.4 \cdot 6 = 374.4$ lbs/ft^2. Since the area of the bottom of the pool is 200 ft^2, the total force on the bottom of the pool is $200 \cdot 374.4 = 74{,}880$ lbs. ∎

In the previous example, we computed the total force on a surface sitting horizontally in the water. Next, we will learn how to determine the total force on a surface that lies vertically in the water. Since pressure varies with depth, this surface experiences different pressures on different parts.

Assume the surface is fully submerged and extends from $y = c$ to $y = d$ along a y-axis that is oriented vertically, and assume the top of the fluid is at $y = k$, as illustrated in Figure 5.16. We partition the interval $[c, d]$ with a regular partition $P = \{y_0, y_1, y_2, \ldots, y_n\}$ where $y_0 = c$ and $y_n = d$ and set $\Delta y = (d - c)/n$. On each subinterval $[y_{i-1}, y_i]$, we approximate the surface with a rectangular strip whose height is Δy and whose width is $w(y_i^*)$ for some y_i^* in the interval $[y_{i-1}, y_i]$. Figure 5.16 illustrates the ith strip in this partition.

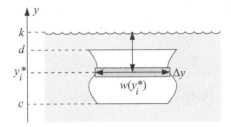

Figure 5.16. Computing force on a vertically submerged surface

For small values of Δy, the pressure on the top of the ith strip is approximately the same as the pressure on the bottom of the strip. We can imagine this strip being rotated about $y = y_i^*$ in such a way that the strip now lies horizontally in the fluid. Once we've done this, we can estimate the total force F_i on the ith strip by computing

$$F_i = \text{density} \cdot \text{depth} \cdot \text{area} = \rho \cdot (k - y_i^*) \cdot w(y_i^*)\Delta y.$$

Thus the total force on the surface can be estimated by the sum of the forces on the horizontal strips:

$$\text{Force} \approx \sum_{i=1}^{n} F_i = \sum_{i=1}^{n} \rho \cdot (k - y_i^*) \cdot w(y_i^*)\Delta y.$$

The expression on the right is a Riemann sum, so we see that the total force on the surface is given by its limit as $n \to \infty$ (or $\Delta y \to 0$):

$$\text{Force} = \lim_{n \to \infty} \sum_{i=1}^{n} F_i = \lim_{n \to \infty} \sum_{i=1}^{n} \rho \cdot (k - y_i^*) \cdot w(y_i^*)\Delta y$$

$$= \int_{c}^{d} \rho \cdot (k - y) \cdot w(y)\,dy. \tag{5.9}$$

Example 5.9 (Hydrostatic force on a submarine viewport) Submarine viewports must be con-structed to withstand large pressures (and hence large forces) from being submerged under perhaps thousands of feet of water.

Figure 5.17. The U.S. Navy DSV Alvin

The U. S. Navy deep sea vehicle Alvin, was originally designed to dive to a depth of 8010 feet. This vehicle had three circular viewports, each 1 foot in diameter. Let's compute the total hydrostatic force (i.e., the force from water pressure) on one of these viewports at a shallow depth and again at the maximum depth of 8010 feet. Placing the origin at the center of the viewport as in Figure 5.18, the viewport extends vertically from $y = -1/2$ to $y = 1/2$. For each

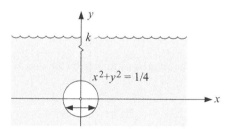

Figure 5.18. A model of a viewport on Alvin

value of y in the interval $[-1/2, 1/2]$ we must determine the width of the viewport. The outside border of the viewport is given by the equation $x^2 + y^2 = 1/4$, and so for a fixed y in the interval $[-1/2, 1/2]$, the width, $w(y)$, is

$$w(y) = 2x = 2\sqrt{\frac{1}{4} - y^2}.$$

When the center of the viewport is at a depth of $k = 20$ feet, we use (5.9) to see that the total force on the viewport is

$$\text{Force} = \rho \int_{-1/2}^{1/2} (20 - y) \cdot 2\sqrt{\frac{1}{4} - y^2}\, dy$$

$$= 40\rho \int_{-1/2}^{1/2} \sqrt{\frac{1}{4} - y^2}\, dy - 2\rho \int_{-1/2}^{1/2} y\sqrt{\frac{1}{4} - y^2}\, dy.$$

The integrand in the second integral is an odd function, and we are integrating over an interval symmetric about 0, therefore this integral is 0. The first integral gives half the area of a circle of radius $1/2$ and therefore

$$\text{Force} = 40\rho \int_{-1/2}^{1/2} \sqrt{\frac{1}{4} - y^2}\, dy = 40\rho \cdot \frac{1}{2}\pi \left(\frac{1}{2}\right)^2 = \rho \cdot 5\pi.$$

Given that the density of saltwater is approximately $\rho = 64$ lbs/ft^3, the total force on the viewport at a depth of 20 feet is approximately $64 \cdot 5\pi \approx 1005$ lbs.

If Alvin's viewport is at a depth of $k = 8010$ feet, similar computations yield

$$\text{Force} = \rho \int_{-1/2}^{1/2} (8010 - y) \cdot 2\sqrt{\frac{1}{4} - y^2}\, dy \approx 402{,}503 \text{ lbs},$$

an enormous force on such a small object! ∎

Example 5.10 (Hydrostatic force on a dam) Dams are subject to many forces including the force from water, force from waves, and force from silt build-up. In this example, we consider the *hydrostatic force* induced by the water behind the dam. The Detroit dam in the Oregon Cascades (see Figure 5.19) has a shape that can be approximated by an isoceles trapezoid whose height is 150 meters, whose width at the base is 100 meters, and whose width at the top is 200 meters.

Figure 5.19. The Detroit dam along the North Santiam River in the Oregon Cascades

Placing the origin at the center of the bottom of the trapezoid, the dam extends vertically from $y = 0$ to $y = 150$ as shown in Figure 5.20. For each value of y in the interval $[0, 150]$, we must determine the width of the dam.

Figure 5.20 makes it apparent that the width $w(y)$ is a linear function of y, and since we know $w(0) = 100$ and $w(150) = 200$, we see that $w(y) = 100 + (2/3)y$. Thus, when the reservoir is completely full, the water reaches the top of the dam and the force from water on the dam is

$$\text{Force} = \rho \int_0^{150} (150 - y) \cdot \left(100 + \frac{2}{3}y\right) dy = \rho \frac{2}{3} \int_0^{150} (150 - y)(150 + y) dy$$

$$= \frac{2}{3}\rho \int_0^{150} (150^2 - y^2) dy = \frac{2}{3}\rho \cdot \frac{2}{3} \cdot 150^3.$$

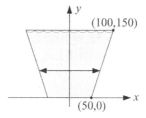

Figure 5.20. A model for the Detroit dam

To obtain the total force, we must replace ρ with the weight-density of water. Since the mass-density of water is $\rho_m = 1000$ kg/m^3, and acceleration due to gravity is $a = 9.8$ m/s^2, we have the weight-density is approximately $\rho = 9{,}800$ n/m^3. Thus, we have

$$\text{Force} = \frac{4}{9} \cdot 150^3 \cdot \rho \approx 14{,}700{,}000{,}000 \text{ n.} \quad \blacksquare$$

5.3.1 Exercises

1. Find the hydrostatic force on the bottom of a circular spa of radius 3 feet that has a constant depth of 2.5 feet.

2. A circular spa tub is 6 feet in diameter. Inside the spa, on the outer edge, there is a circular ledge, 6 inches (1/2 foot) wide, that is 1 foot below the surface of the water. The remainder of the spa is 3 feet deep. Find the hydrostatic force on the entire bottom surface of the spa (including the ledge).

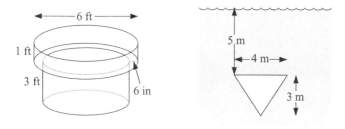

Figure 5.21. Spa in Exercise 2 (left) and plate in Exercise 3 (right)

3. An isosceles triangular plate is 4 meters across the top and 3 meters tall. It is submerged in water so that its top is 5 meters below the surface of the water. Determine the hydrostatic force on the plate.

4. An underwater camera has a circular lens that is 2 inches (1/6 foot) in diameter. The camera is to be used at a depth of 250 feet beneath the surface. When the camera is shooting horizontally, what is the total hydrostatic force on the lens at that depth?

5. A gate in a canal is in the shape of a trapezoid with height 1 meter, width at the top of 5 meters, and width at the bottom of 3 meters. Water rises to the top of the gate. Determine the total force on the gate.

6. A dam is in the shape of a semicircle whose diameter at the top is 80 feet. The reservoir behind the dam is full with water to the top of the dam. Determine the total hydrostatic force on the dam.

7. A dam is in the shape of semi-ellipse, with width 200 feet, and height 50 feet. Water behind the dam is to the top of the dam. Determine the total hydrostatic force on the dam.

8. A swimming pool is in the shape of a rectangle, 12 feet wide and 24 feet long stretching from the shallow end to the deep end. At the shallow end, the water in the pool is 2 feet deep, while at the deep end, the water is 6 feet deep. Assuming the bottom of the pool is a plane, what is the total hydrostatic force on the bottom of the pool?

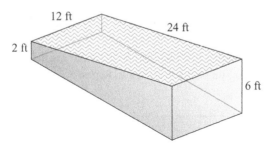

Figure 5.22. Swimming pool in Exercise 8

5.4 Explorations

5.4.1 More on Fluid Force

In Section 5.3 you learned how to find the total force on a submerged object in two special cases: when the object was lying parallel to the surface the water, and when the object was vertical and a Riemann sum could be constructed from horizontal slices. In this Exploration you will extend the technique to cover other situations.

Exercise 1. In this Exploration it will be advantageous to change the orientation of the y-axis so that it becomes a depth axis, that is, its origin is at the water's surface, and the positive direction is downwards. To illustrate, let's calculate the total force on a rectangular plate as shown in Figure 5.23.

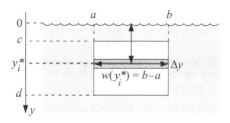

Figure 5.23. Computing force on a vertically submerged rectangular plate

Show that the total force on the plate is

$$\text{Force} = \rho \cdot \frac{c+d}{2} \cdot (b-a)(d-c),$$

that is,

$$\boxed{\text{Force} = \text{density} \cdot \text{average depth} \cdot \text{area}}.$$

If you set up the integral in (5.9) carefully, noting that the y-axis now measures depth, you obtained

$$\text{Force} = \lim_{n \to \infty} \sum_{i=1}^{n} F_i = \sum_{i=1}^{n} \rho \cdot y_i^* \cdot (b-a) \cdot \Delta y$$

$$= \int_c^d (\rho(b-a)y) \, dy = \rho(b-a)\frac{d^2 - c^2}{2} = \rho\frac{c+d}{2}(b-a)(d-c).$$

Exercise 2. Now consider a submerged plate bounded by the graphs of $y = f(x)$, $y = g(x)$ for x in $[a, b]$ where $f(x) \geq g(x)$ on $[a, b]$ as shown in Figure 5.24.

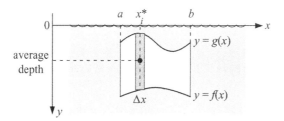

Figure 5.24. General vertical plate

Use the result of Exercise 1 to show that the total force on the plate is given by

$$\text{Force} = \frac{1}{2} \int_a^b \rho \left[(f(x))^2 - (g(x))^2 \right] dx.$$

As indicated in Figure 5.24, a partition of the interval $[a, b]$ on the x-axis yields approximating rectangles, so by the result of Exercise 1 the force F_i on the ith rectangle is given by density · average depth · area, i.e.,

$$F_i = \rho \cdot \frac{f(x_i^*) + g(x_i^*)}{2} \cdot \left[f(x_i^*) - g(x_i^*) \right] \Delta x,$$

and the result follows.

Exercise 3. Under the conditions of Exercise 2, show that the total force on the plate is given by
Force = density · depth of $\bar{y}$ · area where $\bar{y}$ is the y-coordinate of the centroid of the plate. Now
redo Example 5.9.

In Example 5.9, the centroid of Alvin's window is its center, so at a depth of k ft, the force
on the window is $\rho \cdot k \cdot \pi/4$, which agrees with the results in that example.

So far, the surfaces we've considered in fluid force problems have been planar. What about a
curved surface, such as the bottom of a hemispherical basin, or the bottom of the water reservoir
in Example 5.4?

Exercise 4. A hemispherical basin with radius r ft is full of water with density ρ lb/ft^3. Show
that the total force on the bottom of the basin is $\pi \rho r^3$ lb. See Figure 5.25.

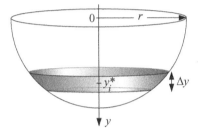

Figure 5.25. Force on a basin

Partitioning the interval $[0, r]$ on the inverted y-axis yields a partition of the surface of the
hemisphere into spherical zones (recall Example 2.16). The area of the ith zone is $2\pi r \, \Delta y$, and
hence the total force is

$$\text{Force} = \int_0^r \rho \cdot y \cdot 2\pi r \, dy = 2\pi \rho r \cdot r^2/2 = \pi \rho r^3 \text{ lb.}$$

Exercise 5. Assume the water reservoir in Example 5.4 is full of water with density ρ. Show
that the total force on the bottom of the reservoir is given by the integral

$$\text{Force} = \int_0^{30} \rho(30 - y) \cdot 10\pi \sqrt{30y + 5625} \, dy.$$

(Hints: use the setup illustrated in Figure 5.5. Partition the interval $[0, 30]$ on the y-axis. The
element of surface area in the ith subinterval is not the lateral area of the approximating disk in
Figure 5.5, but rather the integrand in (2.11) on page 57 where $g(y) = \sqrt{750y}$, i.e., the area of
an approximating conical frustum.)

6

The Hyperbolic Functions

What mathematician has ever pondered over an hyperbola, mangling the unfortunate curve with lines of intersection here and there, in his efforts to prove some property. . . ?

Lewis Carroll

In spite of their different appearances, the circle $x^2 + y^2 = 1$ and the rectangular hyperbola $x^2 - y^2 = 1$ (a hyperbola is rectangular if its asymptotes are perpendicular) have a lot in common. Their equations in cartesian (i.e., rectangular) coordinates are, except for one sign, identical; both exhibit left-right and up-down symmetry; and most importantly for us, both the circle and the hyperbola may be used to define sets of functions useful in calculus and its applications.

In this chapter we will introduce you to the six hyperbolic functions, which mirror the six trigonometric functions (the sine, cosine, tangent, etc.) in many ways, such as their graphs, identities, derivatives, and integrals.

6.1 The six hyperbolic functions

In many precalculus and calculus courses, the sine and cosine are called *circular functions*, since they can be defined as the coordinates of points on the *unit circle* $x^2 + y^2 = 1$. If a ray from the origin makes an angle θ with the positive x-axis, then the coordinates of the point where it intersects the unit circle are $(\cos\theta, \sin\theta)$. Equivalently, the signed arc length (by "signed" we mean that the length is positive if the angle is measured in the counterclockwise direction, and negative in the clockwise direction) on the circle is θ, and the signed area (again, positive for a counterclockwise angle and negative for a clockwise angle) of the circular sector is $\theta/2$, as illustrated in Figure 6.1(a) for θ in $[0, \pi/2)$.

We can now construct *hyperbolic functions* by replacing the unit circle by the right-hand branch of the *unit hyperbola* $x^2 - y^2 = 1$. For a real number u, let a ray from the origin intersect the hyperbola at a point (c, s) so that signed area of the hyperbolic sector is $u/2$. Then we define the *hyperbolic sine* $\sinh u$ and *hyperbolic cosine* $\cosh u$ to be the coordinates of the point on the hyperbola, that is $(c, s) = (\cosh u, \sinh u)$, as illustrated in Figure 6.1(b) for $u > 0$ (the signed area of the sector is positive when it lies above the x-axis and negative when it lies below the x-axis).

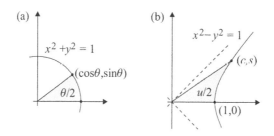

Figure 6.1. Defining the circular and hyperbolic functions

In Example 2.10 we showed that in terms of c and s the signed area of the hyperbolic sector is $(1/2)\ln(c+s)$, and thus we have $u/2 = (1/2)\ln(c+s)$, or equivalently,

$$c + s = e^u. \tag{6.1}$$

But the point (c, s) lies on the hyperbola $x^2 - y^2 = 1$, so that $c^2 - s^2 = 1$ and thus

$$c - s = \frac{c^2 - s^2}{c+s} = \frac{1}{e^u} = e^{-u}. \tag{6.2}$$

Solving (6.1) and (6.2) simultaneously yields

$$\boxed{c = \cosh u = \frac{e^u + e^{-u}}{2} \text{ and } s = \sinh u = \frac{e^u - e^{-u}}{2}.} \tag{6.3}$$

The remaining four hyperbolic functions—the *hyperbolic tangent* $\tanh u$, *hyperbolic cotangent* $\coth u$, *hyperbolic secant* $\operatorname{sech} u$, and *hyperbolic cosecant* $\operatorname{csch} u$—are defined analogously to the corresponding circular function, e.g., $\tanh u = \sinh u / \cosh u$, $\operatorname{sech} u = 1/\cosh u$, etc. Expressions for the six functions and their domains and ranges are given in Table 6.1 (as is customary, we use x for the variable in each function).

Table 6.1. The six hyperbolic functions

Function	Formula	Domain	Range
$\sinh x$	$\dfrac{e^x - e^{-x}}{2}$	$(-\infty, +\infty)$	$(-\infty, +\infty)$
$\cosh x$	$\dfrac{e^x + e^{-x}}{2}$	$(-\infty, +\infty)$	$[1, +\infty)$
$\tanh x$	$\dfrac{e^x - e^{-x}}{e^x + e^{-x}}$	$(-\infty, +\infty)$	$(-1, 1)$
$\operatorname{sech} x$	$\dfrac{2}{e^x + e^{-x}}$	$(-\infty, +\infty)$	$(0, 1]$
$\coth x$	$\dfrac{e^x + e^{-x}}{e^x - e^{-x}}$	$(-\infty, 0) \cup (0, +\infty)$	$(-\infty, -1) \cup (1, +\infty)$
$\operatorname{csch} x$	$\dfrac{2}{e^x - e^{-x}}$	$(-\infty, 0) \cup (0, +\infty)$	$(-\infty, 0) \cup (0, +\infty)$

Graphs of the six hyperbolic functions over the interval $[-2.25, 2.25]$ appear in Figure 6.2. For comparison we have graphs of the six circular functions in Figure 6.2 over the interval $[-\pi/2, \pi/2]$. The similarity of the two graphs is striking, which suggests there may be functional relationships between the two sets of functions. We will investigate these relationships in Section 6.3.

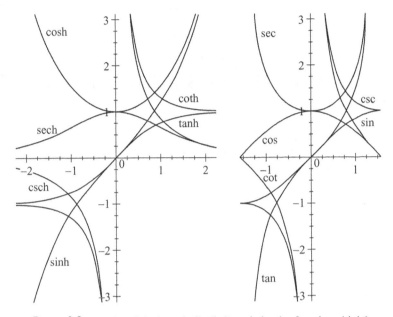

Figure 6.2. Graphs of the hyperbolic (left) and circular functions (right)

You can explore some properties of the hyperbolic functions in the exercises.

Since the hyperbolic functions are expressed in terms of the exponential functions e^x and e^{-x}, it is easy to find their derivatives and antiderivatives. In Exercises 5 and 6 you can verify that

$$\frac{d}{dx} \sinh x = \cosh x \qquad\qquad \frac{d}{dx} \cosh x = \sinh x$$

$$\frac{d}{dx} \tanh x = \operatorname{sech}^2 x \qquad\qquad \frac{d}{dx} \coth x = -\operatorname{csch}^2 x \qquad (6.4)$$

$$\frac{d}{dx} \operatorname{sech} x = -\operatorname{sech} x \tanh x \qquad\qquad \frac{d}{dx} \operatorname{csch} x = -\operatorname{csch} x \coth x$$

and

$$\int \sinh x \, dx = \cosh x + C \qquad\qquad \int \cosh x \, dx = \sinh x + C$$

$$\int \tanh x \, dx = \ln(\cosh x) + C \qquad\qquad \int \coth x \, dx = \ln|\sinh x| + C \qquad (6.5)$$

$$\int \operatorname{sech} x \, dx = \arctan(\sinh x) + C \qquad\qquad \int \operatorname{csch} x \, dx = \ln|\tanh(x/2)| + C$$

6.1.1 Exercises

1. Show that cosh and sech are even functions, and the other four hyperbolic functions are odd functions.

2. Prove the Pythagorean identities for hyperbolic function: for all x,
$$\cosh^2 x - \sinh^2 x = 1,$$
$$\tanh^2 x + \operatorname{sech}^2 x = 1,$$
$$\coth^2 x - \operatorname{csch}^2 x = 1.$$

3. Prove the sum and difference and the double angle identities for hyperbolic sine and cosine: for all x and y,
$$\sinh(x \pm y) = \sinh x \cosh y \pm \cosh x \sinh y,$$
$$\cosh(x \pm y) = \cosh x \cosh y \pm \sinh x \sinh y,$$
$$\sinh(2x) = 2 \sinh x \cosh x,$$
$$\cosh(2x) = \cosh^2 x + \sinh^2 x = 2 \cosh^2 x - 1 = 1 + 2 \sinh^2 x.$$

4. Prove the identities:
   (a) for all real x, $e^x = \dfrac{1 + \tanh(x/2)}{1 - \tanh(x/2)}$,
   (b) for all positive x, $\tanh(\ln x) = \dfrac{x^2 - 1}{x^2 + 1}$,
   (c) for all real x and α, $(\sinh x + \cosh x)^\alpha = \sinh \alpha x + \cosh \alpha x$.

5. Verify (6.4).

6. Verify (6.5). (Hint: use differentiation.)

In Exercises 7 through 12, differentiate each function.

7. $f(x) = \ln \cosh x$ 8. $g(x) = e^x \sinh x$ 9. $h(x) = \ln \coth(x/2)$

10. $f(x) = \arcsin(\tanh x)$ 11. $g(x) = x \cosh x - \sinh x$ 12. $h(x) = (\cosh x - \sinh x)^2$

In Exercises 13 through 21, evaluate the indefinite integral. Be sure to check your work by differentiation.

13. $\int \tanh x \ln(\cosh x)\,dx$ 14. $\int \sinh t \ln(\cosh t)\,dt$ 15. $\int x \sinh x\,dx$

16. $\int x^2 \cosh x\,dx$ 17. $\int x \operatorname{sech}^2 x\,dx$ 18. $\int e^x \cosh x\,dx$

19. $\int \sinh^2 t\,dt$ 20. $\int \operatorname{sech}^3 x\,dx$ 21. $\int \sinh x \cos x\,dx$

22. Find the area of the region bounded by the curve $y = \operatorname{sech} x$, the x-axis, the y-axis, and the line $x = 2$.

23. Find the volume of the solid formed when the region in Exercise 22 is revolved about the x-axis.

24. Find the volume of the solid formed by revolving about the x-axis the region bounded by the curve $y = \cosh x$ and the line $y = 17/8$.

25. Consider the integral $\int e^x \sinh x\, dx$. Integrate by parts twice to obtain

$$\int e^x \sinh x\, dx = e^x \cosh x - \int e^x \cosh x\, dx$$

$$= e^x \cosh x - e^x \sinh x + \int e^x \sinh x\, dx$$

and hence

$$0 = e^x(\cosh x - \sinh x) = e^x \cdot e^{-x} = 1.$$

What went wrong?

26. *Logistic growth and the hyperbolic tangent.* Suppose a population P is growing according to the logistic differential equation (4.4). In Exercise 1 in Section 4.1.1 you showed that the population is growing fastest at the time $t_{1/2}$ that satisfies $P(t_{1/2}) = L/2$, i.e., the time at which the population size is one-half its limiting value.

    (a) Show that

$$\frac{1}{1+e^{-x}} = \frac{1}{2}\left[1 + \tanh(x/2)\right].$$

    (b) Show that the logistic function (4.5) can be written in terms of $t_{1/2}$ and the hyperbolic tangent as

$$P(t) = \frac{L}{2}\left[1 + \tanh\left(\frac{k}{2}(t - t_{1/2})\right)\right].$$

27. Let h be a function given by $h(x) = f(x) + g(x)$ where f is an even function and g is an odd function. In this case, the functions f and g are called the *even and odd parts* of h. For example, if $h(x) = 1/(x^2 - 2x + 2)$, then $f(x) = (x^2 + 2)/(x^4 + 4)$ and $g(x) = 2x/(x^4 + 4)$. See Figure 6.3.

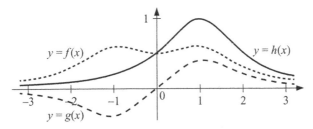

Figure 6.3. Graphs of a function and its even and odd parts

    (a) Verfiy that the functions f and g given above sum to h.
    (b) Show that if the domain of h is symmetric with respect to the origin (i.e., if a is in the domain so is $-a$), then there exists an even function f and an odd function g with the same domain such that $h(x) = f(x) + g(x)$ for all x in the domain of h. (Hint: if $h(x) = f(x) + g(x)$ with f even and g odd, what can you say about $h(-x)$?)
    (c) Find the even and odd parts of the exponential function $h(x) = e^x$.

28. (a) Show that the graphs of $y = \cosh x$ and $y = \operatorname{csch} x$ intersect on the line $y = \sqrt{\varphi}$, where $\varphi = (1 + \sqrt{5})/2$ is the golden ratio (see Section 2.1), and the same is true for the graphs of $y = \sinh x$ and $y = \coth x$. See Figure 6.2.
   (b) Is the same true for the graphs of $y = \sec x$ and $y = \cot x$, and for the graphs of $y = \tan x$ and $y = \csc x$?

6.2 The inverse hyperbolic functions

Computer algebra systems and online integration sites will often produce results such as

$$\int \frac{dx}{\sqrt{x^2 + 1}} = \sinh^{-1} x + C \quad \text{and} \quad \int \frac{dx}{1 - x^2} = \tanh^{-1} x + C.$$

In earlier chapters we evaluated these integrals using trigonometric subtitution and/or partial fractions (see Section 2.2 and 4.2), but it appears that simpler expressions are available using the inverses of the hyperbolic functions. Since the hyperbolic functions are expressible in terms of exponential functions, we anticipate that their inverses will be expressible in terms of logarithms—and that is indeed the case.

The hyperbolic sine is a one-to-one function since it has a positive derivative on $(-\infty, \infty)$, so it has an inverse which we denote by $\sinh^{-1}$. By definition,

$$y = \sinh^{-1} x \quad \text{if and only if} \quad x = \sinh y = (e^y - e^{-y})/2.$$

After clearing fractions and simplifying $x = (e^y - e^{-y})/2$, we have

$$e^{2y} - 2x e^y - 1 = 0.$$

This expression is quadratic in e^y, and the quadratic formula yields

$$e^y = \frac{2x + \sqrt{4x^2 + 4}}{2} = x + \sqrt{x^2 + 1}$$

(we use only the "+" sign in the formula since e^y must be positive). Thus

$$y = \sinh^{-1} x = \ln\left(x + \sqrt{x^2 + 1}\right).$$

The inverses of the other five hyperbolic functions are derived similarly (with the domain restricted to $[0, \infty)$ for cosh and sech, since they are not one-to-one functions). The expressions, domains, and ranges for all six are given in Table 6.2. See Exercise 10 for the graphs of the inverse hyperbolic functions.

Since the inverse hyperbolic functions are expressible in terms of logarithms, it is easy to find their derivatives. For example:

$$\frac{d}{dx}\sinh^{-1} x = \frac{d}{dx}\ln\left(x + \sqrt{x^2 + 1}\right) = \frac{1 + x/\sqrt{x^2 + 1}}{x + \sqrt{x^2 + 1}} = \frac{1}{\sqrt{x^2 + 1}}.$$

Table 6.2. The six inverse hyperbolic functions

Function	Formula	Domain	Range
$\sinh^{-1} x$	$\ln\left(x + \sqrt{x^2 + 1}\right)$	$(-\infty, +\infty)$	$(-\infty, +\infty)$
$\cosh^{-1} x$	$\ln\left(x + \sqrt{x^2 - 1}\right)$	$[1, +\infty)$	$[0, +\infty)$
$\tanh^{-1} x$	$\dfrac{1}{2} \ln\left(\dfrac{1 + x}{1 - x}\right)$	$(-1, 1)$	$(-\infty, +\infty)$
$\operatorname{sech}^{-1} x$	$\ln\left(\dfrac{1 + \sqrt{1 - x^2}}{x}\right)$	$(0, 1]$	$[0, +\infty)$
$\coth x^{-1}$	$\dfrac{1}{2} \ln\left(\dfrac{x + 1}{x - 1}\right)$	$(-\infty, -1) \cup (1, +\infty)$	$(-\infty, 0) \cup (0, +\infty)$
$\operatorname{csch}^{-1} x$	$\ln\left(\dfrac{1 \pm \sqrt{1 + x^2}}{x}\right)$	$(-\infty, 0) \cup (0, +\infty)$	$(-\infty, 0) \cup (0, +\infty)$

The $+$ sign is used for $x > 0$; the $-$ sign for $x < 0$.

Here are the derivatives of all six inverse hyperbolic functions. You will verify them in Exercise 3.

$$\frac{d}{dx} \sinh^{-1} x = \frac{1}{\sqrt{x^2 + 1}} \qquad \frac{d}{dx} \cosh^{-1} x = \frac{1}{\sqrt{x^2 - 1}}$$

$$\frac{d}{dx} \tanh^{-1} x = \frac{1}{1 - x^2} \qquad \frac{d}{dx} \coth^{-1} x = \frac{1}{1 - x^2} \qquad (6.6)$$

$$\frac{d}{dx} \operatorname{sech}^{-1} x = \frac{-1}{x\sqrt{1 - x^2}} \qquad \frac{d}{dx} \operatorname{csch}^{-1} x = \frac{-1}{|x|\sqrt{1 + x^2}}$$

6.2.1 Exercises

1. Verify the *reciprocal identities* for the inverse hyperbolic functions (these are useful for evaluating $\operatorname{sech}^{-1} x$, $\operatorname{csch}^{-1} x$, and $\coth^{-1} x$ on your calculator): for x in the domain of the function,

$$\operatorname{sech}^{-1} x = \cosh^{-1}(1/x)$$
$$\operatorname{csch}^{-1} x = \sinh^{-1}(1/x)$$
$$\coth^{-1} x = \tanh^{-1}(1/x).$$

2. Verify the *composition identities* for the inverse hyperbolic functions: for x in the domain of the function,

$$\sinh\left(\cosh^{-1} x\right) = \sqrt{x^2 - 1} \qquad \cosh\left(\sinh^{-1} x\right) = \sqrt{x^2 + 1}$$
$$\sinh\left(\tanh^{-1} x\right) = x/\sqrt{1 - x^2} \qquad \tanh\left(\sinh^{-1} x\right) = x/\sqrt{1 + x^2}$$
$$\cosh\left(\tanh^{-1} x\right) = 1/\sqrt{1 - x^2} \qquad \tanh\left(\cosh^{-1} x\right) = \sqrt{x^2 - 1}/x.$$

3. Verify the derivative formulas of the inverse hyperbolic functions in (6.6).

For Exercises 4 through 9, differentiate the function.

4. $f(x) = \tanh^{-1}(\sin x)$ 5. $g(x) = \sinh^{-1}(\tan x)$

6. $f(x) = 2\tanh^{-1}(e^x)$ 7. $g(x) = x\sinh^{-1}x - \sqrt{1+x^2}$

8. $f(x) = x\operatorname{sech}^{-1}x + \arcsin x$ 9. $g(x) = x\tanh^{-1}x + \ln\sqrt{1-x^2}$

10. Use your graphing calculator to sketch graphs of the six inverse hyperbolic functions and verify that each graph is the reflection in the line $y = x$ of the graph of the corresponding hyperbolic function (with the restricted domain for cosh and sech).

11. Establish the following integral formulas:

$$\int \sinh^{-1}x\,dx = x\sinh^{-1}x - \sqrt{1+x^2} + C$$

$$\int \tanh^{-1}x\,dx = x\tanh^{-1}x + \frac{1}{2}\ln\left(1-x^2\right) + C$$

$$\int \operatorname{sech}^{-1}x\,dx = x\operatorname{sech}^{-1}x + \arcsin x + C.$$

Hint: see Exercises 7–9.

12. Establish the integral formula $\displaystyle\int \sqrt{x^2+1}\,dx = \frac{1}{2}\left(x\sqrt{x^2+1} + \sinh^{-1}x\right) + C.$

13. Establish the integral formula $\displaystyle\int \frac{\sqrt{1-x^2}}{x}\,dx = \sqrt{1-x^2} - \operatorname{sech}^{-1}x + C.$

In Exercises 14 through 16 evaluate the integral.

14. $\int x\sinh^{-1}x\,dx$ 15. $\int x\tanh^{-1}x\,dx$ 16. $\int x\operatorname{sech}^{-1}x\,dx.$

17. Evaluate $e^{\sinh^{-1}(1/2)}$. Is the value familiar?

6.3 Relationships among circular and hyperbolic functions

In Figure 6.2 we observed a remarkable similarity between the graphs of the six hyperbolic functions and the graphs of the six circular functions. To discover the relationships among these functions we graph the unit circle $x^2 + y^2 = 1$ and the right-hand branch of the unit hyperbola $x^2 - y^2 = 1$ on the same set of axes, as shown in Figure 6.4. Then we draw a ray from the origin making an angle of θ (for θ in $(-\pi/2, \pi/2)$) with the positive x-axis intersecting both the circle and the hyperbola.

The ray intersects the circle at $(\cos\theta, \sin\theta)$ and the hyperbola at $(\cosh u, \sinh u)$ for some real number u. Recall from the first section in this chapter that there is an analogous relationship between the number u and the hyperbolic functions as for the angle θ and the circular functions. The area of the circular sector (in dark gray) in Figure 6.4 is $\theta/2$ and the area of the hyperbolic sector (in two shades of gray) is $u/2$.

We can express u in terms of θ by using similar triangles:

$$\frac{\sinh u}{\cosh u} = \frac{\tan\theta}{1} \quad \text{so that } \tanh u = \tan\theta,$$

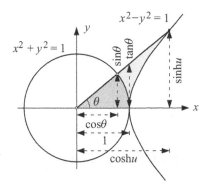

Figure 6.4. Relating circular and hyperbolic functions

and hence $u = \tanh^{-1}(\tan\theta)$. For another relationship between θ and u, see Exercise 11.

To explain the functional relationships evident in Figure 6.2, add a ray from the origin to the point $(1, \sinh u)$ in Figure 6.4 to yield Figure 6.5, and let ϕ denote the angle that this ray makes with the positive x-axis.

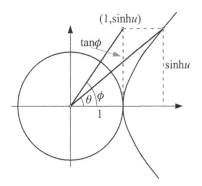

Figure 6.5. Hyperbolic functions of u as circular functions of ϕ

It now follows that $\tan\phi = \sinh u$, and hence we have (see Exercise 1)

$$
\begin{array}{lll}
\sinh u = \tan\phi & \cosh u = \sec\phi & \operatorname{csch} u = \cot\phi \\
\tanh u = \sin\phi & \operatorname{sech} u = \cos\phi & \coth u = \csc\phi
\end{array}
\tag{6.7}
$$

and so each of the hyperbolic functions (of u) is a circular function (of ϕ). The angle ϕ can be expressed in six ways:

$$\phi = \arctan(\sinh u) = \operatorname{arcsec}(\cosh u) = \arcsin(\tanh u)$$
$$= \operatorname{arccot}(\operatorname{csch} u) = \arccos(\operatorname{sech} u) = \operatorname{arccsc}(\coth u).$$

The angle ϕ has a name—it is called the *Gudermannian* of u, written $\phi = \operatorname{gd} u$—named after the German mathematician Christoph Gudermann (1798–1852). The domain and range of $\phi = \operatorname{gd} u$ are $(-\infty, \infty)$ and $(-\pi/2, \pi/2)$, respectively. The Gudermannian is graphed in Figure 6.6 with its horizontal asymptotes. You can explore some the interesting properties of the Gudermannian in the Exercises.

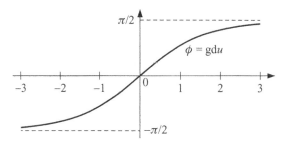

Figure 6.6. The graph of $\phi = \text{gd}u$

The six relationships in (6.7), along with the Pythagorean identities in Exercise 2 of Section 6.1 are illustrated (for $\phi \geq 0$) using the triangles in Figure 6.7.

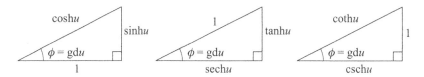

Figure 6.7. Gudermannian right triangles

6.3.1 Exercises

1. Prove (6.7).

In Exercises 2 through 10, prove the property of the Gudermannian.

2. $\text{gd}(-u) = -\text{gd}u$

3. $\lim\limits_{u \to \infty} \text{gd}u = \pi/2$

4. $\lim\limits_{u \to -\infty} \text{gd}u = -\pi/2$

5. $\dfrac{d}{du}\text{gd}u = \text{sech}\,u$

6. $\text{gd}u$ has an inverse $\text{gd}^{-1}\phi = \ln(\sec\phi + \tan\phi) = \ln(\tan(\phi/2 + \pi/4))$

7. $\dfrac{d}{d\phi}\text{gd}^{-1}\phi = \sec\phi$

8. $e^x = \sec(\text{gd}x) + \tan(\text{gd}x)$

9. $\tanh\left(\frac{1}{2}x\right) = \tan\left(\frac{1}{2}\text{gd}x\right)$

10. $\text{gd}x = 2\arctan e^x - \frac{\pi}{2}$

11. Let θ denote the angle at the origin of the hyperbolic sector with signed area $u/2$, as illustrated in Figure 6.4. Show that $2\theta = \text{gd}(2u)$.

12. Let α, where $-\pi/2 < \alpha < \pi/2$, be the angle of inclination of the line tangent to the graph of $y = \cosh x$ at the point $(x_0, \cosh(x_0))$ for $x_0 \neq 0$. Show that $\alpha = \text{gd}x_0$.

13. Evaluate $e^{\text{gd}^{-1}(\arctan(1/2))}$. Is the result familiar?

6.4 Explorations

6.4.1 The catenary and the catenoid

In Exercise 2 of Section 2.1 we introduced the *catenary*, the mathematical curve that describes
the shape of a uniform chain suspended from two points. In the exercise we gave its equation as
$y = (c/2)(e^{x/c} + e^{-x/c})$, where c is a positive constant (the y-intercept of the graph). But now
you recognize the equation as $y = c\cosh(x/c)$. In Figure 6.8 we have graphs of catenaries for
$c = 1, 2, 3$.

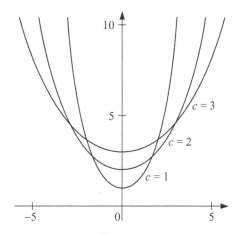

Figure 6.8. Three catenaries

Exercise 1. If you haven't done Exercise 2 in Section 2.1, you should do it now!

In architecture you can find arches large and small whose shapes are approximately inverted
catenaries. In Figure 6.9 we see inverted catenary arches in the attic of the Catalan architect
Antoni Gaudí's Casa Milà in Barcelona, and the Gateway Arch in St. Louis, Missouri.

Figure 6.9. Inverted catenaries in architecture

Although the catenary resembles a parabola, its properties are actually quite different. In
this exploration you will investigate some properties related to the arc length and area under the

catenary, and the volume and surface area of the solid formed by rotating the catenary about the x-axis. We begin by comparing the graph of the catenary to the graph of the parabola.

Exercise 2 Using your graphing calculator or a computer algebra system, graph $y = \cosh x$ and $y = x^2 + 1$ on the same axes. How do the graphs compare?

Your graphs should look something like Figure 6.10, with the curves intersecting at three points: (0, 1) and approximately (± 2.9828, 9.897). The catenary (the solid curve) is flatter than the parabola (the dashed curve) near the bottom, and lies above the parabola for values of x away from the origin.

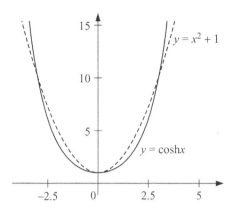

Figure 6.10. Graphs of $y = \cosh x$ and $y = x^2 + 1$

Now let $P(a, c \cosh(a/c))$ and $Q(b, c \cosh(b/c))$, $a < b$, be two points on the graph of the catenary $y = c \cosh(x/c)$; let L be the arc length between P and Q, and let A be the area under the graph over the interval $[a, b]$.

Exercise 3. Show that $A = cL$. Thus the area is a constant multiple (with the same dimension as length) of the arc length, no matter where P and Q are on the catenary.

Is the converse statement true? That is, if the graph of $y = f(x)$ has the property in Exercise 3, must $f(x)$ be $c \cosh(x/c)$ for some constant c? Note that the graph of $y = c$ has the property, so we seek others. To investigate, let $P(a, f(a))$ be a fixed point and $Q(x, f(x))$ a variable point on the graph of $y = f(t)$ for x in $[a, b]$. Also let $A(x)$ denote the area under the graph of $y = f(t)$ over the interval $[a, x]$, let $L(x)$ denote the arc length of the graph between P and Q; and assume $A(x) = cL(x)$ for all x in $[a, b]$.

Exercise 4. Show that $f(x) = c\sqrt{1 + [f'(x)]^2}$. (Hint: use the fundamental theorem of calculus.)

Exercise 5. Letting $y = f(x)$, show that $\frac{dy}{dx} = \frac{\sqrt{y^2 - c^2}}{c}$.

Exercise 6. Solve the differential equation in Exercise 5. (Hint: make the substitution $y = c \cosh u$ before separating the variables.)

Did you obtain $\cosh^{-1}(y/c) = (x/c) + C$ as a solution? This is equivalent to $y = c \cosh[(x/c) + C]$, which is a catenary with a horizontal shift if $C \neq 0$. Thus you have shown that $A = cL$ is a necessary and sufficient condition for a curve $y = f(x)$ to be a catenary.

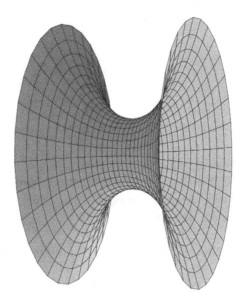

Figure 6.11. A catenoid

If you rotate the catenary $y = c \cosh(x/c)$ about the x-axis, the resulting object is called a *catenoid*, shown in Figure 6.11. Again let $P(a, c \cosh(a/c))$ and $Q(b, c \cosh(b/c))$, $a < b$, be two points on the graph of the catenary, and let V and S be the resulting volume and surface area respectively of the catenoid obtained when the part of the graph of the catenary between P and Q is revolved about the x-axis.

Exercise 7. Show that $cS = 2V$ for all points P and Q on the catenary.

If you wish, you can show that the property in Exercise 7 leads to the same differential equation as in Exercise 5, and forces the graph of the function rotated to be a catenary. The work parallels Exercises 4, 5, and 6.

6.4.2 The Gudermannian substitution for trigonometric integrals

Do you recall the difficulty we faced in Chapters 1 and 2 evaluating integrals such as $\int \sec x \, dx$ and $\int \sec^3 x \, dx$ (Example 1.9), $\int \tan^2 x \sec x \, dx$ (Example 2.11), and $\int dx/(\cos x \sin^2 x)$? In this Exploration you will discover that a substitution illustrated with the Gudermannian triangle in Figure 6.7 can be used to integrate an expression equivalent to $\int \tan^m x \sec^n x \, dx$ when m and n are integers with m even and n odd. (A similar substitution can be used for $\int \cot^m x \csc^n x \, dx$.)

Exercise 1. Verify that each of the first four integrals in the above paragraph has the form $\int \tan^m x \sec^n x \, dx$ with m even and n odd.

You should discover that for the four integrals we have $(m, n) = (0, 1), (0, 3), (2, 1)$, and $(-2, 3)$. The substitution is $x = gdu$, hence $\tan x = \sinh u$, $\sec x = \cosh u$, $dx = du/\cosh u$ (since $\sec^2 x \, dx = \cosh u \, du$), and $u = \ln(\cosh u + \sinh u) = \ln|\sec x + \tan x|$. In Figure 6.12 we illustrate the substitution (for x in $[0, \pi/2)$) with a Gudermannian triangle.

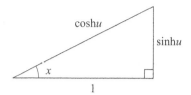

Figure 6.12. A hyperbolic function trigonometric substitution

Since the range of the Gudermannian is $(-\pi/2, \pi/2)$ the substitution is valid only for values of x in this interval. However, it produces the correct antiderivative outside that interval as long as absolute values are used with logarithms, as we have done in writing $u = \ln |\sec x + \tan x|$.

Let's try the substitution on $\int \sec x \, dx$:

$$\int \sec x \, dx = \int \cosh u \, \frac{du}{\cosh u} = \int du = u + C$$
$$= \ln |\sec x + \tan x| + C,$$

which (of course!) agrees with (1.10). Now you try some.

Exercise 2. Evaluate $\int \sec^3 x \, dx$.

Exercise 3. Evaluate $\int \tan^2 x \sec x \, dx$. (Hint: recall that $\cosh 2u = 2 \sinh^2 u + 1$ and $\sinh 2u = 2 \sinh u \cosh u$ from Exercise 3 in Section 6.1.1.)

Exercise 4. Evaluate $\int dx/(\cos x \sin^2 x)$. (Hint: recall that $\coth^2 u - \operatorname{csch}^2 u = 1$ from Exercise 2 in Section 6.1.1.)

6.5 Acknowledgments

1. Exercise 25 in Section 6.1.1 is from E.J. Barbeau, *Mathematical Fallacies, Flaws, and Flimflam*, Mathematical Association of America, Washington, DC (2000).
2. The material in Section 6.3 is adapted from J. M. H. Peters, The Gudermannian, *The Mathematical Gazette*, **68** (1984), pp. 192–196.
3. Exploration 6.4.1 is adapted from F. Chorlton, Some geometrical properties of the catenary, *The Mathematical Gazette*, **83** (1999), pp. 121–123.
4. Exploration 6.4.2 is adapted from W. K. Viertel, Use of hyperbolic substitution for certain trigonometric integrals, *Mathematics Magazine*, **38** (1965), pp. 141–144.

7

Numerical Integration: What to Do When You Can't Antidifferentiate

Far better an approximate answer to the right question, than the exact answer to the wrong question.

John Tukey

There are many situations that call for the computation of a definite integral, but making use of the fundamental theorem of calculus is either impossible or impractical. In these circumstances, one course of action is to use an approximation method to estimate the integral's value.

In your PCE, you studied several methods for approximating definite integrals including left endpoint sums, right endpoint sums, midpoint sums, and trapezoidal sums. In this chapter, we review these methods and introduce some more efficient methods. We also consider their accuracy.

7.1 A review of methods from your PCE

The integral $\int_a^b f(x)\,dx$ is defined by

$$\int_a^b f(x)\,dx = \lim_{\Delta x \to 0} \sum_{i=1}^n f(x_i^*)\Delta x, \tag{7.1}$$

where we have partitioned the interval $[a, b]$ by $a = x_0 < x_1 < \cdots < x_n = b$, $\Delta x = \frac{b-a}{n}$, $x_i = a + i\,\Delta x$, and for each i, $x_{i-1} \le x_i^* \le x_i$ (see Section 0.2). The right-hand side of (7.1) contains a sum in which each term is the signed area of a rectangle. This sum provides a simple way to estimate the value of the integral. We select either a specific value for n or for Δx, make selections for x_i^*, and compute the finite sum. It helps to be systematic in our choice for x_i^*, and that is precisely what you learned to do in constructing left endpoint sums (L_n), right endpoint sums (R_n), and midpoint sums (M_n).

The other method you learned in your PCE is the trapezoidal sum (T_n). In this case, rather than having each term in the sum represent the signed area of a rectangle, we use the signed area of an approximating trapezoid for each term.

Integral Approximations. Let f be defined on $[a, b]$ and $a = x_0 < x_1 < \cdots < x_n = b$ be a partition of $[a, b]$ with $\Delta x = \frac{b-a}{n}$ and $x_i = a + i \Delta x$. The left endpoint, right endpoint, midpoint and trapezoidal estimates of $\int_a^b f(x)\, dx$ are given by

$$L_n = \sum_{i=1}^{n} f(x_{i-1}) \Delta x, \qquad\qquad R_n = \sum_{i=1}^{n} f(x_i) \Delta x,$$

$$M_n = \sum_{i=1}^{n} f\left(\frac{x_{i-1} + x_i}{2}\right) \Delta x, \qquad T_n = \sum_{i=1}^{n} \frac{f(x_{i-1}) + f(x_i)}{2} \Delta x.$$

From these formulas we see that $T_n = \dfrac{L_n + R_n}{2}$.

Example 7.1 Suppose that the velocity of an automobile headed down a highway is recorded every 5 seconds. These values of velocity, $v(t)$, in feet per second over 30 seconds, are given in the table below. We wish to determine the distance the car travels during the 30 seconds.

t (seconds)	0	5	10	15	20	25	30
$v(t)$ (feet/second)	27	37	59	62	70	82	88

We know that if we had an analytic description of $v(t)$, that is a formula for $v(t)$ for all time t in the interval $[0, 30]$, we would be able to find the total distance the car was driven by computing $\int_0^{30} v(t)\, dt$. We will estimate the integral using the data and our approximation methods.

Figure 7.1 shows three of our four estimates. The area shaded dark gray is L_6 while the area in two shades of gray is R_6. The area below the polygonal line joining the white dots is T_6. Can you identify M_3?

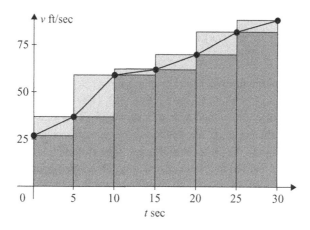

Figure 7.1. Three estimates of distance traveled

For the left endpoint, right endpoint and trapezoidal estimates, we find that $\int_0^{30} v(t)\,dt$ is approximated by

$$L_6 = (27 + 37 + 59 + 62 + 70 + 82) \cdot 5 = 1685 \text{ feet,}$$

$$R_6 = (37 + 59 + 62 + 70 + 82 + 88) \cdot 5 = 1990 \text{ feet,}$$

$$T_6 = \frac{1}{2}(L_6 + R_6) = 1837.5 \text{ feet.}$$

Using our data, we recognize that $t = 5, 15,$ and 25 are the midpoints of the intervals $[0, 10], [10, 20], [20, 30]$. So, we can compute M_3, and estimate $\int_0^{30} v(t)\,dt$ by

$$M_3 = (37 + 62 + 82) \cdot 10 = 1810 \text{ feet.} \ \blacksquare$$

Example 7.2 The integral $\int_a^b e^{-x^2/2}\,dx$ is very important in statistics. However, the integrand $e^{-x^2/2}$ does not have an antiderivative that is expressible in terms of ordinary (elementary) functions. So, rather than using the fundamental theorem of calculus, we use approximation methods to estimate its value.

Let's consider $\int_0^1 e^{-x^2/2}\,dx$. Selecting $n = 10$, we have

$$L_{10} = \left(1 + e^{-(0.1)^2/2} + e^{-(0.2)^2/2} + \cdots + e^{-(0.9)^2/2}\right) \cdot \frac{1}{10} \approx 0.8748,$$

$$R_{10} = \left(e^{-(0.1)^2/2} + e^{-(0.2)^2/2} + \cdots + e^{-(0.9)^2/2} + e^{-1/2}\right) \cdot \frac{1}{10} \approx 0.8354,$$

$$T_{10} = \frac{L_{10} + R_{10}}{2} \approx 0.8551,$$

$$M_{10} = \left(e^{-(0.05)^2/2} + e^{-(0.15)^2/2} + \cdots + e^{-(0.85)^2/2} + e^{-(0.95)^2/2}\right) \cdot \frac{1}{10} \approx 0.8559.$$

To four digits, the actual value of the integral is 0.8556. We can see that with this example, the trapezoidal and midpoint estimates are best. This is often the case. $\blacksquare$

Usually we calculate a trapezoidal estimate T_n without calculating L_n and R_n. Setting $y_i = f(x_i)$, we have

$$T_n = \frac{1}{2}\left(\underbrace{(y_0 + y_1) + (y_1}_{2y_1} + \underbrace{y_2) + (y_2}_{2y_2} + y_3) + \cdots + (y_{n-2} + \underbrace{y_{n-1}) + (y_{n-1}}_{2y_{n-1}} + y_n)\right)\Delta x$$

$$= \frac{\Delta x}{2}(y_0 + 2y_1 + 2y_2 + \cdots + 2y_{n-1} + y_n). \tag{7.2}$$

7.1.1 Understanding error

Each time we estimate the value of an integral, we should try to understand how our approximation relates to the actual value of the integral. For that, we investigate the error in our approximation. The word "error" refers to the difference between our approximation and the true but unknown value of the integral. It does not refer to a mistake as it does in everyday language. Is our estimate too small or too large? How many digits of accuracy do we have in our estimate? For certain integrals, we are able to answer these questions.

Monotone functions Suppose that f is an increasing function on $[a, b]$. When we partition the interval $[a, b]$ by $a = x_0 < x_1 < \cdots < x_n = b$, we have $f(x_{i-1}) \leq f(x_i)$ for each i. Thus $f(x_{i-1})\Delta x \leq \int_{x_{i-1}}^{x_i} f(x)\,dx \leq f(x_i)\Delta x$. This means that the left endpoint sum will underestimate our integral while the right endpoint sum will overestimate our integral. If f is a decreasing function, these behaviors reverse. Figure 7.2 illustrates this.

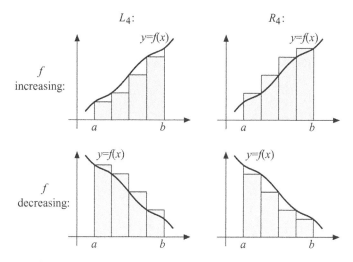

Figure 7.2. Left and right estimates for monotone functions

Theorem 7.1 (Monotonicity and integral estimates) *Let f be a continuous function on $[a, b]$.*

1. If f is increasing on the interval $[a, b]$, then for all $n \geq 1$,

$$L_n \leq \int_a^b f(x)\,dx \leq R_n.$$

2. If f is decreasing on the interval $[a, b]$, then for all $n \geq 1$,

$$R_n \leq \int_a^b f(x)\,dx \leq L_n.$$

3. If f changes from increasing to decreasing or vice versa in the interval, then no conclusion can be drawn.

A nice consequence of this theorem is that when f is monotone, the integral is sandwiched between the left endpoint and right endpoint sums.

Concave down or concave up functions Suppose the graph of $y = f(x)$ is concave down and positive on the interval (a, b). We see in the left side of Figure 7.3 that the area of the trapezoid constructed on this interval underestimates the area under the graph of $y = f(x)$ on the interval.

In the right side of Figure 7.3, we see that the two shaded triangles have the same area. So the midpoint rectangle (the region below the horizontal line at height $f\left(\frac{a+b}{2}\right)$) has the same area as the trapezoidal region below the dashed line. Thus the midpoint rectangle overestimates

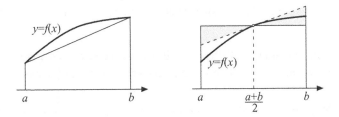

Figure 7.3. Concavity, trapezoidal (left) and midpoint (right) estimates

the area under the graph of $y = f(x)$ on this interval. This result also holds when f is concave down and negative on the interval (a, b) (see Exercise 15). The concave up case is similar (see Exercise 16).

We summarize these observations in the following theorem.

Theorem 7.2 (Concavity and integral estimates) *Let f be continuous on $[a, b]$ and differentiable on (a, b).*

1. *If the graph of f is concave up on the interval (a, b), then for all $n \geq 1$,*

$$M_n \leq \int_a^b f(x)\,dx \leq T_n.$$

2. *If the graph of f is concave down on the interval (a, b), then for all $n \geq 1$,*

$$T_n \leq \int_a^b f(x)\,dx \leq M_n.$$

3. *If f changes concavity within the interval, then no conclusion can be drawn.*

Theorem 7.2 reveals that for a function whose concavity does not change on the interval (a, b), the integral $\int_a^b f(x)\,dx$ lies between the midpoint and trapezoidal estimates.

We can say even more about functions that do not change concavity. Figure 7.4 is a visual argument showing that the midpoint rule is better than the trapezoidal rule for functions that are concave down. In Figure 7.4, the gray region on the far left represents the error in the midpoint rule approximation and the gray region on the far right represents the error in the trapezoidal

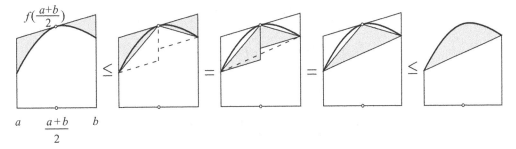

Figure 7.4. Midpoint is better than trapezoid for concave down functions

rule approximation. A similar diagram shows that the same is true for functions that are concave up.

Error bounds If our integrand is differentiable and we have certain information about the derivatives, we can specify with certainty a maximal error for each of our estimates.

> **Theorem 7.3** *Suppose f is continuous on $[a, b]$. Also suppose f has as many derivatives necessary on $[a, b]$ to be able to write the right-hand side of each of the following inequalities. Then*
>
> $$\left| \int_a^b f(t)\, dt - L_n \right| \le \frac{(b-a)^2}{2n} \max_{a \le x \le b} |f'(x)|, \tag{7.3}$$
>
> $$\left| \int_a^b f(t)\, dt - R_n \right| \le \frac{(b-a)^2}{2n} \max_{a \le x \le b} |f'(x)|, \tag{7.4}$$
>
> $$\left| \int_a^b f(t)\, dt - T_n \right| \le \frac{(b-a)^3}{12n^2} \max_{a \le x \le b} |f''(x)|, \tag{7.5}$$
>
> $$\left| \int_a^b f(t)\, dt - M_n \right| \le \frac{(b-a)^3}{24n^2} \max_{a \le x \le b} |f''(x)|. \tag{7.6}$$

In each of the inequalities, the left-hand side is the distance between our estimate and the actual value of the integral. The right-hand side then gives us an upper bound on this distance. Each right-hand side involves the length of the interval over which we are integrating, $b - a$, and the maximum absolute value of a derivative of f over the interval. These two numbers are fixed. It's the n in the denominator that gives us the power in estimation. Since n appears in the denominator, the larger we make n, the smaller the error bound is, and the more certain we are about the accuracy of our estimate.

Let's prove (7.3). The proof of (7.4) is similar. Inequalities (7.5) and (7.6) are proved in a course in numerical analysis.

Proof of (7.3). We begin by partitioning $[a, b]$ by $a = x_0 < x_1 < \cdots < x_{n-1} < x_n = b$. Consider the left endpoint approximation on one subinterval $[x_{i-1}, x_i]$. On this interval, we use the mean value theorem (Theorem 0.5) to conclude that for each x in $[x_{i-1}, x_i]$, there is a c_i in (x_{i-1}, x_i) such that $f'(c_i)(x - x_{i-1}) = f(x) - f(x_{i-1})$. Integrating both sides of this equation over the interval $[x_{i-1}, x_i]$ and taking absolute values yields

$$\left| \int_{x_{i-1}}^{x_i} f(x) - f(x_{i-1})\, dx \right| = \left| \int_{x_{i-1}}^{x_i} f'(c_i)(x - x_{i-1})\, dx \right|$$

$$\le \max_{x_{i-1} \le x \le x_i} |f'(x)| \int_{x_{i-1}}^{x_i} (x - x_{i-1})\, dx$$

$$= \max_{x_{i-1} \le x \le x_i} |f'(x)| \cdot \frac{(x_i - x_{i-1})^2}{2}$$

$$= \max_{x_{i-1} \le x \le x_i} |f'(x)| \cdot \frac{\big((b-a)/n\big)^2}{2}$$

$$= \frac{(b-a)^2}{2n^2} \cdot \max_{x_{i-1} \le x \le x_i} |f'(x)| \le \frac{(b-a)^2}{2n^2} \max_{a \le x \le b} |f'(x)|. \tag{7.7}$$

Therefore, we have

$$
\left| \int_a^b f(x)\,dx - L_n \right| = \left| \int_a^b f(x)\,dx - \sum_{i=1}^n f(x_{i-1})(x_i - x_{i-1}) \right|
$$

$$
= \left| \sum_{i=1}^n \int_{x_{i-1}}^{x_i} f(x)\,dx - f(x_{i-1})(x_i - x_{i-1}) \right|
$$

$$
= \left| \sum_{i=1}^n \int_{x_{i-1}}^{x_i} (f(x) - f(x_{i-1}))\,dx \right|
$$

$$
\leq \sum_{i=1}^n \left| \int_{x_{i-1}}^{x_i} (f(x) - f(x_{i-1}))\,dx \right|
$$

$$
\leq \sum_{i=1}^n \frac{(b-a)^2}{2n^2} \max_{a \leq x \leq b} |f'(x)| = \frac{(b-a)^2}{2n} \max_{a \leq x \leq b} |f'(x)|.
$$

The last inequality comes from (7.7). $\square$

Example 7.3 Consider the integral $\int_0^2 \sin(e^x)\,dx$. If we are to estimate it using each of our methods, what should n be to guarantee the estimate is within 10^{-3} of the actual value of the integral?

In other words, we must find the values for n that make the right sides of (7.4), (7.3), (7.5), and (7.6) less than 0.001. To do so, we first determine upper bounds on the maximum value of the absolute values of the first and second derivatives of $f(x) = \sin(e^x)$ on the interval $0 \leq x \leq 2$.

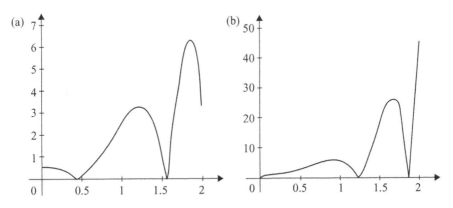

Figure 7.5. Graphs of (a) $|f'(x)|$ and (b) $|f''(x)|$ of $f(x) = \sin(e^x)$

The graphs of $y = |f'(x)|$ and $y = |f''(x)|$ in Figure 7.5 indicate that $|f'(x)| \leq 7$ on $[0, 2]$ and $|f''(x)| \leq 60$ on $[0, 2]$. In estimating the maximum value of the derivatives, it is okay to overestimate, so long as we use reasonable approaches.

Now, for both the left endpoint and right endpoint sums, we want to find n so that

$$
\frac{|2 - 0|^2}{2n} \max_{0 \leq x \leq 2} |f'(x)| < \frac{2}{n} \cdot 7 \leq 0.001.
$$

So we solve

$$\frac{2}{n} \cdot 7 \leq 0.001 \text{ so that } n \geq 14,000.$$

It appears that to guarantee a reasonable accuracy with left or right endpoint sums, we must use a large number of subintervals.

To determine n for the trapezoidal sum, we have

$$\frac{|2-0|^3}{12n^2} \max_{0 \leq x \leq 2} |f''(x)| \leq \frac{8}{12n^2} \cdot 60 \leq 0.001 \text{ so that } n \geq 200.$$

And finally, for the midpoint sum

$$\frac{|2-0|^3}{24n^2} \max_{0 \leq x \leq 2} |f''(x)| \leq 0.001 \text{ so that } n \geq 142.$$

It is impressive how many fewer subintervals are needed to guarantee good accuracy when using a trapezoidal or midpoint sum rather than an endpoint sum! However, remember that the computations are based on what the maximum error might be. The following table shows the estimates based on these values for n. The actual value of the integral to ten decimal places is 0.5509351737. Each of our estimates is within the desired tolerance of 0.001. For each method the actual error is substantially less than the error bound of 0.001. This is often the case.

Method	n	Estimate	Error
left	14,000	0.5509314397	-0.0000037340
right	14,000	0.5509389231	$+0.0000037494$
trapezoidal	200	0.5509582830	$+0.0000231093$
midpoint	142	0.5509122410	-0.0000229327

Symbolically finding error bounds The graphs in Figure 7.5 are helpful to estimate bounds on the maximum values of $|f'(x)|$ and $|f''(x)|$. We can also use a symbolic approach to determine bounds. With $f(x) = \sin(e^x)$, we have $f'(x) = e^x \cos(e^x)$ and $f''(x) = e^x \cos(e^x) - e^x \cdot e^x \sin(e^x) = e^x \cos(e^x) - e^{2x} \sin(e^x)$. Since for all x, $|\sin(x)| \leq 1$ and $|\cos(x)| \leq 1$, and $2 < e < 3$, we can see that

$$\max_{0 \leq x \leq 2} |f'(x)| = \max_{0 \leq x \leq 2} |e^x \sin(e^x)| \leq \max_{0 \leq x \leq 2} e^x = e^2 < 7.5$$

and

$$\max_{0 \leq x \leq 2} |f''(x)| = \max_{0 \leq x \leq 2} |e^x \cos(e^x) - e^x \cdot e^x \sin(e^x)|$$

$$\leq \max_{0 \leq x \leq 2} e^x + \max_{0 \leq x \leq 2} e^{2x} = e^2 + e^4 < 62.$$

These bounds are not the tightest bounds (our graphical estimates were better), but they are bounds nevertheless and can be used in estimating error. ∎

7.1.2 Exercises

In Exercises 1 through 6 estimate the value of each integral using a (i) left endpoint sum (ii) right endpoint sum (iii) midpoint sum (iv) trapezoidal sum with the given value of n. Compute the integral exactly and determine the error in each of your estimates.

1. $\displaystyle\int_0^1 t^4\, dt, n = 8$
2. $\displaystyle\int_0^4 t^4\, dt, n = 8$
3. $\displaystyle\int_1^3 \frac{1}{1+x^2}\, dx, n = 4$

4. $\displaystyle\int_2^4 \ln x\, dx, n = 4$
5. $\displaystyle\int_0^6 \sin(\pi x), n = 6$
6. $\displaystyle\int_0^{1/2} \frac{1}{\sqrt{1-x^2}}\, dx, n = 4$

7. Compute the trapezoidal sum with one subinterval for the integral $\int_0^1 5x\, dx$. Determine the exact value of the integral as well. Can you explain your results? What happens when you use a midpoint sum with one subinterval? Can you explain why?

In Exercises 8–10, for each of the four methods (left, right, trapezoid, and midpoint), we wish to determine a number of subintervals, n, that will guarantee that the estimate obtained will be within 0.00001 of the exact value of the integral. Use Theorem 7.3 to determine such values of n for each of the four methods.

8. $\displaystyle\int_0^1 e^{-t^2}\, dt$
9. $\displaystyle\int_{-1}^1 \sqrt{9 - x^3}\, dx$
10. $\displaystyle\int_1^3 \frac{1}{1+s^6}\, ds$

In Exercises 11–14, use technology to find the indicated approximation. Then, use Theorem 7.3 to find a bound on the error in your approximation.

11. $\displaystyle\int_0^1 e^{-x^2}\, dx, \quad R_{100}$.

12. $\displaystyle\int_0^1 e^{-x^2}\, dx, \quad T_{100}$.

13. $\displaystyle\int_1^3 e^{-x}\sin(x^2)\, dx, \quad M_{100}$.

14. $\displaystyle\int_{-2}^2 \frac{1}{\sqrt{1+x^3+x^6}}\, dx, \quad T_{50}$.

15. Use graphs similar those in Figure 7.3 to show that for a continuous function f on $[a, b]$, if $f(x) < 0$ for all x in $[a, b]$ and if f is concave down on $[a, b]$, then $T_n \le \int_a^b f(x)\, dx \le M_n$ for all n.

16. Use graphs similar those in Figure 7.3 to show that for a continuous function f on $[a, b]$, if $f(x) > 0$ for all x in $[a, b]$ and if f is concave up on $[a, b]$, then $T_n \le \int_a^b f(x)\, dx \le M_n$ for all n. Do the same for a function f such that $f(x) < 0$ for all x in $[a, b]$.

17. Consider the integral $\displaystyle\int_0^{10} \frac{1}{1+e^{-2t}}\, dt$.
    (a) Determine R_{10} and L_{10}.
    (b) Determine upper bounds on the errors in the estimates using Theorem 7.3.

(c) Show that the function $f(t) = \frac{1}{1+e^{-2t}}$ is an increasing function on the interval $[0, 10]$. What does Theorem 7.1 tell you about the exact value of the integral?

(d) How does your answer to part (b) compare with your answer to part (c)? Why are the answers compatible?

18. Consider the integral $\int_{-5}^{5} \ln\left(\sqrt{e^{2x} + 1}\right) dx$.

(a) Determine T_{10} and M_{10}, the trapezoidal and midpoint sums with ten subintervals of equal length.

(b) Determine upper bounds on the errors in the estimates using Theorem 7.3.

(c) Show that the graph of $y = \ln\left(\sqrt{e^{2x} + 1}\right)$ is concave up on the interval $[-5, 5]$. What does Theorem 7.2 tell you about the exact value of the integral?

(d) How does your answer to part (b) compare with your answer to part (c)? Why are the answers compatible?

7.2 Simpson's rule

We have seen that if a function does not change its concavity in the interval (a, b), then the integral, $\int_a^b f(x)\,dx$, is sandwiched between the midpoint estimate and the trapzoidal estimate. It makes sense, therefore, to create a new estimate that is an average of the estimates. We know that the midpoint estimate is better than the trapezoidal estimate. *But how much better?*

Example 7.4 Consider the integral $\int_1^5 \frac{1}{x}\,dx$. We use the fundamental theorem of calculus to find $\int_1^5 \frac{1}{x}\,dx = \ln 5$. We can use trapezoidal or midpoints approximations for the integral to estimate $\ln 5$. Let's examine the estimates and their errors for various choices of n. In the following table, the error in each approximation is the estimate minus $\ln 5$, and the percent relative error (the error divided by $\ln 5$ expressed as a percent) is given in parentheses.

n	Trapezoidal	Error	Midpoint	Error
2	1.866666667	0.25722876 (15.98%)	1.500000000	−0.109437912 (−6.80%)
4	1.683333333	0.073895421 (4.59%)	1.574603175	−0.034834737 (−2.16%)
8	1.628968254	0.019530342 (1.21%)	1.599844394	−0.009593518 (−0.60%)
16	1.614406324	0.004968412 (0.31%)	1.606965468	−0.002472444 (−0.15%)
32	1.610685896	0.001247984 (0.08%)	1.608814675	−0.000623237 (−0.04%)

Evident in this table is that the error in the midpoint estimate is approximately half that of the trapezoidal estimate and the errors have opposite signs. We also see that each time we double the size of n, the error in both the trapezoidal and the midpoint estimates is reduced by a factor of approximately 4. This is consistent with n^2 appearing in the denominator of their error bounds (see Theorem 7.3). ∎

For many integrals, the errors given by the trapezoidal and midpoint rules compare similarly to Example 7.4 and so it makes sense to construct a new integral approximation that is a weighted average of the trapezoidal and midpoint estimates. Given the relationship of their errors, we will use the weighted average $\frac{1}{3}T_n + \frac{2}{3}M_n$. This new estimate is called *Simpson's rule*, named after the English mathematician Thomas Simpson (1710–1761).

To accomplish this averaging, we will use some of the points in the partition to evaluate T_n and the others to evaluate M_n. We illustrate with an example using ten subintervals so that $\Delta x = \frac{b-a}{10}$. The points x_1, x_3, x_5, x_7, x_9 will be used for the midpoint estimate, M_5. They are the midpoints of the intervals $[x_0, x_2], [x_2, x_4], \ldots, [x_8, x_{10}]$. Each of the five intervals has width $2\Delta x$ and so

$$M_5 = 2\Delta x \left(y_1 + y_3 + y_5 + y_7 + y_9 \right).$$

The points $x_0, x_2, x_4, x_6, x_8, x_{10}$ will be used for the trapezoidal estimate. They are endpoints of the five intervals $[x_0, x_2], [x_2, x_4], \ldots, [x_8, x_{10}]$, each with width $2\Delta x$, and so

$$T_5 = \frac{2\Delta x}{2} \left(y_0 + 2y_2 + 2y_4 + 2y_6 + 2y_8 + y_{10} \right).$$

So, since we have used ten subintervals, we have

$$S_{10} = \frac{1}{3}T_5 + \frac{2}{3}M_5 = \frac{\Delta x}{3} \left(y_0 + 4y_1 + 2y_2 + 4y_3 + \cdots + 2y_8 + 4y_9 + y_{10} \right).$$

This method applies for any partition $a = x_0 < x_1 < x_2 < \cdots < x_{n-1} < x_n = b$ where n is even.

Simpson's rule Let n be even, and let $a = x_1 < x_2 < \cdots < n_{n-1} < x_n = b$ be a partition of the interval $[a, b]$. With $\Delta x = \frac{b-a}{n}$, let

$$S_n = \frac{\Delta x}{3} \left(f(x_0) + 4f(x_1) + 2f(x_2) + 4f(x_3) + \cdots + 2f(x_{n-2}) + 4f(x_{n-1}) + f(x_n) \right).$$

(7.8)

S_n is known as a Simpson's rule estimate for the integral $\int_a^b f(x)\,dx$.

Back to Example 7.4 We use Simpson's rule to estimate $\int_1^5 \frac{1}{x}\,dx$. The table below shows the Simpson's rule estimates for the same values of n we used for the trapezoidal and midpoint estimates.

n	Simpson's	Error
2	1.688888889	0.0794509769 (4.94%)
4	1.622222222	0.0127843100 (0.79%)
8	1.610846561	0.0014086490 (0.09%)
16	1.609552347	0.0001144351 (0.007%)
32	1.609445753	0.0000078412 (0.0005%)

In terms of relative error, the Simpson's rule estimate with $n = 32$ is nearly 100 times better than the midpoint estimate! For many integrals, this is the case. ■

Of course, we would like to know an upper bound on the error in Simpson's rule. It is known and has a form not unlike those we've seen for our other methods of integral approximation.

Theorem 7.4 (Simpson's rule error bound) *Let f be a function that is at least four times differentiable on (a, b) and continuous on $[a, b]$. Then*

$$\left| \int_a^b f(x)\,dx - S_n \right| \le \frac{(b-a)^5}{180n^4} \max_{a \le x \le b} |f^{(4)}(x)|.$$

Back to example 7.3 Let's use Simpson's rule to estimate $\int_0^2 \sin(e^x)\,dx$ to within 0.001.

The graph of the absolute value of the fourth derivative of $\sin(e^x)$ on the interval $[0, 2]$ is shown in Figure 7.6. We can see that $\max_{0 \le x \le 2} |f^{(4)}(x)| \le 1550$. Thus,

$$\left| \int_0^2 \sin(e^x)\,dx - S_n \right| \le \frac{2^5}{180n^4} 1550 = \frac{2480}{9n^4}.$$

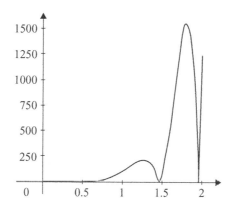

Figure 7.6. Graph of $|f^{(4)}(x)|$

To obtain an estimate that is within 0.001 of the actual value of the integral, it is sufficient that $2480/9n^4 \le 0.001$. In other words, $n \ge 23$. Since n must be even to use Simpson's rule, we select $n = 24$, compute $S_{24} = 0.5509296818$ and conclude that this is within 0.001 of the actual value of the integral. ∎

7.2.1 An approximation based on parabolas

To approximate the integral $\int_a^b f(x)\,dx$, the trapezoidal estimate works by approximating the function $f(x)$ by a piecewise linear function. Can we do better with a piecewise quadratic function, using parabolic arcs to approximate the graph of $f(x)$?

Figure 7.7 illustrates both a trapezoidal and parabolic approach to estimating an integral. Just as two points determine a line segment, three points determine a parabolic arc, so we begin at the left endpoint (our first in the partition), and use the next two points for the first parabolic arc. For the next parabolic arc, we begin at the third point in our partition and use the next two, and so on.

For the general case, begin by considering a partition $a = x_0 < x_1 < x_2 < \cdots < x_{n-2} < x_{n-1} < x_n = b$ where n is even. We work with the first three points in our partition, x_0, x_1, x_2.

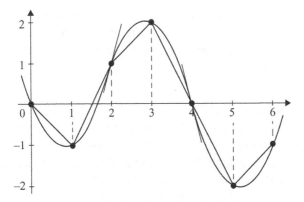

Figure 7.7. Approximation by trapezoids and parabolas

Let $y_i = f(x_i)$. Our goal is to find the signed area between the x-axis and the parabolic arc passing through (x_0, y_1), (x_1, y_1), (x_2, y_2).

We can simplify our work by translating horizontally so that x_1 moves to $x = 0$. Letting $h = \Delta x$, our task now is to find the signed area between the x-axis and the parabolic arc passing through $(-h, y_0)$, $(0, y_1)$, (h, y_2).

Let $Ax^2 + Bx + C$ be the quadratic that passes through these three points. Then we have

$$Ah^2 - Bh + C = y_0$$
$$C = y_1$$
$$Ah^2 + Bh + C = y_2.$$

Adding the first and third equations and substituting for C, we obtain

$$2Ah^2 = y_0 - 2y_1 + y_2.$$

We can then compute the signed area in question as

$$\int_{-h}^{h} \left(Ax^2 + Bx + C \right) dx = \frac{2}{3}Ah^3 + 2Ch$$
$$= \frac{h}{3}(y_0 - 2y_1 + y_2) + 2hy_1 = \frac{h}{3}(y_0 + 4y_1 + y_2).$$

This same method works for each triplet of consecutive points x_i, x_{i+1}, x_{i+2} where i is even. To get our integral estimate, we add the approximations together and see that

$$\int_a^b f(x)\, dx \approx \frac{h}{3}(y_0 + 4y_1 + 2y_2 + \cdots + 2y_{n-2} + 4y_{n-1} + y_n)$$
$$= \frac{\Delta x}{3}(f(x_0) + 4f(x_1) + 2f(x_2) + \cdots + 2f(x_{n-2}) + 4f(x_{n-1}) + f(x_n)).$$

But this is Simpson's rule! See (7.8). So, we have two approaches to Simpson's rule: one a weighted average of the midpoint and trapezoidal estimates, and another the estimate obtained from approximations by parabolic arcs.

Approximations by cubics and higher degree polynomials lead to additional approximation methods. These were important before the advent of computers when computation was hard and mathematicians strived to obtain as much accuracy as possible with a minimum of computation.

Why is Simpson's rule exact for cubics?

One of the great freebies of calculus is the fact that Simpson's rule, guaranteed to be exact for quadratics, is also exact for cubics. Of course this follows from the error bound, but the following explanation is more direct. It suffices to consider the interval $[-h, h]$ for arbitrary positive h. Let $f(x)$ be an cubic polynomial, and let $g(x)$ be the unique quadratic polynomial that agrees with f at $-h, 0$, and h. Now let $p(x) = f(x) - g(x)$.

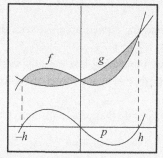

A quadratic and a cubic through the same three points.

Since $p(-h) = p(0) = p(h) = 0$, the three zeros of p are $-h, 0$, and h. Therefore $p(x) = ax(x + h)(x - h) = ax^3 - ah^2 x$ (where a is the coefficient of x^3 in f). Hence p is an odd function, so that $\int_{-h}^{h} p(x)\,dx = 0$. Thus $\int_{-h}^{h} f(x)\,dx = \int_{-h}^{h} g(x)\,dx$, that is, the two gray regions in the figure have the same area. Hence, since Simpson's rule is exact for g, it is also exact for f.

Example 7.5 Since the area of a circle of radius r is πr^2, the area of one-quarter of the circle $x^2 + y^2 = 4$ is π. That is, $\int_0^2 \sqrt{4 - x^2}\,dx = \pi$, as illustrated in Figure 7.8(a).

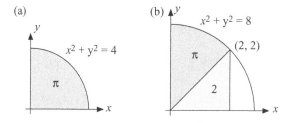

Figure 7.8. Estimating π

Let's use Simpson's rule to approximate this integral. Here are some results (S_n denotes the Simpson's rule approximation with n subintervals):

n	S_n
10	3.1364470643
20	3.1397753524
40	3.1409504859
100	3.1414302492

The disappointing results (e.g., only three correct decimals with $n = 100$) are not surprising, since the first and higher order derivatives of the integrand are unbounded at $x = 2$ (recall the formula for the error bound for Simpson's rule).

But we can dramatically improve the results by using one-eighth of the circle $x^2 + y^2 = 8$ (See Figure 7.8(b)), and evaluating $\int_0^2 \sqrt{8 - x^2}\, dx = \pi + 2$ ($\pi + 2$ and $\pi + 2$ look exactly the same to the right of the decimal point). The derivatives of this integrand are bounded on $[0, 2]$, and here are the results from Simpson's rule:

n	$S_n - 2$
10	3.1415918322
20	3.1415926017
40	3.1415926503
100	3.1415926535

Here we get five decimals correct with $n = 10$ and ten decimals correct with $n = 100$. ∎

7.2.2 Exercises

In Exercises 1 through 6 estimate the value of each integral using Simpson's rule with the given value of n. Compute the integral exactly and determine the error in each of your estimates.

1. $\displaystyle\int_0^1 t^4\, dt, n = 8$ 2. $\displaystyle\int_0^4 t^4\, dt, n = 8$ 3. $\displaystyle\int_1^3 \frac{1}{1+x^2}\, dx, n = 4$

4. $\displaystyle\int_2^4 \ln(x)\, dx, n = 4$ 5. $\displaystyle\int_0^6 \sin(\pi x)\, dx, n = 6$ 6. $\displaystyle\int_0^{1/2} \frac{1}{\sqrt{1 - x^2}}\, dx, n = 4$

In Exercises 7–9, we wish to determine a number of subintervals, n, that will guarantee that the Simpson's rule estimate will be within 0.00001 of the exact value of the integral. Use Theorem 7.4 to determine such a value of n in each exercise.

7. $\displaystyle\int_0^1 e^{-t^2}\, dt$ 8. $\displaystyle\int_{-1}^1 \sqrt{9 - x^3}\, dx$ 9. $\displaystyle\int_1^3 \frac{1}{1+s^6}\, ds$

In Exercises 10–13, use technology to find S_n for the given value of n. Then, use Theorem 7.4 to find a bound on the error in your approximation.

10. $\displaystyle\int_0^1 e^{-x^2}\, dx, n = 50.$

11. $\displaystyle\int_0^1 e^{-x^2}\, dx, n = 100.$

12. $\int_1^3 e^{-x} \sin(x^2)\,dx$, $n = 100$.

13. $\int_{-2}^2 \dfrac{1}{\sqrt{1 + x^3 + x^6}}\,dx$, $n = 50$.

14. Consider the integral $\int_0^2 e^{x^2}\,dx$.
    (a) Use technology to compute this integral.
    (b) Use Theorem 7.4 to estimate an upper bound on the error in Simpson's rule with 10 subintervals.
    (c) Compute, using technology, the Simpson's rule estimate of the integral with 10 subintervals. Compute the *actual* error in this estimate (using your answer to part (a)) and compare it to the error bound computed in part (b).

15. Show that
$$\pi = \int_0^1 \frac{6}{\sqrt{4 - x^2}}\,dx = \int_0^1 \frac{4}{1 + x^2}\,dx,$$
and use Simpson's rule with these integrals to approximate π.

7.3 Explorations

7.3.1 Symmetry and definite integrals

Consider the integral
$$\int_0^{2\pi} \frac{1}{1 + e^{\sin x}}\,dx. \tag{7.9}$$

None of the techniques you have studied for finding antiderivatives seem to help here. But before resorting to one of the numerical methods in this chapter, take a look at the graph of the integrand. Since it is positive on $[0, 2\pi]$, the integral represents the area of the shaded region in Figure 7.9

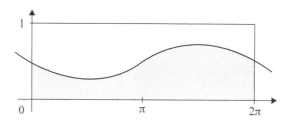

Figure 7.9. Graph of $y = 1/(1 + e^{\sin x})$

The function has symmetry similar to the symmetry of an odd function, but the center of symmetry is not the origin but rather the point $(\pi, 1/2)$. Hence we suspect that the value of the integral in (7.9) should equal one-half the area of the enclosing rectangle with width 2π and height 1, that is, π.

To show that this is indeed the case, we make a symmetry substitution. For an integral of the form $\int_a^b f(x)\,dx$ where the graph of f has the symmetry described above, we proceed as follows:

1. Split the interval of integration at its midpoint to obtain two integrals.
2. Substitute $x = a + b - t$ in the second integral, then replace t by x.
3. Combine the two integrals, simplify the integrand, and integrate.

If the apparent symmetry is indeed present, then the integrand should now be a constant (equal to $f(a) + f(b)$). Let's illustrate with the integral in (7.9):

$$\int_0^{2\pi} \frac{1}{1 + e^{\sin x}}\,dx = \int_0^{\pi} \frac{1}{1 + e^{\sin x}}\,dx + \int_{\pi}^{2\pi} \frac{1}{1 + e^{\sin x}}\,dx$$

$$= \int_0^{\pi} \frac{1}{1 + e^{\sin x}}\,dx - \int_{\pi}^{0} \frac{1}{1 + e^{\sin(2\pi - t)}}\,dt$$

$$= \int_0^{\pi} \frac{1}{1 + e^{\sin x}}\,dx + \int_0^{\pi} \frac{1}{1 + e^{-\sin x}}\,dx$$

$$= \int_0^{\pi} \left(\frac{1}{1 + e^{\sin x}} + \frac{e^{\sin x}}{1 + e^{\sin x}} \right)\,dx = \int_0^{\pi} 1\,dx = \pi.$$

Here are some for you to try (answers are in parentheses). We recommend using your graphing calculator first to observe the symmetry of the integrand in each exercise.

Exercise 1. Evaluate $\int_0^4 \frac{dx}{4+2^x}$. $(1/2)$

Exercise 2. Evaluate $\int_{-1}^{1} \arctan(e^x)\,dx$. $(\pi/2)$

Exercise 3. Evaluate $\int_0^{\pi/4} \ln(1 + \tan x)\,dx$. $((\pi \ln 2)/8)$

Exercise 4. Evaluate $\int_0^2 \frac{dx}{x+\sqrt{x^2-2x+2}}$. (1)

The William Lowell Putnam Mathematical Competition is an annual mathematics contest for college and university students in the United States and Canada. An article in *TIME* magazine (December 16, 2002) called it the "world's toughest math test." Here are two problems (with answers) from that competition that you can now solve:

Exercise 5 [Problem A3, 1980]. Evaluate $\int_0^{\pi/2} \frac{1}{1+(\tan x)^{\sqrt{2}}}\,dx$. $(\pi/4)$

Exercise 6 [Problem B1, 1987]. Evaluate $\int_2^4 \frac{\sqrt{\ln(9-x)}}{\sqrt{\ln(9-x)}+\sqrt{\ln(x+3)}}\,dx$. (1)

Note: Graphing calculators are *not permitted* on the Putnam Competition!

7.3.2   Means for two positive numbers and the Hermite-Hadamard inequality

There are many ways to compute the average of two positive numbers a and b. These averages are called *means* (from the French *moyen*, "medium" or "middle"). Six of the most common means are listed below:

- the *arithmetic mean*: $\dfrac{a+b}{2}$,

- the *geometric mean*: $\sqrt{ab}$,

- the *harmonic mean*: $\dfrac{2ab}{a+b}$,

- the *root mean square*: $\sqrt{\dfrac{a^2+b^2}{2}}$,

- the *contraharmonic mean*: $\dfrac{a^2+b^2}{a+b}$, and

- the *logarithmic mean*: $\dfrac{b-a}{\ln b - \ln a}$ (for $a \neq b$), a (for $a = b$).

The arithmetic mean is certainly familiar, it is the usual average of, for example, exam scores in one of your courses. The origin of the geometric mean is of course geometry: it is the length of the side of a square equal in area to a rectangle with side lengths a and b. The harmonic mean is often used for rates: if you drive from Portland to Eugene (two cities in Oregon about 180 km apart) at 90 km/h and return at 60 km/h, then the average speed for the round trip 72 km/h (since the 360 km round trip takes 5 hours), the harmonic mean of 90 and 60. The root mean square (the square *root* of the arithmetic *mean* of the *squares*) is sometimes used to average quantities that can be both positive and negative. The contraharmonic mean complements the harmonic mean, as it is as much larger than the arithmetic mean as the harmonic mean is smaller. Finally, the logarithmic mean is used in certain applications in physics and engineering.

In this exploration you will establish the following inequalities between the means: If $0 < a \leq b$, then

$$a \leq \frac{2ab}{a+b} \leq \sqrt{ab} \leq \frac{b-a}{\ln b - \ln a} \leq \frac{a+b}{2} \leq \sqrt{\frac{a^2+b^2}{2}} \leq \frac{a^2+b^2}{a+b} \leq b. \tag{7.10}$$

Exercise 1. As an example, show that the inequalities between the means in (7.10) for $a = 2$ and $b = 8$ are

$$2 < 3.2 < 4 < \frac{3}{\ln 2} < 5 < \sqrt{34} < 6.8 < 8.$$

Exercise 2. Assuming (7.10) has been established, show that equality holds throughout (7.10) when $a = b$. (Hint: review the squeeze theorem, Theorem 0.2.)

Means and the apportionment of the House of Representatives
Article 1, Section 2 of the Constitution of the United States reads in part "Representatives and direct Taxes shall be apportioned among the several States which may be included within this Union, according to their respective Numbers. . . . " However, the Constitution is silent on how the apportionment is to be done, and since 1790 several different methods have been used. When one divides a state's population by a proposed House district size, the result is generally a non-integer that must be rounded up or down. The apportionment methods that have been used or proposed for use differ on the rounding rule. Some of the methods (and the rounding rule) are Jefferson's (always round down), Adams' (always round up), Webster's (round up or down at the arithmetic mean), Dean's (round up or down at the geometric mean), and Huntington-Hill's (round up or down at the harmonic mean). The method in current use is Huntington-Hill. For further details, see the book by Balinski and Young mentioned in the Acknowledgments at the end of this chapter.

Several of the inequalities in (7.10) can be established with simple algebra. However, we will need to approximate definite integrals with the trapezoidal and midpoint rules to establish some of them.

Exercise 3. Your first task is to show that the geometric mean is less than or equal to the arithmetic mean (we will insert the logarithmic mean between them in Exercise 6). Here are two ways to do it, one algebraic, one geometric: (i) show that $0 \leq (\sqrt{a} - \sqrt{b})^2$ implies $\sqrt{ab} \leq (a + b)/2$; and (ii) use Figure 7.10. (Hint: compare the sum of the areas of the two isosceles right triangles to the area of the shaded rectangle.)

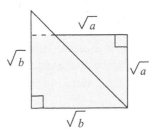

Figure 7.10. A geometric proof of the arithmetic mean-geometric mean inequality

Exercise 4. Show that the harmonic mean is less than or equal to the geometric mean. (Hint: either apply the inequality between the geometric and arithmetic means to the numbers $1/a$ and $1/b$, or multiply both sides of the inequality between the geometric and arithmetic means by $2\sqrt{ab}/(a + b)$.)

Exercise 5. Use algebra (with $0 < a \leq b$) to establish the two outer inequalities in (7.10): $a \leq 2ab/(a + b)$ and $(a^2 + b^2)/(a + b) \leq b$.

To insert the logarithmic mean between the geometric and arithmetic means requires calculus. The tool that you will use to do this is the *Hermite-Hadamard inequality*, named for the French mathematicians Charles Hermite (1822–1901) who first published it in 1883, and Jacques Hadamard (1865–1963) who rediscovered it ten years later.

Theorem 7.5 (The Hermite-Hadamard inequality) *If f is continuous and concave up on $[a, b]$, then*

$$f\left(\frac{a+b}{2}\right) \le \frac{1}{b-a} \int_a^b f(x)\,dx \le \frac{f(a)+f(b)}{2},$$

and if f is continuous and concave down on $[a, b]$, then

$$\frac{f(a)+f(b)}{2} \le \frac{1}{b-a} \int_a^b f(x)\,dx \le f\left(\frac{a+b}{2}\right).$$

The middle term in the Hermite-Hadamard inequality is f_{ave}, the *average value* of f on $[a, b]$, and the inequality states that if f is either concave up or concave down on $[a, b]$ then f_{ave} lies between f evaluated at the average of a and b and the average of f at a and f at b.

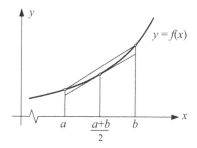

Figure 7.11. Proving the Hermite-Hadamard inequality

Exercise 6. Prove the Hermite-Hadamard inequality. (Hint: consider the integral $\int_a^b f(x)\,dx$ and its approximations T_1 (the trapezoidal rule with one trapezoid) and M_1 (the midpoint rule with just one midpoint). See Figure 7.11 for the concave up case.)

Exercise 7. Show that

$$\sqrt{ab} \le \frac{b-a}{\ln b - \ln a} \le \frac{a+b}{2}.$$

(Hint: apply the Hermite-Hadamard inequality to $f(x) = e^x$ on the interval $[\ln a, \ln b]$.)

Exercise 8. Show that the arithmetic mean is less than or equal to the root mean square. (Hint: apply the Hermite-Hadamard inequality to $f(x) = \sqrt{x}$ on $[a^2, b^2]$ to establish

$$\frac{a+b}{2} \le \frac{2(a^2+ab+b^2)}{3(a+b)} \le \sqrt{\frac{a^2+b^2}{2}}.$$

Note that f is concave down. The middle term in the above inequality is known as the *centroidal mean* of a and b.)

Exercise 9. To conclude the proof of 7.10, use the inequality between the arithmetic mean and the root mean square to establish the inequality between the root mean square and the contraharmonic mean. (Hint: multiply both sides of the inequality between the arithmetic mean and the root mean square by $(2/(a + b)) \cdot \sqrt{(a^2 + b^2)/2}$.)

Exercise 10. By now you realize that there are many means of two positive numbers a and b. The final one we consider is the *identric mean* $(1/e) \cdot (b^b/a^a)^{1/(b-a)}$. Show that the identric mean lies between the geometric mean and the arithmetic mean by applying the Hermite-Hadamard inequality to the function $f(x) = \ln x$ on $[a, b]$. Note that f is concave down on $[a, b]$.

Exercise 11. Show that $(\pi/e) + (e/\pi) > 2$. (Hint: more generally, show that the sum of a positive number and its reciprocal is always at least 2. This is a special case of one of the inequalities between means in this Exploration.)

Exercise 12. In your PCE you may have shown that the number e satisfies the following double inequality: For every positive integer n,

$$\left(1 + \frac{1}{n}\right)^n < e < \left(1 + \frac{1}{n}\right)^{n+1}.$$

Use the result of Exercise 7 to derive the much better inequality: for every positive integer n,

$$\left(1 + \frac{1}{n}\right)^{\sqrt{n(n+1)}} < e < \left(1 + \frac{1}{n}\right)^{n+\frac{1}{2}}.$$

(Hint: let $[a, b] = [n, n + 1]$.) For example, when $n = 50$, the first inequality yields $2.69159 < e < 2.74542$ while the second one yields $2.71824 < e < 2.71837$.

7.3.3 Using substitutions to improve numerical integration

Simpson's rule is the most powerful of the numerical integration techniques presented in this chapter, as seen by comparing the error formulas for the techniques we've studied. But if the integrand fails to have a bounded fourth derivative on the interval of integration, the results can be disappointing.

Exercise 1. Use S_{20} to approximate (a) $\int_1^4 \sqrt{x}\, dx = 14/3$ and (b) $\int_0^1 \sqrt{x}\, dx = 2/3$. Since you know the exact values, in which integral is Simpson's rule more accurate? Why?

Using technology you should get (rounded to eight decimal places) in (a) $S_{20} \approx 4.66666567$ and in (b) $S_{20} \approx 0.66565901$. Simpson's rule is much more accurate in (a) because in (b) the first and higher-order derivatives are all unbounded at 0.

Exercise 2. To improve the performance of Simpson's rule in (b) in the preceding exercise, consider a substitution that will have bounded derivatives on $[0, 1]$. Clearly one such substitution is $u = \sqrt{x}$. How does Simpson's rule perform on the u-integral?

Since $dx = 2u\, du$, the integral becomes $\int_0^1 2u^2\, du$, and Simpson's rule with *any* n is exact (do you see why?), a dramatic improvement in accuracy! So the lesson learned is this: if the integrand has unbounded derivatives on the interval of integration, consider a substitution to yield an integrand with bounded derivatives.

In probability and statistics, the family of *chi-square distributions* is defined in terms of the functions

$$f_n(x) = c_n x^{(n/2)-1} e^{-x/2}$$

where n is a positive integer and c_n is a constant that depends on n. Probabilities associated with chi-square distributions are computed by integration. When n is even this can be done using integration by parts, but for odd n we must do the integration numerically.

Exercise 3. Consider the case $n = 3$ and the integral $\int_0^1 f_3(x) \, dx = c_3 \int_0^1 x^{1/2} e^{-x/2} \, dx$. The integrand has unbounded derivatives. Find a substitution with bounded derivatives and evaluate S_{20} for the integral $\int_0^1 x^{1/2} e^{-x/2} \, dx$.

If you used the substitution $u = \sqrt{x}$ as in Exercise 2, the integral becomes

$$\int_0^1 2u^2 e^{-u^2/2} \, du.$$

The new integrand has bounded derivatives, and $S_{20} \approx 0.4981872957$, which is accurate to six decimal places.

Exercise 4. Approximate the area A under the graph of $y = \sqrt{\tan x}$ over the interval $[0, \pi/4]$.

Since $A = \int_0^{\pi/4} \sqrt{\tan x} \, dx$, we approximate A with $S_{20} \approx 0.4868655764$. However, since the derivatives of the integrand are unbounded at 0, we suspect that this value is not very accurate. The substitution $u^2 = \tan x$ yields $A = \int_0^1 \left(2u^2/(1 + u^4) \right) du$, and for this integral $S_{20} \approx 0.4874959173$, which is accurate to 6 decimal places.

The same advice about substitution holds for integrals with unbounded integrands. Such integrals are called *improper integrals*, which we will study in Chapter 9. Often a substitution can make an improper integral proper, with an integrand that has bounded derivatives, so the integral can be evaluated numerically. See Exploration 9.6.3.

7.3.4 The prismoidal formula

A *prismatoid* is a polyhedron all of whose vertices lie in two parallel planes, as illustrated in Figure 7.12.

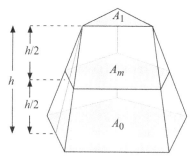

Figure 7.12. A prismatoid

The volume V of a prismatoid can be found using the *prismoidal formula*. Let A_0 and A_1 be the areas of the faces in the two parallel planes, and let A_m be the area of the intersection of the prismatoid with the plane parallel to the two planes and midway between them. If h denotes the distance between the planes, then

$$V = \frac{h}{6}\left(A_0 + 4A_m + A_1\right).$$

For example, a pyramid with a square base is a prismatoid. If the length of a side of the base is s and the height h, then $A_0 = s^2$, $A_1 = 0$, and $A_m = (s/2)^2$, so that $V = (h/6)(0 + s^2 + s^2) = (1/3)s^2 h$, the familiar formula for the volume of a pyramid.

In this Exploration we'll examine whether the prismoidal formula applies to solids other than prismatoids.

Exercise 1. Does the prismoidal formula work for a sphere? (Hint: let the two planes be tangent to the sphere at the north and south poles so that $A_0 = A_1 = 0$, and then A_m is the area of the circle at the equator of the sphere.)

Exercise 2. Does the prismoidal formula work for a cylinder? For a cone?

You may have been surprised to learn that the prismoidal formula gives the exact volume for spheres, cylinders, and cones. It also gives the exact volumes for many other solids. Let's see why.

Exercise 3. Using the known cross-sectional area approach, set up an integral of the form $\int_a^b A(x)\,dx$ for the volume of a sphere (where $A(x)$ is the area of the cross-section at the point x between a and b). Now evaluate the Simpson's rule approximation S_2 to the volume. How does it compare to the prismoidal formula? Why is S_2 exact?

Exercise 4. What must be true about the cross-sectional area function $A(x)$ of a solid so that the prismoidal formula yields the exact volume of the solid?

7.4 Acknowledgments

1. The visual argument after Theorem 7.2 comparing the midpoint and trapezoidal rules is from F. Burk, Behold! The midpoint rule is better than the trapezoidal rule for concave functions, *College Mathematics Journal*, **16** (1985), p. 56.
2. The subsection showing the Simpson's rule is exact for cubics is adapted from R. N. Greenwell, Why Simpson's rule gives exact answers for cubics, *The Mathematical Gazette*, **83** (1999), p. 508.
3. For more on means and the apportionment of the House of Representatives, see M. Balinski and H. Young, *Fair Representation: Meeting the Ideal of One Man, One Vote*, 2nd ed., Brookings Institution Press, Washington, 2001.
4. Exploration 7.3.3 is adapted from C. W. Avery and F. P. Soler, Applications of transformations to numerical integration, *College Mathematics Journal*, **19** (1988), pp. 166–168.

8

Parametric Equations and Polar Coordinates

To this point, we have considered the calculus of functions of the form $y = f(x)$. While our techniques and results have proven powerful, there are real problems in physics and elsewhere that require more information. For example, consider a baseball moving in a vertical plane, from the batter's bat to the outfielder's glove. At any given time, the baseball will have a position (x, y) where x is the horizontal distance from the batter, and y is the height above the ground. The calculus we have studied so far will not allow us to get detailed information that involves time—how fast is the baseball moving and in what direction is the ball moving at a particular time?

In this chapter, we will study the calculus of parametric functions and calculus in polar coordinates. These techniques will prove powerful, especially in gaining a deeper understanding of objects in motion, and in finding beautiful properties of curves that could not be described in a simple manner in standard Cartesian coordinates as graphs of functions.

8.1 Parametric equations and parametric curves

When an object is in motion over time in an xy-plane, we often use *parametric equations* to describe the position of the object.

Example 8.1 A baseball hit by a batter at home plate travels toward right field. Suppose the height, in feet, of the baseball at time t is given by the function $y(t) = 3 + 75t - 16t^2$ where t

is measured in seconds. Furthermore, suppose the horizontal distance from home plate is given by $x(t) = 130t$. (This is a very simplified model of actual baseball motion.)

Figure 8.1. The path of a baseball

The functions $x(t)$ and $y(t)$ are called *coordinate functions* and the equations $x = x(t)$, $y = y(t)$ are known as *parametric equations* involving the *parameter t*. You may be used to parameters being constants, but in this case, the parameter is a variable and all other variables depend on it. In the context of particle (or baseball) motion, parametric equations are sometimes given implicitly in the form $(x(t), y(t))$.

In this example, we can specify the position of the baseball with the parametric equations

$$x = 130t, \quad y = 3 + 75t - 16t^2, \quad t \geq 0.$$

or more succinctly in the parametric form

$$(130t, 3 + 75t - 16t^2), \quad t \geq 0.$$

From either of these forms, we can determine the speed of the baseball and its direction of motion at any given time. ∎

Example 8.2 (The unit circle: the standard trigonometric parametrization) Consider a particle whose position in the plane is given by

$$x = \cos(t), \quad y = \sin(t), \quad 0 \leq t \leq 2\pi. \tag{8.1}$$

By squaring and adding the coordinate functions, we see that this particle's motion is restricted to the unit circle:

$$x^2 + y^2 = \cos^2(t) + \sin^2(t) = 1.$$

We were able to *eliminate the parameter* to obtain an equation in x and y that describes a curve along which the particle moves.

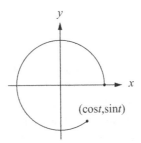

Figure 8.2. The standard trigonometric parametrization of the unit circle

In our case, given that the parameter t satisfies $0 \le t \le 2\pi$, we know that the particle travels once completely around the circle in a counter-clockwise fashion, beginning and ending at the point $(1, 0)$. The unit circle equation, $x^2 + y^2 = 1$, although useful, does not have any information about time and therefore encodes less information about the motion of the particle than do the parametric equations. ■

Example 8.3 When we convert from parametric coordinates to Cartesian coordinates, we must carefully consider the domain and range of the functions. For example, consider the parametric equations

$$x(t) = \cos t + 1, \quad y(t) = \cos 2t, \quad t \ge 0. \tag{8.2}$$

A particle with these coordinates moves along the curve in Figure 8.3, tracing out part of a parabola, starting at $(2, 1)$, moving to $(0, 1)$, then returning to $(2, 1)$ and repeating the path.

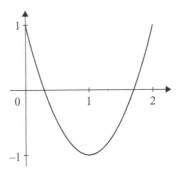

Figure 8.3. Graph of (8.2)

The path looks like a portion of a parabola, and indeed it is—eliminating t yields the Cartesian function given by the equation $y = 2(x-1)^2 - 1$ with domain $0 \le x \le 2$. ■

Example 8.4 (Converting ordinary functions to parametric form) Sometimes it is convenient to parametrize a curve that is the graph of an ordinary function $y = f(x)$. A standard way to do this is to write $x(t) = t$, $y(t) = f(t)$. For example, the graph of $y = x^2$, where $-1 \le x \le 1$ can be parametrized by (t, t^2) where $-1 \le t \le 1$. ■

Integration by parts revisited. Theorem 1.3 in Chapter 1 tells us that if f and g are functions with continuous derivatives, then

$$\int_a^b f(x)g'(x)\,dx = f(x)g(x)\Big|_{x=a}^b - \int_a^b g(x)f'(x)\,dx. \tag{8.3}$$

A parametric representation of a curve in the uv-plane yields a nice geometric illustration of this formula when f and g are increasing functions in the first quadrant. Let the curve be given parametrically with parameter x by $u = f(x)$ and $v = g(x)$ for $a \le x \le b$, as show in the figure below.

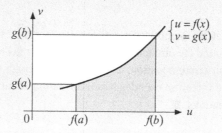

The area of the white region is $\int_{g(a)}^{g(b)} u\,dv = \int_a^b f(x)g'(x)\,dx$, the area of the gray region is $\int_{f(a)}^{f(b)} v\,du = \int_a^b g(x)f'(x)\,dx$, and the sum of the areas equals the difference of the areas of the two rectangles:

$$f(b)g(b) - f(a)g(a) = f(x)g(x)\Big|_{x=a}^b$$

and hence (8.3) follows.

8.1.1 A collection of interesting parametrized curves

Many interesting and useful curves can be described using parametric equations. We give examples of several historically important curves that are often generated geometrically, and then described analytically using parametric equations.

Example 8.5 (The unit circle: a rational parametrization) In Example 8.2, we considered the standard trigonometric parametrization of the unit circle. There are other ways to parametrize the unit circle, some quite surprising.

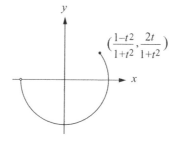

Figure 8.4. A rational parametrization of the unit circle

Consider a particle whose position at time t is given by

$$x = \frac{1-t^2}{1+t^2}, \quad y = \frac{2t}{1+t^2}, \quad -\infty < t < \infty. \tag{8.4}$$

As in Example 8.2, you can show that the coordinate functions satisfy $x^2 + y^2 = 1$. There-fore, the particle is moving along the unit circle. However, the particle does not hit every point on the unit circle (there is one point missed; can you find it?). These parametric equations, consisting of rational functions, give a *rational parametrization* of the unit circle. We will return to this example later. ∎

Example 8.6 (The semicubical parabola) In Chapter 2, we considered the curve given by the equation $y^2 = a^2 x^3$, where a is a constant. This curve can be given parametrically by the equations

$$x(t) = t^2, \quad y(t) = at^3, \quad -\infty < t < \infty. \tag{8.5}$$

You should check that the functions $x(t)$, $y(t)$ satisfy the equation $y^2 = a^2 x^3$. ∎

Example 8.7 (The astroid) Consider a circle of radius a centered at the origin. Take another circle of radius $a/4$ centered at $(3a/4, 0)$. Roll the smaller circle around the inside of the larger circle. Let P be the point on the smaller circle that begins at $(a, 0)$. The point P traces out a curve known as an *astroid*.

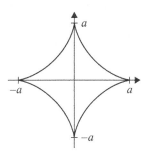

Figure 8.5. Graph of the astroid as given in (8.6)

We can show that the astroid is given by the parametric equations

$$x(t) = a \cos^3 t, \quad y(t) = a \sin^3 t, \quad 0 \le t \le 2\pi. \tag{8.6}$$

We can further check that the coordinate functions satisfy the equation $x^{2/3} + y^{2/3} = a^{2/3}$. (See Exercise 19 in Section 2.1.1.) ∎

Example 8.8 (The cissoid of Diocles) Consider the circle C of radius a centered at the point $(0, a)$ and the horizontal line T tangent to the circle at the point $(0, 2a)$. Let O denote the origin. Pick a point A on the circle, and draw the segment from O through A to T. Label the endpoint of the segment B. Now, pick a point P on the segment such that $AB = OP$. As A moves around the circle, the point P traces out a path called the *cissoid of Diocles*.

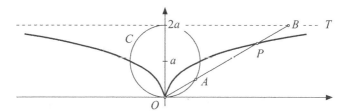

Figure 8.6. Generating a cissoid

It can be shown that parametric equations for the cissoid are

$$x(t) = \frac{2at^3}{1+t^2}, \quad y(t) = \frac{2at^2}{1+t^2}, \quad -\infty < t < \infty, \tag{8.7}$$

where a is a positive constant (the radius of the circle). A particle moving according to them traces out the path in Figure 8.6. The coordinate functions for the cissoid satisfy the equation $y^3 = (2a - y)x^2$. ▨

Example 8.9 (The cardioid) A cardioid is the path traced out by a point on a circle that is rolling around the outside of another circle. More specifically, let C_1 be the circle of radius a centered at the point $(-a, 0)$. Let C_2 be the circle of radius a centered at $(a, 0)$. Consider the point P on C_2 that begins at the origin. Now, roll C_2 around the outside of C_1 and follow the point P. The point P traces out a curve known as a *cardioid*. The word cardioid comes from the Greek χαρδια meaning "heart", and indeed a cardioid is (sort of) heart shaped. Figure 8.7 illustrates this construction in the case where the circles have radius 1. We will determine a parametrization for this cardioid, assuming the outside circle C_1 is moving at a constant rate and completes its journey once around the inside circle C_1 over the interval $0 \le t \le 2\pi$.

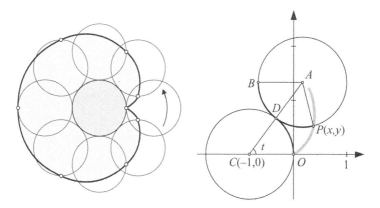

Figure 8.7. Construction of the cardioid

The arcs $\widehat{DP}$ and $\widehat{DO}$ are the same length t, and since the circles have the same radius 1, $\angle DAP = \angle DCO$. Both these angles have measure t. Choose B on C_2 so that $\overline{BA}$ is parallel to the x-axis and $\angle BAD = \angle DCO$. Then $\angle BAP$ has measure $2t$.

Point D, the point of tangency of the two circles, has coordinates $(\cos(t) - 1, \sin(t))$ (do you see why?). Thus, $A = (2\cos(t) - 1, 2\sin(t))$. If A were the origin, point P would have coordinates $(-\cos(2t), -\sin(2t))$. Thus, P actually has coordinates

$$P(x,y) = (x(t), y(t)) = (2\cos(t) - \cos(2t) - 1, 2\sin(t) - \sin(2t)).$$

Using the identities $\cos 2t = 2\cos^2 t - 1$ and $\sin 2t = 2\cos t \sin t$, we arrive at the parametric equations

$$x(t) = 2(1 - \cos t)\cos t, \quad y(t) = 2(1 - \cos t)\sin t, \quad 0 \le t \le 2\pi.$$

When the circles have radius a, an equivalent argument gives the parametric equations for the cardiod

$$x(t) = 2a\,(1 - \cos t)\cos t, \quad y(t) = 2a\,(1 - \cos t)\sin t, \quad 0 \le t \le 2\pi. \ \blacksquare \qquad (8.8)$$

The Mandelbrot Cardioid. The *Mandelbrot set* (the dark region in the figure below), named for the French-American mathematician Benoit Mandelbrot (1924–2010), is the set of complex numbers c such that the limit of the sequence $\{c, c^2 + c, (c^2 + c)^2 + c, \dots\}$ is *not* ∞. The boundary of the central bulb of the Mandelbrot set appears to be a cardioid. And so it is—for a proof, see the article by B. Branner in the Acknowledgments.

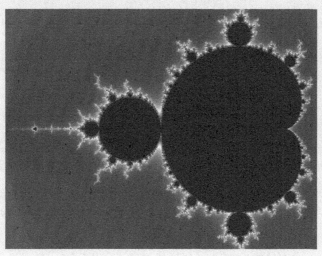

8.1.2 Tangent lines to curves given parametrically

When a curve is the graph of a differentiable function $y = f(x)$, the slope of the tangent line to the curve at a point $(x_0, f(x_0))$ is $f'(x_0)$ or $\frac{dy}{dx}|_{x=x_0}$. For curves that are given parametrically we also can describe the slope of the tangent line if it exists.

Suppose a curve is given parametrically by $(x(t), y(t))$. The slope of the tangent line to this curve at the point $(x_0, y_0) = (x(t_0), y(t_0))$ should still be given by $\frac{dy}{dx}|_{(x_0,y_0)}$. Since we do not necessarily have an equation relating x and y directly, we must look for a way to compute $\frac{dy}{dx}$ from the functions $x(t)$ and $y(t)$.

We know by the chain rule that when all quantities exist

$$\frac{dy}{dt} = \frac{dy}{dx} \cdot \frac{dx}{dt} \quad \text{and so} \quad \frac{dy}{dx} = \frac{(dy/dt)}{(dx/dt)} = \frac{y'(t)}{x'(t)}.$$

Note that $\frac{dy}{dx}$ is undefined whenever either of the quantities $y'(t)$ or $x'(t)$ is undefined or when $x'(t) = 0$. Otherwise, $\frac{dy}{dx}$ exists and represents the slope of the tangent line to the path.

Theorem 8.1 (Tangent line to a parametric curve) *Suppose a curve is given by $(x(t), y(t))$ for some interval of the parameter t. If $\frac{dy}{dx} = \frac{(dy/dt)}{(dx/dt)}$ is defined at $t = t_0$, then the tangent line to the curve at the point $(x(t_0), y(t_0))$ has slope $y'(t_0)/x'(t_0)$ and has equation*

$$y = y(t_0) + \frac{y'(t_0)}{x'(t_0)}(x - x(t_0)).$$

Back to Example 8.2 Recall that a standard trigonometric parametrization of the unit circle is

$$x = \cos(t), \quad y = \sin(t), \qquad 0 \le t \le 2\pi.$$

At time t, $0 < t < 2\pi$, the slope of the tangent line to this curve is

$$\frac{dy}{dx} = \frac{y'(t)}{x'(t)} = \frac{\cos(t)}{-\sin(t)} = -\cot(t).$$

Alternatively, we can use implicit differentiation on the equation of the unit circle:

$$x^2 + y^2 = 1,$$

$$2x + 2y\frac{dy}{dx} = 0,$$

$$\frac{dy}{dx} = -\frac{x}{y} = -\frac{\cos(t)}{\sin(t)} = -\cot(t).$$

This slope, $\frac{dy}{dx}$, exists for all values of t except $t = 0, \pi$, and 2π, where $\sin(t) = 0$. These values of t correspond to the points $(1, 0)$, $(-1, 0)$ and again $(1, 0)$, all points at which the tangent line to the circle is vertical. ■

Back to Example 8.7 We compute the slope of the tangent line to the astroid given parametrically by

$$x(t) = a\cos^3 t, \quad y(t) = a\sin^3 t, \quad 0 \le t \le 2\pi.$$

At any point where it exists,

$$\frac{dy}{dx} = \frac{dy/dt}{dx/dt} = \frac{3a\sin^2 t \cdot \cos t}{-3a\cos^2 t \cdot \sin t}.$$

For t values such that $\sin t = 0$ or $\cos t = 0$, this expression is undefined. Thus $\frac{dy}{dx}$ is undefined for $t = 0, \pi/2, \pi, 3\pi/2$, and 2π. For all other values of t, we can simplify the expression to

$$\frac{dy}{dx} = -\frac{\sin t}{\cos t} = -\tan t. ■$$

Back to Example 8.9 For the cardioid given in (8.8), we see

$$\frac{dy}{dx} = \frac{dy/dt}{dx/dt} = \frac{2a(\sin t \sin t + (1 - \cos t)\cos t)}{2a(\sin t \cos t - (1 - \cos t)\sin t)} = \frac{\cos t - \cos 2t}{\sin t \cdot (2\cos t - 1)}.$$

Notice that $\cos t - \cos 2t = 0$ when $t = 0, 2\pi/3, 4\pi/3$, and $t = 2\pi$ and $\sin t(2\cos t - 1) = 0$ when $t = 0, \pi/3, \pi, 5\pi/3$, and $t = 2\pi$. So, when $t = 2\pi/3$ and $4\pi/3$, that is, at the points $(-a/8, 3a\sqrt{3}/8)$ and $(-a/8, -3a\sqrt{3}/8)$, the tangent lines to the cardioid are horizontal. When $t = \pi/3, \pi$, and $5\pi/3$, that is, at the points $(a/8, 3a\sqrt{3}/8)$, $(-a, 0)$, and $(a/8, -3a\sqrt{3}/8)$, the tangent lines to the cardioid are vertical. When $t = 0$, and $t = 2\pi$ both $\frac{dx}{dt}$ and $\frac{dy}{dt}$ are 0. Can you see what is happening on the cardioid at these t values? ∎

8.1.3 Area and parametric curves

Let f be a continuous function on the interval $[a, b]$. In your PCE you learned that the *signed area* of the region bounded by the x-axis, the graph of $y = f(x)$, and the lines $x = a$ and $x = b$ is given by $\int_a^b f(x)\,dx$.

Suppose that the graph of $y = f(x)$ on $[a, b]$ can be parametrized by

$$x = x(t), \quad y = y(t), \quad \alpha \le t \le \beta,$$

where the parametrization yields the graph exactly once (that is, the curve is not retraced) and $x(\alpha) = a$ and $x(\beta) = b$. Here we have $y(t) = f(x(t))$.

In the area integral, we make the substitution $x = x(t)$ so that $dx = x'(t)\,dt$. This transforms the integral for signed area to

$$\int_a^b f(x)\,dx = \int_\alpha^\beta f(x(t))\,x'(t)\,dt = \int_\alpha^\beta y(t)x'(t)\,dt.$$

In a similar way, if a curve is the graph of $x = g(y)$ and can be parametrized by $(x(t), y(t))$ where $\gamma \le t \le \delta$ and $y(\gamma) = c$ and $y(\delta) = d$, then the signed area between the y-axis and the curve is given by

$$\int_c^d g(y)\,dy = \int_\gamma^\delta g(y(t))\,y'(t)\,dt = \int_\gamma^\delta x(t)y'(t)\,dt.$$

Example 8.10 Consider the curve given by $x(t) = 2t^3$, $y(t) = t^2$, $0 \le t \le 1$, as in Figure 8.8.

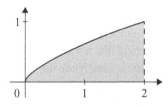

Figure 8.8. Area using parametric equations

This curve represents y as a function of x. The curve is traced out beginning at $(x(0), y(0)) = (0, 0)$ and ending at $(x(1), y(1)) = (2, 1)$. The area of the region bounded by this curve, the x-axis, and the line $x = 2$ is given by

$$\int_0^1 y(t)x'(t)\,dt = \int_0^1 t^2(6t^2)\,dt = \int_0^1 6t^4\,dt = \frac{6}{5}. \quad ∎$$

Back to Example 8.7 Let's compute the area of the region bounded by the astroid. To do so, we begin by computing the area of the region, R, in the first quadrant with boundary given by the astroid, as in Figure 8.9. The equations

$$x(t) = a\cos^3 t, \quad y(t) = a\sin^3 t, \quad 0 \le t \le \frac{\pi}{2}$$

trace out this portion of the astroid beginning with the point $(a, 0)$ and ending at $(0, a)$. This orientation of the boundary of our region is backwards from the one we use in an ordinary integral to compute area. To account for this, we will reverse the limits of integration, with t starting at $\pi/2$ and ending at 0.

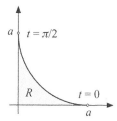

Figure 8.9. Determining the area of the astroid

Since $x'(t) = -3a\sin^2 t \cdot \cos t$, we have that the area of R is

$$\int_{\pi/2}^{0} \left(a\sin^3 t\right)\left(-3a\cos^2 t \cdot \sin t\right) dt = \int_0^{\pi/2} 3a^2 \sin^4 t \cdot \cos^2 t \, dt.$$

We can rewrite this integral as

$$3a^2 \int_0^{\pi/2} \sin^4 t(1 - \sin^2 t) \, dt = 3a^2 \left(\int_0^{\pi/2} \sin^4 t \, dt - \int_0^{\pi/2} \sin^6 t \, dt \right).$$

We evaluate these integrals using the result from Exercise 35 in Section 1.2.1 to obtain an area of $\frac{3}{32}a^2\pi$. The astroid encloses four such regions, each with the same area. Therefore, the astroid bounds an area of $\frac{3}{8}a^2\pi$. ∎

Back to Example 8.9 The cardioid given by the parametric equations in (8.8) is illustrated in Figure 8.10. We can see from the graph that in this case, the top half of the cardioid is not the graph of y as a function of x—the vertical line test fails in the first quadrant. It appears we must break up the graph into two pieces that separately are the graphs of functions.

We have already determined that the cardioid has a vertical tangent line when $t = \frac{\pi}{3}$ at the point $(a/8, 3a\sqrt{3}/8)$, and we find that on the interval $0 \le t \le \frac{\pi}{3}$, the graph is the graph of y as a function of x. Similarly on the interval $\frac{\pi}{3} \le t \le \pi$, we have y as a function of x. However, as t increases on this interval, the cardioid is being traced out to the left. This means that to compute the area of the region bounded by the x-axis and the curve over the interval $\frac{\pi}{3} \le t \le \pi$, we must calculate

$$\int_\pi^{\pi/3} y(t)x'(t) \, dt = \int_\pi^{\pi/3} (2a(1 - \cos t)\sin t) \cdot (2a(2\cos t - 1)\sin t) \, dt.$$

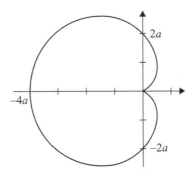

Figure 8.10. Standard cardioid

To find the area of the top half of the cardioid, we compute

$$\int_{\pi}^{\pi/3} y(t)x'(t)\,dt - \int_{0}^{\pi/3} y(t)x'(t)\,dt = -\left(\int_{\pi/3}^{\pi} y(t)x'(t)\,dt + \int_{0}^{\pi/3} y(t)x'(t)\,dt\right)$$

$$= -\int_{0}^{\pi} y(t)x'(t)\,dt.$$

The reminder of the calculation of the area is left as an exercise. ∎

8.1.4 Exercises

In Exercises 1–15, the parameter t can be eliminated to yield an equation with just x and y. Eliminate the parameter and sketch the resulting curve.

1. $x = 3t - 5, y = 2t + 2, -1 \le t \le 2$

2. $x = 4t + 2, y = 3t + 1, -\infty < t < \infty$

3. $x = \sqrt{t}, y = t^{3/2}, -2 \le t \le 2$

4. $x = \frac{1}{1+t}, y = t + 1, -1 < t \le 4$

5. $x = \sqrt{t}, y = t^2 - 4, 0 \le t \le 5$

6. $x = t^2, y = \sin(t^2), 0 \le t \le 2\pi$

7. $x = \csc t, y = \sin t, 0 < t < \pi$

8. $x = 2\cos t, y = 3\sin t, 0 \le t < 2\pi$

9. $x = \cos t, y = 2\sin t, 0 \le t \le 2\pi$

10. $x = 3\cos t + 1, y = 3\sin t - 4, -\pi \le t \le \pi$

11. $x = \cos^2 t, y = 3\cos^4 t, 0 \le t \le \pi/2$

12. $x = \sec t, y = \tan t, 0 < t < \pi/2$

13. $x = \cosh t$, $y = \sinh t$, $-2 \le t \le 2$

14. $x = \cos 2t$, $y = 4 \sin t$, $-\pi/2 \le t \le \pi/2$

15. $x = \tan^2 t$, $y = \tan^4 t$, $0 \le t < \pi/2$

In Exercises 16–20 determine expressions for $\frac{dy}{dx}$.

16. $x = 1 - t$, $y = 3t + 4$

17. $x = \cos t$, $y = \sin 2t$

18. $x = \cosh t$ $y = \sinh t$

19. $x = e^t$, $y = \ln t$, $t > 0$

20. $x = 1 + \sqrt{t}$, $y = t^2 - 4t$

In Exercises 21–27 find an equation for the tangent line, if it exists, to the graph at the indicated value of t. If there is no tangent line, explain why.

21. $x = \cos t$, $y = \sin 2t$, $t = \pi/3$

22. $x = \cosh t$ $y = \sinh t$, $t = -1$

23. $x = t^3 - 2t^2 - 4t + 1$, $y = t - \sqrt{t}$, $t = 4$

24. $x = \frac{1}{2} - \cos t$, $y = \frac{1}{2} \tan t + \sin t$, $t = \pi/3$

25. $x = \frac{1}{2} - \cos t$, $y = \frac{1}{2} \tan t + \sin t$, $t = \pi$

26. $x = 4 \cos t - \cos 5t$, $y = 4 \sin t - \sin 5t$, $t = \pi$

27. $x = 4 \cos t - \cos 5t$, $y = 4 \sin t - \sin 5t$, $t = \pi/2$

28. A curve is given by the parametric equations

$$x = 4 \cos t + \cos 4t, \qquad y = 4 \sin t - \sin 4t.$$

This curve is an example of a *hypocycloid*.
(a) Find all values of t, $0 \le t < 2\pi$, for which the tangent line does not exist.
(b) Find all values, if any, of t, $0 \le t \le 2\pi$, at which the tangent line is vertical.

29. Find the area of the cardioid given in Example 8.9.

30. Consider a circle of radius r rolling along on top of the x-axis. Assume at time $t = 0$, the point P on the circle is at the origin. As the circle rolls, the path followed by the point P is called a *cycloid*. The cycloid can be given by the parametric equations

$$x = r(t - \sin t), \qquad\qquad y = r(1 - \cos t).$$

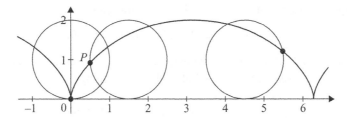

Figure 8.11. The cycloid generated when $r = 1$

(a) For all values of $t > 0$ at which the tangent line to the cycloid exists, find the slope of the tangent line to the cycloid.
(b) Determine the area between the x-axis and one arch of the cycloid.

8.2 Parametrized motion and arc length

It is common to describe the position of an object (mathematicians usually call them particles!) with parametric equations. From these equations, we investigate the speed of the particle and the total distance the particle travels.

8.2.1 Speed

We can determine a general expression for the speed of an object under parametric motion. Consider the position of the object at two times: t and $t + h$ where h is a small increment of time. These positions are $(x(t), y(t))$ and $(x(t + h), y(t + h))$, so the displacement of the object is approximately $\sqrt{(x(t + h) - x(t))^2 + (y(t + h) - y(t))^2}$. Dividing by the elapsed time, we see that the speed of the object can be approximated by

$$\text{speed} \approx \frac{1}{h}\sqrt{(x(t + h) - x(t))^2 + (y(t + h) - y(t))^2}$$

$$= \sqrt{\left(\frac{x(t + h) - x(t)}{h}\right)^2 + \left(\frac{y(t + h) - y(t)}{h}\right)^2}.$$

This expression involves some very familiar quotients. In fact, if we take the limit as h approaches 0, we obtain the *instantaneous speed* of the object at time t:

$$\text{speed} = \lim_{h \to 0}\sqrt{\left(\frac{x(t + h) - x(t)}{h}\right)^2 + \left(\frac{y(t + h) - y(t)}{h}\right)^2} = \sqrt{(x'(t))^2 + (y'(t))^2}.$$

Theorem 8.2 *Suppose a particle is moving over time in such a way that its position at time t is given by $(x(t), y(t))$ where x and y are differentiable functions of t. Then the instantaneous speed of the particle at time t is*

$$\text{speed} = \sqrt{\left(x'(t)\right)^2 + \left(y'(t)\right)^2}.$$

Back to Example 8.1. Recall our simple model for the trajectory of a baseball:

$$x = 130t, \quad y = 3 + 75t - 16t^2, \quad t \geq 0,$$

where x is the horizontal distance in feet from the batter and y is the height in feet of the ball off the ground. At time t seconds, the speed of the ball is given by

$$\text{speed} = \sqrt{\left(x'(t)\right)^2 + \left(y'(t)\right)^2} = \sqrt{130^2 + (75 - 32t)^2} \text{ ft/sec}.$$

We can use this expression to find how fast the ball is moving as it hits the ground (well, just before it hits the ground). We determine the time when the ball hits the ground, that is, when $y = 0$. To do so, we solve $3 + 75t - 16t^2 = 0$ to get $t = \frac{75 + \sqrt{5817}}{32} \approx 4.73$ seconds. So, the speed of the ball just before it hits the ground is approximately

$$\text{speed} \approx \sqrt{130^2 + (75 - 32(4.73))^2} \approx 150.72 \text{ ft/sec}. \quad \blacksquare$$

Back to Example 8.2. For the particle moving along the unit circle according to the standard trigonometric parametrization, we find the speed of the particle at any given time t. From Theorem 8.2 we see

$$\text{speed} = \sqrt{\left(x'(t)\right)^2 + (y'(t))^2} = \sqrt{\sin^2(t) + \cos^2(t)} = 1.$$

Therefore, a particle moving according to these parametric equations has a constant speed of 1. $\blacksquare$

Back to Example 8.5. The particle moving along the unit circle according to the rational parametrization of the circle

$$x(t) = \frac{1 - t^2}{1 + t^2}, \quad y(t) = \frac{2t}{1 + t^2}$$

is moving in a fundamentally different way from the particle moving according to Example 8.2. We can check (and you should!) that

$$x'(t) = -\frac{4t}{\left(1 + t^2\right)^2} \quad \text{and} \quad y'(t) = \frac{2(1 - t^2)}{\left(1 + t^2\right)^2}.$$

So, at time t, the particle is moving along the path in such a way that the tangent line has slope

$$\text{slope} = \frac{y'(t)}{x'(t)} = -\frac{(1 - t^2)}{2t} = -\frac{\left(\dfrac{1 - t^2}{1 + t^2}\right)}{\left(\dfrac{2t}{1 + t^2}\right)} = \frac{-x(t)}{y(t)}.$$

This is the same expression for slope in terms of x and y as in Example 8.2 in Section 8.1.2 but in terms of t, it is different.

The speed of the particle in this example is given by

$$\text{speed} = \sqrt{\left(x'(t)\right)^2 + (y'(t))^2} = \frac{2}{1+t^2}.$$

Unlike the preceding example, the speed is not constant. ■

8.2.2 Velocity and acceleration

The velocity of a particle whose position is given by $(x(t), y(t))$ is the rate of change of position. Unlike speed, velocity has direction and can be given by the ordered pair $v(t) = (x'(t), y'(t))$. Viewing $v(t)$ as a vector (that is, a quantity with direction and magnitude), we have that speed is the magnitude of velocity. Therefore, *speed*, being the length of the vector $v(t)$, is given by $||v(t)|| = \sqrt{(x'(t))^2 + (y'(t))^2}$.

In a similar fashion, acceleration, the rate of change of velocity, is a vector quantity given by $a(t) = v'(t) = (x''(t), y''(t))$.

Back to Example 8.1 Returning to our baseball example, we see that

$$v(t) = (120, 75 - 32t) \qquad \text{and} \qquad a(t) = (0, -32).$$

The first component of acceleration is 0, not surprising given that the simplified model we use here assumes that gravity, a vertical force, is the only force acting on the baseball once it is hit. ■

Back to Example 8.5 We return to the particle moving according to the rational parametrization of the circle. We have that

$$v(t) = \left(-\frac{4t}{(1+t^2)^2}, \frac{2(1-t^2)}{(1+t^2)^2}\right) \qquad \text{and} \qquad a(t) = \left(\frac{4(3t^2-1)}{(1+t^2)^3}, \frac{4t(t^2-3)}{(1+t^2)^3}\right).$$

If we consider what is happening to this particle as $t \to \infty$, we see that both the first and second coordinates of velocity are tending to 0. We investigate this further in the next section.

8.2.3 Arc length

In Figure 8.12 on the left we see a *figure eight knot*, familiar to scouts and sailors around the world. Such a knot always shortens the length of the line in which it is tied — but by how much? To answer, measure the length of the line with and without the knot, but in the absence of a line we'll consider a mathematical model of the knot in Figure 8.12.

Figure 8.12. The figure eight knot and a Lissajous curve

A mathematical model of the knot is the graph of a curve given parametrically as

$$x(t) = 2\sin 3t, \quad y(t) = \sin 5t, \quad \text{where} \quad -\frac{\pi}{2} \leq t \leq \frac{\pi}{2} \tag{8.9}$$

and x and y are measured in centimeters. This curve is an example of a *Bowditch* or *Lissajous* curve, first studied by the American mathematician Nathaniel Bowditch (1773–1838) in 1815 and later by the French mathematician Jules Antoine Lissajous (1822–1880) in 1857.

In Chapter 2, we developed a method for computing the length of a curve that is the graph of a function. In this section we will use a similar procedure to determine the length of a curve given with parametric equations, such as the Lissajous curve.

Let C be a curve parametrized by $(x(t), y(t))$, $a \leq t \leq b$, where both x and y have continuous first derivatives. Let $P = \{t_0, t_1, \ldots, t_{n-1}, t_n\}$ be a regular partition of the interval $[a, b]$ where $t_0 = a$ and $t_n = b$, so that $\Delta t = \frac{b-a}{n}$. The partition, P, induces a partition of the curve into n pieces. The ith portion has endpoints $(x(t_{i-1}), y(t_{i-1}))$ and $(x(t_i), y(t_i))$. Proceeding as in Chapter 2 we construct the line segment between them and use the Pythagorean theorem to find its length L_i:

$$L_i = \sqrt{\left(x(t_i) - x(t_{i-1})\right)^2 + \left(y(t_i) - y(t_{i-1})\right)^2}.$$

The total length of the curve C is approximately the sum

$$\sum_{i=1}^{n} L_i = \sum_{i=1}^{n} \sqrt{\left(x(t_i) - x(t_{i-1})\right)^2 + \left(y(t_i) - y(t_{i-1})\right)^2}.$$

Next, we multiply and divide each L_i by Δt:

$$L_i = \sqrt{\left(x(t_i) - x(t_{i-1})\right)^2 + \left(y(t_i) - y(t_{i-1})\right)^2} \, \frac{\Delta t}{\Delta t}$$

$$= \sqrt{\left(\frac{x(t_i) - x(t_{i-1})}{\Delta t}\right)^2 + \left(\frac{y(t_i) - y(t_{i-1})}{\Delta t}\right)^2} \, \Delta t.$$

We apply the mean value theorem to the functions $x(t)$ and $y(t)$ on the interval to obtain

$$L_i = \sqrt{\left(x'(t_i^*)\right)^2 + \left(y'(t_i^{**})\right)^2} \, \Delta t$$

for some t_i^* and t_i^{**} in the interval (t_{i-1}, t_i) so that the total length of the curve C is approximately

$$\sum_{i=1}^{n} L_i = \sum_{i=1}^{n} \sqrt{\left(x'(t_i^*)\right)^2 + \left(y'(t_i^{**})\right)^2} \, \Delta t. \tag{8.10}$$

On each subinterval, the mean value theorem gives us what are most likely two different points, t_i^* and t_i^{**}, one for x' and one for y'. Therefore, the right-hand side of (8.10) is not in the form of a Riemann sum, for we would need our expression involving x' and y' to be evaluated at a single point in each subinterval. It is reasonable to conjecture that this sum approaches an integral, just as a Riemann sum does, and this conjecture is true. The following theorem summarizes this fact. The proof of the theorem is rather technical and can be found in a text on advanced calculus.

Theorem 8.3 (Length of a parametrized curve) *Let C be a curve parametrized by* $(x(t), y(t))$, $a \leq t \leq b$, *where x and y are continuously differentiable functions. Assume as well that for all points P on C (except possibly one point) there is a unique t in the interval* $a \leq t \leq b$ *such that* $(x(t), y(t)) = P$. *Then the length L of the curve C is given by*

$$L = \int_a^b \sqrt{\left(x'(t)\right)^2 + (y'(t))^2}\, dt. \tag{8.11}$$

In the case where $(x(t), y(t))$ *describes the position of a particle in motion we have that the distance D traveled by the particle over the time interval* $a \leq t \leq b$ *is*

$$D = \int_a^b \|v(t)\|\, dt = \int_a^b \text{speed}(t)\, dt. \tag{8.12}$$

Example 8.11 Let's try to determine the length of the Lissajous curve given by

$$x(t) = 2\sin 3t, \quad y(t) = \sin 5t, \text{ where } -\frac{\pi}{2} \leq t \leq \frac{\pi}{2}. \tag{8.13}$$

We have $x'(t) = 6\cos 3t$ and $y'(t) = 5\cos 5t$, so the length L is given by

$$L = \int_{-\pi/2}^{\pi/2} \sqrt{36\cos^2 3t + 25\cos^2 5t}\, dt.$$

There are trigonometric identities we can attempt to use here, but given the complexity of expressions we obtain, we might be best served approximating this integral to obtain that the length is approximately 16.583. ∎

Back to Example 8.5 We return once again to the particle moving according to the rational parametrization of the circle

$$x(t) = \frac{1 - t^2}{1 + t^2}, \quad y(t) = \frac{2t}{1 + t^2}, \quad a \leq t \leq b.$$

We saw that $v(t) = (x'(t), y'(t)) = \left(\frac{-4t}{(1+t^2)^2}, 2\frac{1-t^2}{(1+t^2)^2}\right)$. From (8.12), the distance the particle travels is given by

$$\int_a^b \|v(t)\|\, dt = \int_a^b \frac{2}{1 + t^2} = 2\arctan(t)\Big|_{t=a}^b = 2\left[\arctan(b) - \arctan(a)\right].$$

If the particle is moving for all t in the interval $(-\infty, \infty)$, the total distance it travels can be computed by letting $b \to \infty$ and $a \to -\infty$. This distance is

$$2\left[\lim_{b\to\infty} \arctan(b) - \lim_{a\to-\infty} \arctan(a)\right] = 2\pi.$$

Since the particle travels once around the unit circle, this is not surprising! ∎

8.2.4 Exercises

For Exercises 1 through 8, a parametrization for the position of a particle is given. In each case, at time t, find (i) the velocity of the particle at time t, (ii) the acceleration of the particle at time t, (iii) the distance traveled by the particle over the given interval. For most of the distance integrals, you will need to use technology to approximate the value.

1. $x = \cos 2t$, $y = \sin t$, $0 \le t \le 2\pi$

2. $x = t^2$, $y = \sqrt{t}$, $0 \le t \le 1$

3. $x = t^3 - 4t$, $y = t^2 - 3t$, $0 \le t \le 2$

4. $x = 1 + t^2$, $y = 3 + t^3$, $0 \le t \le 1$

5. $x = \cosh t$, $y = \sinh t$, $-1 \le t \le 1$

6. $x = \tan t$, $y = \sec t$, $-\pi/4 \le t \le \pi/4$

7. $x = t \cos t$, $y = t \sin t$, $0 \le t \le \pi$

8. $x = e^t \cos t$, $y = e^t \sin t$, $0 \le t \le 1$

9. $x = \cos t + t \sin t$, $y = \sin t - t \cos t$, $0 \le t \le \pi$

10. The parametric equations $x = a \cos t$, $y = b \sin t$, $0 \le t \le 2\pi$ describe an ellipse centered at the origin with horizontal radius a and vertical radius b. Write an integral that gives the perimeter of this ellipse. (Note: the integral you obtain is an example of an *elliptic integral* and in general cannot be computed exactly without the use of infinite series, which we cover in Chapters 10 and 11.)

11. There are two functions in the field of optics known as *Fresnel integrals*, named for the French engineer Augustin-Jean Fresnel (1788-1827): for any real t,

$$S(t) = \int_0^t \sin(u^2)\, du \quad \text{and} \quad C(t) = \int_0^t \cos(u^2)\, du.$$

The beautiful *Euler spiral* is defined parametrically by $x = C(t)$ and $y = S(t)$, and is illustrated in Figure 8.13 for t in $[-5, 5]$.

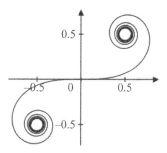

Figure 8.13. The Euler spiral

Find the arc length of the Euler spiral from the origin to a point $(C(t), S(t))$.

12. Determine the length of one arch of a cycloid given by the parametric equations

$$x = r(u - \sin u), \quad y = r(\cos u + 1) \quad 0 \le u \le 2\pi.$$

(Hint: $2 - 2\cos u = 4\sin^2(u/2)$.)

13. The *deltoid* (or *tricuspoid*) is given parametrically by the equations $x = a(2\cos t + \cos 2t)$ and $y = a(2\sin t - \sin 2t)$ for $0 \le t \le 2\pi$ and a positive number a. It is illustrated for $a = 1$ in Figure 8.14, along with its circumscribing circle (the dashed curve) of radius $3a$. Find the arc length of the deltoid. (Hints: you may need to use a sum formula and a half-angle formula from Appendix B.)

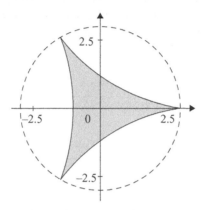

Figure 8.14. A deltoid with $a = 1$

14. Let C be the curve defined parametrically by $x = t^2$, $y = (1 - t)^2$, $-\infty < t < \infty$.
    (a) Find the Cartesian equation of C.
    (b) Show that the curve is a parabola.
    (Hints: the obvious answer $\sqrt{x} + \sqrt{y} = 1$ yields only part of the curve. The correct answer is $2(x + y) - 1 = (x - y)^2$ or anything equivalent. Technology helps here. You may need to recall rotation of axes from precalculus for part (b).)

8.3 Polar coordinates

Throughout our studies, we have dealt with describing points and curves in Cartesian coordinates that measure the horizontal and vertical distances from a fixed point, the origin. But there are situations in which curves are more conveniently described in terms of the direction and distance from the origin.

Definition 8.1 Let P be a point in the plane. Draw the segment connecting P to a fixed point O called the *pole* (a synonym for the origin which is why we use the letter O for the pole). We say (r, θ) are *polar coordinates* for P if $|r|$ is the length of the segment $\overline{PO}$ and angle θ is an angle (in radians!) that this segment makes with a fixed *polar axis* (usually the positive x-axis). The pole has polar coordinates $(0, 0)$ but it can also be given by $(0, \theta)$ for any θ.

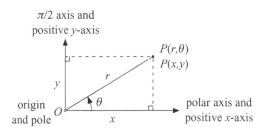

Figure 8.15. Polar and Cartesian coordinates

You may have used Cartesian graph paper in previous courses. This paper has horizontal and vertical lines on it, which are graphs of the lines where each coordinate equals a constant, i.e., $x = a$ and $y = b$ for appropriate values of a and b.

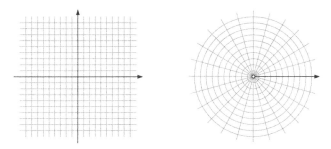

Figure 8.16. Cartesian and polar graph paper

The same is true for polar graph paper. The points where r equals a constant all lie on a circle whose radius is that constant (i.e., the graph of $r = k$ is a circle centered at the pole with radius $|k|$) and the points where θ equals a constant all lie on the line through the pole whose angle of inclination is that constant (i.e., the graph of $\theta = \alpha$ is a line through the pole making an angle of α with the polar axis). Polar graph paper will be useful when you work the exercises in this chapter.

In Definition 8.1, we stated that the pole has coordinates $(0, \theta)$ for any θ. All other points in the plane have multiple representations as well. For example, the point $(6, \pi/4)$ also can be represented by $(6, 9\pi/4)$, and by $(-6, 5\pi/4)$, a distance of $+6$ in the direction opposite to $\theta = 5\pi/4$. In general, if a point has polar coordinates (r, θ), then it can also be represented by $(r, \theta + 2\pi)$ or by $(-r, \theta + \pi)$.

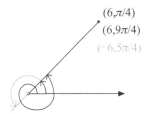

Figure 8.17. Multiple Representations

8.3.1 Converting between Cartesian and polar coordinates

Since each point in the plane has both Cartesian and polar coordinates, we should find a way to convert between them. Figure 8.15 shows that the following theorem holds.

> **Theorem 8.4 (Converting between Cartesian and polar coordinates)** *Let P be a point with polar coordinates (r, θ) and with Cartesian coordinates (x, y). Then the following equations hold.*
>
> $$r^2 = x^2 + y^2 \qquad\qquad x = r\cos\theta$$
> $$\tan\theta = \frac{y}{x} \quad (x \neq 0) \qquad\qquad y = r\sin\theta$$

Example 8.12 Let P be a point with polar coordinates $(3, 2\pi/3)$. The Cartesian coordinates for P are

$$x = 3\cos\left(\frac{2\pi}{3}\right) = -\frac{3}{2}, \quad y = 3\sin\left(\frac{2\pi}{3}\right) = \frac{3\sqrt{3}}{2}. \quad\blacksquare$$

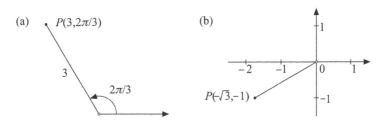

Figure 8.18. Converting between polar and Cartesian coordinates

Example 8.13 Let P be a point with Cartesian coordinates $(-\sqrt{3}, -1)$. There are many polar representations for P. With each, however, we must have

$$r^2 = (-\sqrt{3})^2 + (-1)^2 = 4 \quad \text{and} \quad \tan\theta = \frac{-1}{\sqrt{3}}.$$

Given that P is in the third quadrant, we can see that $(2, 7\pi/6)$ is one valid representation in polar coordinates. Some others are $(-2, \pi/6)$ and $(2, -5\pi/6)$. $\blacksquare$

Now, consider a curve given in polar coordinates by the equation $r = f(\theta)$. We can give parametric equations with parameter θ for this curve by substituting for r in the conversion equations to obtain

$$x = r\cos\theta = f(\theta)\cos\theta, \tag{8.14}$$

$$y = r\sin\theta = f(\theta)\sin\theta. \tag{8.15}$$

We will use this form later in this chapter to do calculus in polar coordinates.

8.3.2 A collection of polar curves

Many interesting graphs are best worked with in polar form. We begin by examining several simple equations.

Example 8.14 (Circle centered at the pole) A circle is the collection of points in the plane that are equidistant from a fixed point, the center of the circle. Consider the polar equation $r = a$. This equation has no restriction on the polar angle θ, and so is satisfied by every point whose distance from the pole is the fixed value $|a|$. Thus, the graph of $r = a$ is the circle of radius $|a|$ centered at O. ■

Example 8.15 (Line through the pole) The polar equation $\theta = a$ is satisfied by all points with a fixed polar angle a. Thus, the equation, without restriction on r, gives a line through the origin. For values of a other than odd multiples of $\pi/2$, the slope of the line is given by $\tan a$. When a is an odd multiple of $\pi/2$, the line is vertical. We can also see this by converting the equation to Cartesian form. Since $\theta = a$, we have $\tan\theta = \tan a$, whenever a is not an odd multiple of $\pi/2$. So, $\frac{y}{x} = \tan a$ or $y = x \tan a$. ■

Example 8.16 (Other circles) Consider the polar equation $r = \sin\theta$, $0 \leq \theta \leq 2\pi$. Mutiplying by r on both sides we obtain $r^2 = r \sin\theta$. In Cartesian coordinates, the equation then becomes $x^2 + y^2 = y$. We complete the square on y to obtain the equation

$$x^2 + (y - 1/2)^2 = 1/4.$$

Thus, the equation yields a circle of radius $1/2$ centered at $(0, 1/2)$.

We can obtain other circles as well. Consider $r = 4\cos\theta$. Proceeding as with the previous equation, this equation in Cartesian coordinates becomes

$$(x - 2)^2 + y^2 = 4.$$

Therefore, our graph is a circle of radius 2, whose center, in Cartesian coordinates, is $(2, 0)$. ■

⊘ Be careful! When you multiply both sides of a polar equation by r (as we did in this example), you *add* the pole to the graph if it is not already on it! So before doing so, it's best to check that the pole is already a point on the graph.

Example 8.17 (Spirals) Spirals are ubiquitous in nature. They appear as defining shapes in hurricanes, galaxies, shells, and plants (see Figure 8.19). Spirals centered at the pole have a

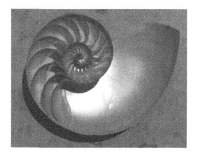

Figure 8.19. Spirals in nature (left to right) a nautilus shell, the M74 spiral galaxy, and hurricane Irene from August 2011

Figure 8.20. An Archimedean spiral given by $r = \theta$, $-8\pi \leq \theta \leq 8\pi$

common feature that the distance from from the pole, $|r|$, increases as we monotonically increase (or perhaps as we monotonically decrease) the angle θ from the polar axis.

The most straightforward polar equation yielding a spiral is $r = \theta$. The graph of this polar equation is an example of an *Archimedean spiral*. Archimedean spirals are used in the design of scroll compressors or scroll pumps. These efficient compressors are used in air conditioners, automobiles, and even certain soft-serve ice cream machines. A *logarithmic spiral* is given by the polar equation $r = ae^{b\theta}$ where a and b are fixed constants. The spiral arms appearing in Figure 8.19 are approximately logarithmic spirals.

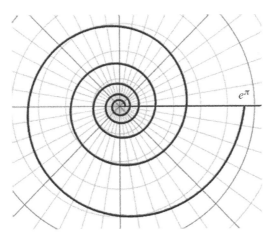

Figure 8.21. The graph of $r = e^{0.1\theta}$, $0 \leq \theta \leq 10\pi$

Logarithmic spirals have the property that as one moves along a radial axis (that is, a ray with constant polar angle), the radial coordinates of the points of intersection with the spiral grow in geometric progression. Indeed, for a fixed θ, two successive points of intersection are given by $(ae^{b\theta}, \theta)$ and $(ae^{b(\theta+2\pi)}, \theta + 2\pi) = (ae^{b(\theta+2\pi)}, \theta)$. The ratio of the radial coordinates of these points is

$$\frac{ae^{b(\theta+2\pi)}}{ae^{b\theta}} = e^{2\pi b},$$

a fixed constant that does not depend upon θ. ∎

Example 8.18 (A cardioid in polar coordinates) Recall from Example 8.9 that a cardioid can be given by the parametric equations

$$x(t) = 2a\,(1 - \cos t)\cos t, \quad y(t) = 2a\,(1 - \cos t)\sin t \quad 0 \leq t \leq 2\pi.$$

From Theorem 8.4, we see that with $t = \theta$, we have

$$\begin{aligned}
r^2 &= x^2 + y^2 \\
&= [2a(1 - \cos\theta)\cos\theta]^2 + [2a(1 - \cos\theta)\sin\theta]^2 \\
&= 4a^2(1 - \cos\theta)^2
\end{aligned}$$

and so $r = 2a\,(1 - \cos\theta)$, $0 \leq \theta \leq 2\pi$. ∎

8.3.3 Calculus in polar coordinates

The meaning of $\frac{dr}{d\theta}$

In Cartesian coordinates, we often have functions $y = f(x)$, and we know that the derivative $\frac{dy}{dx}$ is the slope of the tangent line to the graph of $y = f(x)$. In polar coordinates, many of the curves of interest are given as $r = f(\theta)$. So, what does $\frac{dr}{d\theta}$ represent? Conceptually, $\frac{dr}{d\theta}$ is the rate of change of distance to the origin with respect to angle measure (in radians!).

Slope of the tangent line to a polar curve

We saw in (8.14) and (8.15) that when $r = f(\theta)$, we have $x = f(\theta)\cos\theta$ and $y = f(\theta)\sin\theta$. If f is a differentiable function, we can use these equations to determine the slope of a tangent line to the curve $r = f(\theta)$:

$$\text{slope} = \frac{dy}{dx} = \frac{dy/d\theta}{dx/d\theta} = \frac{f'(\theta)\sin\theta + f(\theta)\cos\theta}{f'(\theta)\cos\theta - f(\theta)\sin\theta}. \tag{8.16}$$

This expression exists as long as the denominator $f'(\theta)\cos\theta - f(\theta)\sin\theta \neq 0$. When the denominator os 0, it may be the case that the tangent line to the curve is vertical. This occurs if the numerator, $f'(\theta)\sin\theta + f(\theta)\cos\theta \neq 0$. If both the numerator and denominator are 0 for a particular value of θ, then more analysis is needed.

Example 8.19 For the logarithmic spiral given by $r = ae^{b\theta}$, we have

$$\begin{aligned}
\frac{dy}{dx} = \frac{dy/d\theta}{dx/d\theta} &= \frac{abe^{b\theta}\sin\theta + ae^{b\theta}\cos\theta}{abe^{b\theta}\cos\theta - ae^{b\theta}\sin\theta} \\
&= \frac{b\sin\theta + \cos\theta}{b\cos\theta - \sin\theta} \\
&= \frac{b\tan\theta + 1}{b - \tan\theta}.
\end{aligned}$$

From this form, we can see that when $\tan\theta = b$, the tangent line to the logarithmic spiral is vertical. ■

Area of polar regions

We move now to finding the area of a region R in the plane bounded by the rays $\theta = a$ and $\theta = b$, and the curve $r = f(\theta)$. Our procedure is exactly the same as the procedure you used in your PCE to find the area of a region R in the plane bounded by the lines $y = 0$, $x = a$, and $x = b$, and the curve $y = f(x)$. See Figure 8.22. The calculus we employ—setting up a Riemann sum and taking the limit to obtain a definite integral—is independent of the coordinate system (Cartesian or polar) used.

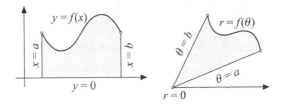

Figure 8.22. A Cartesian region and a polar sector

Just as we did with area in Cartesian coordinates, we begin by partitioning the interval $[a, b]$ by a regular partition $P = \{\theta_0, \theta_1, \ldots, \theta_{n-1}, \theta_n\}$ with $\theta_0 = a$ and $\Delta\theta = (b - a)/n$. Consider the polar sector bounded by $\theta = \theta_{i-1}, \theta = \theta_i$, and $r = f(\theta)$, as illustrated in Figure 8.23.

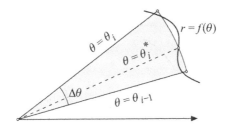

Figure 8.23. A polar sector

A circular sector whose angle measure is θ and whose radius is r has area $\frac{\theta}{2}r^2$ (see Appendix B). We can approximate the area of our polar sector with the area of a circular sector with angle measure $\Delta\theta$ and radius $f(\theta_i^*)$ where θ_i^* is a point in the interval $[\theta_{i-1}, \theta_i]$. That is, our polar sector has area

$$\text{Area}_i \approx \frac{1}{2}f(\theta_i^*)^2\,\Delta\theta.$$

Therefore the area of the region bounded by the rays $\theta = a, \theta = b$, and $r = f(\theta)$ is approximated by

$$\text{Area} = \sum_{i=1}^{n}\text{Area}_i \approx \sum_{i=1}^{n}\frac{1}{2}f(\theta_i^*)^2\,\Delta\theta. \tag{8.17}$$

This approximation is a Riemann sum whose limit as $n \to \infty$ gives us the area of the sector.

Theorem 8.5 (Area of a polar region) *Let R be the region bounded by the rays $\theta = a$, $\theta = b$, and the curve $r = f(\theta)$ where f is continuous and nonnegative on $a \leq \theta \leq b$. The area of R is given by*

$$Area = \frac{1}{2} \int_a^b (f(\theta))^2 \, d\theta. \tag{8.18}$$

Example 8.20 (Area of a cardioid) We saw in Example 8.18 that a cardioid can be given by $r = 2a(1 - \cos\theta)$, where $0 \leq \theta \leq 2\pi$. Thus, the area enclosed by the cardioid is

$$
\begin{aligned}
\text{Area} &= \frac{1}{2} \int_0^{2\pi} 4a^2 (1 - \cos\theta)^2 \, d\theta \\
&= 2a^2 \int_0^{2\pi} 1 - 2\cos\theta + \cos^2\theta \, d\theta \\
&= 2a^2 \left(\frac{3}{2}\theta - 2\sin\theta + \frac{1}{2}\sin\theta\cos\theta \right)\Bigg|_{\theta=0}^{2\pi} = 6\pi a^2.
\end{aligned}
$$

We've just shown that the area of the cardioid is *exactly* 6 times the area of the generating circle! ∎

Example 8.21 (Limaçon) Consider the curve given by $r = 1 - 2\cos\theta$, $0 \leq \theta \leq 2\pi$. This curve, illustrated in Figure 8.24, is an example of a limaçon.

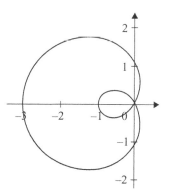

Figure 8.24. A limaçon of Pascal

Limaçons were first studied by Etienne Pascal (1588–1651), the son of a wealthy tax collector in France. His son, Blaise Pascal, was a child prodigy who went on to become one of history's most famous mathematicians.

Let us compute the area enclosed by the outer loop, but outside the inner loop of the limaçon. The outer loop of the curve is formed over the interval $\pi/3 \leq \theta \leq 5\pi/3$ while the inner loop is formed by the two intervals $0 \leq \theta \leq \pi/3$ and $5\pi/3 \leq \theta \leq 2\pi$. With the symmetry in the figure, it will suffice to compute the area of the region bounded by the outer loop on the interval

$\pi/3 \leq \theta \leq \pi$ and subtract the area of the region bounded by the inner loop on the interval $0 \leq \theta \leq \pi/3$. We double this number to obtain the area of our region.

We begin by finding an antiderivative of $(1 - 2\cos\theta)^2$. To do so, we use the trigonometric identity $\cos^2\theta = \frac{1+\cos 2\theta}{2}$ to obtain

$$\int (1 - 2\cos\theta)^2 \, d\theta = \int 1 - 4\cos\theta + 4\cos^2\theta \, d\theta$$

$$= 3\theta - 4\sin\theta + \sin 2\theta + C.$$

Using this expression with $C = 0$, the area corresponding to the outer loop is given by

$$\text{Area}_{\text{outer}} = \frac{1}{2} \int_{\pi/3}^{\pi} (1 - 2\cos\theta)^2 \, d\theta$$

$$= \frac{1}{2} \left[3\theta - 4\sin\theta + \sin 2\theta \right]_{\theta=\pi/3}^{\pi} = \pi + \frac{3\sqrt{3}}{4}.$$

The area corresponding to the inner loop is given by

$$\text{Area}_{\text{inner}} = \frac{1}{2} \int_{0}^{\pi/3} (1 - 2\cos\theta)^2 \, d\theta$$

$$= \frac{1}{2} \left[3\theta - 4\sin\theta + \sin 2\theta \right]_{\theta=0}^{\pi/3} = \pi/2 - \frac{3\sqrt{3}}{4}.$$

Therefore, the region between the outer loop and inner loop has area

$$2(\text{Area}_{\text{outer}} - \text{Area}_{\text{inner}}) = \pi + 3\sqrt{3} \approx 8.3377. \ \blacksquare$$

Arc length in polar coordinates

(8.14) and (8.15) gave us a way to produce parametric equations for a curve that is given in polar coordinates by $r = f(\theta)$. We can use these equations, together with the arc length formula (8.11) from Theorem 8.3 to determine an expression for arc length in polar coordinates

First, note that since $x = f(\theta)\cos(\theta)$, we have $\frac{dx}{d\theta} = f'(\theta)\cos\theta - f(\theta)\sin\theta$. Similarly, $\frac{dy}{d\theta} = f'(\theta)\sin(\theta) + f(\theta)\cos\theta$. We use these expressions to simplify the integrand from the integral in (8.11).

$$\sqrt{(x'(\theta))^2 + (y'(\theta))^2} = \sqrt{(f'(\theta)\cos\theta - f(\theta)\sin\theta)^2 + (f'(\theta)\sin(\theta) + f(\theta)\cos\theta)^2}$$

$$= \sqrt{(f'(\theta))^2 (\cos^2\theta + \sin^2\theta) + (f(\theta))^2 (\sin^2\theta + \cos^2\theta)}$$

$$= \sqrt{(f'(\theta))^2 + (f(\theta))^2}.$$

Thus, we arrive at the following theorem.

Theorem 8.6 (Length of a polar curve) *Let C be a curve given by the polar equation $r = f(\theta)$, $\alpha \le \theta \le \beta$ where f is continuously differentiable on $[\alpha, \beta]$. Assume as well that for all points P on C (except possibly one point) there is a unique θ in the interval $\alpha \le \theta \le \beta$ such that $(f(\theta), \theta) = P$. Then the length L of the curve C is given by*

$$L = \int_\alpha^\beta \sqrt{(f'(\theta))^2 + (f(\theta))^2} \, d\theta. \tag{8.19}$$

8.3.4 Exercises

In Exercises 1 through 8, a point is given in Cartesian coordinates. Find polar coordinates for the point.

1. $(1, 0)$ 2. $(0, 1)$ 3. $(1, \sqrt{3})$ 4. $(-1, \sqrt{3})$

5. $(-2, 2)$ 6. $(3, -3)$ 7. $(1, 3)$ 8. $(-1, 3)$

In Exercises 9 through 16, a point is given in polar coordinates. Find Cartesian coordinates for the point.

9. $(1, 0)$ 10. $(0, 1)$ 11. $(2, \pi/3)$ 12. $(2, -\pi/3)$

13. $(-2, \pi/3)$ 14. $(3, 5\pi/2)$ 15. $(-2, 3\pi/4)$ 16. $(1, 17\pi/3)$

17. For a positive integer n, the graph of the polar equation $r = \sin n\theta$ over an appropriate interval of θ is called a *rose*.
    (a) Sketch the graph of $r = \sin(2\theta)$ for $0 \le \theta \le 2\pi$.
    (b) Sketch the graph of $r = \sin(3\theta)$ for $0 \le \theta \le \pi$.
    (c) Sketch the graph of $r = \sin(4\theta)$ for $0 \le \theta \le 2\pi$.
    (d) Does the total area enclosed by the rose $r = \sin n\theta$ (n a positive integer) depend on the value of n? If so, how?

18. Find the area of the hyperbolic sector in Example 2.10 using integration in polar coordinates. (Hint: a polar representation of $x^2 - y^2 = 1$ is $r^2 \cos 2\theta = 1$, and the limits of integration are $\theta = 0$ and $\theta = \arctan(s/c)$.)

In Exercises 19 through 22, let C be the curve given by the equation. Sketch the curve and determine its length.

19. $r = e^\theta$, $0 \le \theta \le 2\pi$

20. $r = e^{2\theta}$, $0 \le \theta \le 2\pi$

21. $r = \theta$, $0 \le \theta \le 4\pi$

22. $r = \theta^2$, $0 \le \theta \le 2\pi$

23. Determine the perimeter of a cardioid as given in Example 8.18.

24. The *folium of Descartes* is a curve with a loop. Its parametric equations are $x = 3at/(1 + t^3)$ and $y = 3at^2/(1 + t^3)$ for $-\infty < t < \infty$ where a is a nonzero real number; its polar equation is $r(\sin^3 \theta + \cos^3 \theta) = 3a \sin \theta \cos \theta$. It is illustrated for $a = 1$ in Figure 8.25

along with its asymptote. Find the area of the loop (shaded). (Hint: first find the interval of t or θ values for which the graph lies in the first quadrant.)

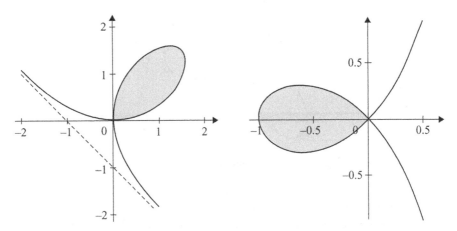

Figure 8.25. The folium of Descartes (left) and the right strophoid (right)

25. Another curve with a loop is the *right strophoid*, which has the polar equation $r = a(\sec \theta - 2 \cos \theta)$ where a is a nonzero real number. It is illustrated for $a = 1$ in Figure 8.25. Find the area of the loop.

26. In Example 8.8 we presented the *cissoid of Diocles*. Find the area of the region between the cissoid and its horizontal asymptote (the line $y = 2a$ in Figure 8.6).

8.4 Explorations

8.4.1 Rhodonea curves and double roses

A *rhodonea curve* (rhodonea is a fancy name for rose) is a simple sine or cosine function plotted in polar coordinates. They were studied and named by the Italian mathematician Guido Grandi (1672–1742) in the early eighteenth century. The polar equation of a rhodonea is $r = a \sin(b\theta)$ or $r = a \cos(b\theta)$, where b is a rational number, i.e., $b = n/d$ for positive integers n and d (n for numerator, d for denominator). When $d = 1$ we have the roses you encountered in Exercise 17 in Section 8.3.4. In this Exploration you will investigate these curves for $d > 1$. We will set $a = 1$ and consider the curves $r = \sin(n\theta/d)$ (the curves $r = \cos(n\theta/d)$ are similar).

Exercise 1. Use technology to graph $r = \sin(n\theta/d)$ for n and d between 1 and 7 inclusive. You will need to let θ range from 0 to $d\pi$ or $2d\pi$ to get a complete graph.

Exercise 2. Consider now the graphs when $n = 1$. Describe the differences you see in the graphs for d even versus d odd.

Exercise 3. In this chapter you learned that the graph of $r = \sin(n\theta)$ is an n-leaf rose when n is odd. How would you describe the graph of $r = \sin(n\theta/2)$ for odd n?

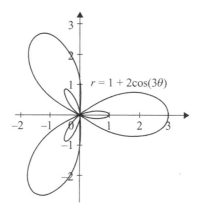

Figure 8.26. The double rose $r = 1 + 2\cos 3\theta$

Exercise 4. Assume that the fraction n/d is in reduced form, that is, n and d have no common integer factors, and both n and d are at least 2. These rhodonea all have either n or $2n$ petals. When are there n petals, and when are there $2n$ petals?

Exercise 5. The graphs of $r = a + b\sin(n\theta)$ and $r = a + b\cos(n\theta)$ for positive integer values of a, b, and n with $b > a$ (these polar functions generalize the limaçons from this chapter) are known as *double roses*. To see why, use technology to graph several. (Hint: a good choice of b is a value close to $3a$.) Now graph some with $a > b$. Is the name still appropriate? What happens at $a = b$?

Exercise 6. In Figure 8.26 we see a graph of the double rose $r = 1 + 2\cos 3\theta$. Show that the area of the region inside the three large petals but outside the three small petals is $\pi + 3\sqrt{3} \approx 9.3377$.

Figure 8.27 illustrates the graphs of rhodonea curves from Exercise 1.

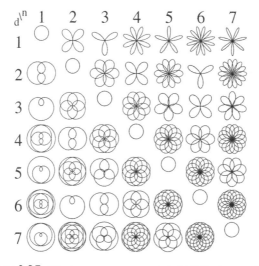

Figure 8.27. Rhodonea curves $r = \sin(n\theta/d)$ for $1 \le n, d \le 7$

8.4.2 Polar area as a source of antiderivatives

In this Exploration you will learn that interpreting a definite integral as the area of a region in polar coordinates may help you find an antiderivative of the integrand. The idea is based on the fundamental theorem: if you can use geometry to evaluate $\int_a^x f(\theta)d\theta$, then you have found an antiderivative of f since $\frac{d}{dx}\int_a^x f(\theta)\,d\theta = f(x)$.

We begin with a simple example. In Exercise 28(b) of Section 1.2.1 you used integration by parts to show that $\int \cos^2 x\,dx = (1/2)(x + \sin x \cos x) + C$. Now consider a definite integral in polar coordinates with the same integrand, i.e., $\int_0^\alpha \cos^2 \theta\,d\theta$. This integral looks suspiciously like a polar area integral since the integrand is a square. Since the graph of $r = 2\cos\theta$ is a circle, let's write $\int_0^\alpha \cos^2 \theta\,d\theta = (1/2)\cdot(1/2)\int_0^\alpha (2\cos\theta)^2\,d\theta$ so that $\int_0^\alpha \cos^2 \theta\,d\theta$ represents half the area of the region in the plane bounded by $r = 2\cos\theta$ and the rays $\theta = 0$ and $\theta = \alpha$, as illustrated in Figure 8.28 for α in $(0, \pi/2)$.

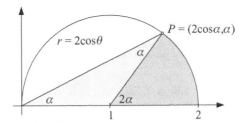

Figure 8.28. An area interpretation of $\int_0^\alpha \cos^2 \theta\,d\theta$

We can now compute the area of the region using simple geometry. The circular sector (in dark gray) has angle 2α and radius 1, so its area is $(1/2)(2\alpha) = \alpha$. The light gray triangle has base 1 and altitude $2\cos\alpha \sin\alpha$ (the y-coordinate of P) so its area is $\sin\alpha \cos\alpha$. Hence one antiderivative (in the variable α) is

$$\int_0^\alpha \cos^2 \theta\,d\theta = \frac{1}{2}(\alpha + \sin\alpha \cos\alpha),$$

and hence

$$\int \cos^2 \theta\,d\theta = \frac{1}{2}(\theta + \sin\theta \cos\theta) + C.$$

Exercise 1. Use the polar area method to integrate the square of the secant by showing that $\int_0^\alpha \sec^2 \theta\,d\theta = \tan\theta$. (Hint: consider the graph of $r = \sec\theta$.)

Exercise 2. Show that

$$\int \frac{d\theta}{(a\cos\theta + b\sin\theta)^2} = \frac{1}{a(a\cos\theta + b)} + C.$$

(Hint: the graph of $r = 1/(a\cos\theta + b\sin\theta)$ is a straight line! Sketch a picture for a and b both positive.)

Exercise 3. Show that

$$\int \frac{d\theta}{(1 + \cos\theta)^2} = \frac{\sin\theta(1 + \cos\theta)}{2(1 + \cos\theta)^2} + C.$$

(Hint: the graph of $r = 1/(1 + \cos\theta)$ is a parabola, so you will need to use geometry and integration in Cartesian coordinates to find the area of the region.)

8.4.3 The tautochrone

Consider a track in the shape of an inverted cycloid given by the parametric equations

$$x(u) = r(u - \sin u), \quad y(u) = r(\cos u + 1), \quad 0 \le u \le \pi.$$

In this exploration, we consider an object sliding without friction on this track and influenced only by the force of gravity.

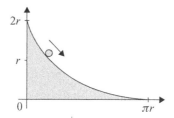

Figure 8.29. Object sliding along an inverted cycloid

We begin by determining the velocity of an object on the track at some point after it has been released. We assume the track is frictionless so that the only force on the object at any moment is the force of gravity.

The kinetic energy of an object of mass m and velocity v is given by $\frac{1}{2}mv^2$. The potential energy of the object is the product of gravitational force on the object (its mass times the gravitational constant) and its height h, that is, mgh.

A well-known conservation law in physics is the law of conservation of energy. In our context, this law tells us that the potential energy of the object on the track added to the kinetic energy of the object should equal the original potential energy when the object was at its starting position, at rest.

Exercise 1. If the object is released from rest from the point $(x(a), y(a))$, determine the potential energy of the object at its starting point $(x(a), y(a))$ and the potential energy and kinetic energy of the object at the point $((x(u), y(u))$.

Exercise 2. Use your result from Exercise 1 and the law of conservation of energy to determine the velocity of the object, $v(u)$, when it is at position $((x(u), y(u))$.

You should have found $v(u) = \sqrt{2g}\sqrt{y(a) - y(u)}$.

When velocity, v, is constant, we know the time t to move a distance d is given by $t = \frac{d}{v}$. We can use this to approximate the time it takes our object to move along a small portion of the track, from position $(x(u_1), y(u_1))$ to $(x(u_2), y(u_2))$.

Exercise 3. Show that the distance the object travels along the track as it moves from $(x(u_1), y(u_1))$ to $(x(u_2), y(u_2))$ is approximately $\sqrt{(x'(u_1))^2 + (y'(u_1))^2}\,\Delta u$ where $\Delta u = u_2 - u_1$. Use this result and the approximation $v(u) \approx \sqrt{2g}\sqrt{y(a) - y(u_1)}$ to approximate the time it takes the object to move along the track between the two points.

Exercise 4. Show using the limit of Riemann sums that the total time to move from $(x(a), y(a))$ to $(x(b), y(b))$ is

$$\text{Time} = \frac{1}{\sqrt{2g}} \int_a^b \frac{\sqrt{(x'(u))^2 + (y'(u))^2}}{\sqrt{y(a) - y(u)}} \, du. \tag{8.20}$$

Now consider an object moving in free fall along a frictionless track in the shape of the inverted cycloid given by

$$x = r(u - \sin u), \quad y = r(\cos u + 1), \quad 0 \le u \le \pi.$$

Exercise 5. If the object starts at rest and is released from the point $(x(a), y(a))$, show that the time to get to the point $(x(b), y(b))$ is given by

$$\text{Time} = \sqrt{\frac{r}{g}} \int_a^b \frac{\sqrt{1 - \cos u}}{\sqrt{\cos a - \cos u}} \, du.$$

Exercise 6. Use the trigonometric identities

$$\sin^2 \theta = \frac{1 - \cos 2\theta}{2} \quad \text{and} \quad \cos^2 \theta = \frac{1 + \cos 2\theta}{2}$$

and the integral in Exercise 5 to show

$$\text{Time} = \sqrt{\frac{r}{g}} \int_a^b \frac{\sin(u/2)}{\sqrt{\cos^2(a/2) - \cos^2(u/2)}} \, du.$$

Exercise 7. Assume $b = \pi$ so that the object moves to the bottom of the track. Use the substitution

$$w = \frac{\cos(u/2)}{\cos(a/2)}$$

to evaluate the integral in Exercise 6.

You should have gotten that the time it takes the object to reach the bottom of the track is $\pi \sqrt{r/g}$. Amazingly, this time does not depend on where the object starts on the track!

It was Christiaan Huygens (1629–1695) in 1659 who solved the problem of determining the curve with the property that the time taken for an object sliding without friction to the lowest point does not depend on the object's starting point. Such a curve is called a *tautochrone* curve and as you've just shown, the curve is a cycloid.

The cycloid has another startling property. It is also the *brachistochrone* curve. That is, the inverted cycloid with a cusp at A decreasing to a minimum at B is the path along which the time it takes for an object to move from A to B only under the force of gravity is minimized.

8.4.4 Polar area and Cartesian area

Suppose we have a region in the plane for which it is possible to find its area using either Cartesian (i.e., xy) coordinates or polar (i.e., $r\theta$) coordinates. We certainly hope that the two procedures yield the same answer for the area, and in this Exploration you will show that they indeed do.

We consider only a simple case—when the region lies in the first quadrant and its curved boundary is the graph of a function in both Cartesian and polar coordinates, as illustrated in Figure 8.30. Other cases can be dealt with similarly.

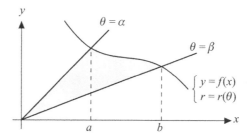

Figure 8.30. A simple region

Let A_{cart} and A_{polar} denote the area of the shaded region when computed in Cartesian and polar coordinates, respectively.

Exercise 1. Show that

$$A_{\text{cart}} = \int_a^b f(x)\,dx + \frac{1}{2}af(a) - \frac{1}{2}bf(b).$$

(Hint: Express the area of the shaded region in terms of the area under the graph of $y = f(x)$ and the areas of two right triangles.)

Exercise 2. Show that

$$A_{\text{cart}} = \frac{1}{2}\left(xf(x)\Big|_{x=b}^a - 2\int_b^a f(x)\,dx \right).$$

(Hint: notice the change in the order of the limits of integration a and b in the integral.)

Exercise 3. Show that

$$A_{\text{polar}} = \frac{1}{2}\int_\beta^\alpha [r(\theta)]^2\,d\theta = \frac{1}{2}\int_\beta^\alpha [r(\theta)\cos\theta]^2\,d(\tan\theta).$$

Exercise 4. Show that

$$A_{\text{polar}} = \frac{1}{2}\int_b^a x^2\,d\left(\frac{f(x)}{x}\right).$$

(Hint: use the change of variables $x = r(\theta)\cos\theta$ and $f(x) = y = r(\theta)\sin\theta$.)

Exercise 5. Show that

$$A_{\text{polar}} = \frac{1}{2}\left[x^2\,\frac{f(x)}{x}\Big|_{x=b}^a - \int_b^a \frac{f(x)}{x}\cdot 2x\,dx \right]$$

to conclude that $A_{\text{polar}} = A_{\text{cart}}$. (Hint: integrate the result in Exercise 4 by parts and then compare to Exercise 2.)

For the final exercise, consider the case where the function f has an inverse g, as illustrated in Figure 8.31, and set $A = f(a)$ and $B = f(b)$.

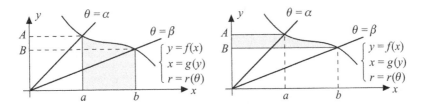

Figure 8.31. Two more simple regions

Exercise 6. Derive the somewhat unexpected result that A_{polar} is equal to the average of the areas of the two shaded regions in Figure 8.31, i.e.,

$$A_{polar} = \frac{1}{2} \int_{\beta}^{\alpha} [r(\theta)]^2 \, d\theta = \frac{1}{2} \left(\int_{a}^{b} f(x) \, dx + \int_{B}^{A} g(y) \, dy \right).$$

(Hint: use the results in Exercises 1 and 5 to show that

$$A_{polar} = \int_{B}^{A} g(y) \, dy + \frac{1}{2} bf(b) - \frac{1}{2} af(a).)$$

Exercise 7. Are there any curves in the first quadrant for which the three shaded regions in Figures 8.30 and 8.31 have identical areas for every choice of (positive) a and b? (Hint: the answer is yes. Equating any two of the three yields $af(a) = bf(b)$, so that $xf(x)$ must be a (positive) constant.)

9

Improper Integrals, L'Hôpital's Rule, and Probability

The theory of probabilities is at bottom nothing but common sense reduced to calculus.

Pierre-Simon Laplace

In the preceding chapters, nearly all the definite integrals we encountered were proper integrals. A definite integral $\int_a^b f(x)\, dx$ is *proper* if

i. $[a, b]$ is an interval of finite length and

ii. $f(x)$ is defined at every x in $[a, b]$, is bounded on $[a, b]$, and has at most a finite number of discontinuities in $[a, b]$.

Conditions (i) and (ii) both must hold in order to construct Riemann sums and to take their limits. Furthermore, the use of the fundamental theorem requires that $f(x)$ is continuous at every x in $[a, b]$.

When condition (i), condition (ii), or both fail to hold for an integral, we call the integral *improper*.

Condition (i) fails when the interval of integration has infinite length, and this can happen in exactly one of three ways, which we write as $\int_a^\infty f(x)\, dx$, $\int_{-\infty}^b f(x)\, dx$, and $\int_{-\infty}^\infty f(x)\, dx$ when the interval of integration is $[a, \infty)$, $(-\infty, b]$, or $(-\infty, \infty)$, respectively. These integrals are rather uncreatively known as *type I improper integrals*.

One way condition (ii) fails is when $f(x)$ has an infinite discontinuity for at least one x in $[a, b]$. These integrals are known as *type II improper integrals*. Some improper integrals are both type I and type II—the interval of integration has infinite length and contains at least one infinite discontinuity of the integrand.

In this chapter we will extend our definition of the definite integral to include some improper integrals. In many instances we will need a new technique for evaluating limits known as *L'Hôpital's rule* (pronounced "low-pea-tahl's" rule), named for the seventeenth century French mathematician Guillaume François Antoine Marquis de l'Hôpital (1661–1704), who wrote the first calculus textbook. We conclude this chapter with a major application of improper integrals, continuous probability models.

9.1 Type I improper integrals

Why can't we use the definition of the definite integral as the limit of a Riemann sum to evaluate $\int_a^\infty f(x)\,dx$? We can't construct a Riemann sum since Δx is undefined when the interval has infinite length. Nor can we use the fundemantal theorem of calculus: if we try to do so with an antiderivative $F(x)$, then we have $\int_a^\infty f(x)\,dx = F(x)\big|_{x=1}^\infty = F(\infty) - F(a)$, which is nonsense since ∞ is *not* a real number, and so clearly not in the domain of $F(x)$!

Our solution to this situation is motivated by considering the following example.

Example 9.1 Evaluate $\int_1^\infty \frac{1}{x^2}\,dx$.

Such an integral arises by trying to compute the area under the graph of $y = 1/x$ over the interval $[1, \infty)$. See Figure 9.1.

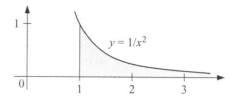

Figure 9.1. A graph of $y = 1/x^2$

We can approximate this area with integrals like

$$\int_1^{10} \frac{1}{x^2}\,dx, \qquad \int_1^{100} \frac{1}{x^2}\,dx, \qquad \int_1^{1000} \frac{1}{x^2}\,dx, \qquad \int_1^{10,000} \frac{1}{x^2}\,dx,$$

and so on, which motivates the following definition:

Definition 9.1 If $f(x)$ is continuous on $[a, \infty)$, then

$$\int_a^\infty f(x)\,dx = \lim_{b \to \infty} \int_a^b f(x)\,dx$$

if the limit exists as a real number L, in which case we say that $\int_a^\infty f(x)\,dx$ *converges to* L. If the limit does not exist, we say that $\int_a^\infty f(x)\,dx$ *diverges*.

Thus for Example 9.1 we have

$$\int_1^\infty \frac{1}{x^2}\,dx = \lim_{b \to \infty} \int_1^b \frac{1}{x^2}\,dx = \lim_{b \to \infty} \left[-\frac{1}{x} \right]_{x=1}^b = \lim_{b \to \infty} \left[-\frac{1}{b} + 1 \right] = 0 + 1 = 1.$$

Consequently, the integral $\int_1^\infty \frac{1}{x^2}\,dx$ converges to 1. ∎

The example illustrates the following three-step procedure for using Definition 9.1 to evaluate $\int_a^\infty f(x)\,dx$:

1. Evaluate the indefinite integral: $\int f(x)\,dx = F(x)$.

2. Use the fundamental theorem of calculus: $\int_a^b f(x)\,dx = F(b) - F(a)$.

3. Take the limit as $b \to \infty$: $\int_a^\infty f(x)\,dx = \lim_{b \to \infty} [F(b) - F(a)]$.

Example 9.2 Evaluate $\int_0^\infty \frac{x}{x^2+1}\,dx$.

See Figure 9.2 for a graph of $y = x/(x^2 + 1)$.

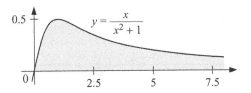

Figure 9.2. A graph of $y = x/(x^2 + 1)$

Following the three-step procedure, we have

$$\int_0^\infty \frac{x}{x^2+1}\,dx = \lim_{b\to\infty} \int_0^b \frac{x}{x^2+1}\,dx = \lim_{b\to\infty} \left[\frac{1}{2}\ln(x^2+1)\right]_{x=0}^b$$

$$= \lim_{b\to\infty}\left[\frac{1}{2}\ln(b^2+1) - 0\right]$$

$$= \infty,$$

and hence $\int_0^\infty \frac{x}{x^2+1}\,dx$ diverges. ∎

The following theorem generalizes Example 9.1, and will be useful when we consider infinite series in Chapters 10 and 11.

Theorem 9.1 *Let $a > 0$. Then the improper integral $\int_a^\infty \frac{1}{x^p}\,dx$ converges for $p > 1$ and diverges for $p \le 1$.*

Proof. We first note that the integrand is continuous on $[a, \infty)$ for all p, and thus it suffices to prove the theorem for $a = 1$ (do you see why?). Now consider the case $p = 1$:

$$\int_1^\infty \frac{1}{x}\,dx = \lim_{b\to\infty} \int_1^b \frac{1}{x}\,dx = \lim_{b\to\infty} \ln x \Big|_{x=1}^b = \lim_{b\to\infty}[\ln b - 0] = \infty,$$

and hence $\int_1^\infty 1/x^p\,dx$ diverges for $p = 1$. When $p \ne 1$ we have

$$\int_1^\infty \frac{1}{x^p}\,dx = \lim_{b\to\infty} \int_1^b x^{-p}\,dx = \lim_{b\to\infty} \frac{x^{1-p}}{1-p}\Big|_{x=1}^b = \lim_{b\to\infty}\left[\frac{b^{1-p}}{1-p} - \frac{1}{1-p}\right] = \begin{cases} \infty, & p < 1 \\ \frac{1}{p-1}, & p > 1 \end{cases}$$

so that $\int_1^\infty 1/x^p\,dx$ (and consequently $\int_a^\infty 1/x^p\,dx$) converges for $p > 1$ and diverges for $p \le 1$. □

Example 9.3 Evaluate $\int_0^\infty e^{-x} \cos(\pi x)\, dx$.

See Figure 9.3 for a graph of the integrand.

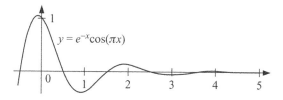

Figure 9.3. A graph of $y = e^{-x} \cos(\pi x)$

In Example 1.8 we evaluated $\int e^{-x} \cos x\, dx$ by parts. The evaluation of $\int e^{-x} \cos(\pi x)\, dx$ is similar and yields

$$\int e^{-x} \cos(\pi x)\, dx = \frac{\pi \sin \pi x - \cos \pi x}{(1 + \pi^2)e^x} + C.$$

Hence

$$\int_0^\infty e^{-x} \cos(\pi x)\, dx = \lim_{b \to \infty} \int_0^b e^{-x} \cos(\pi x)\, dx$$

$$= \lim_{b \to \infty} \left[\frac{\pi \sin \pi x - \cos \pi x}{(1 + \pi^2)e^x} \right]_{x=0}^b$$

$$= \lim_{b \to \infty} \left[\frac{\pi \sin \pi b - \cos \pi b}{(1 + \pi^2)e^b} - \frac{-1}{1 + \pi^2} \right].$$

Since both $\sin \pi b$ and $\cos \pi b$ lie in the interval $[-1, 1]$, $\pi \sin \pi x - \cos \pi x$ lies in $[-1 - \pi, 1 + \pi]$ so that

$$\frac{-1 - \pi}{(1 + \pi^2)e^b} \leq \frac{\pi \sin \pi b - \cos \pi b}{(1 + \pi^2)e^b} \leq \frac{1 + \pi}{(1 + \pi^2)e^b},$$

the squeeze theorem yields

$$\lim_{b \to \infty} \frac{\pi \sin \pi b - \cos \pi b}{(1 + \pi^2)e^b} = 0,$$

and thus $\int_0^\infty e^{-x} \cos(\pi x)\, dx$ converges to $1/(1 + \pi^2)$. ■

The procedure for $\int_{-\infty}^b f(x)\, dx$ is similar, and is given in the following definition.

Definition 9.2 If $f(x)$ is continuous on $(-\infty, b]$, then

$$\int_{-\infty}^b f(x)\, dx = \lim_{a \to -\infty} \int_a^b f(x)\, dx$$

if the limit exists as a real number L, in which case we say that $\int_{-\infty}^b f(x)\, dx$ *converges to L*. If the limit does not exist, we say that $\int_{-\infty}^b f(x)\, dx$ *diverges*.

Example 9.4 Evaluate $\int_{-\infty}^1 e^x\, dx$.

The steps to evaluate this integral are similar to the three steps we used following Definition 9.1. So

$$\int_{-\infty}^{1} e^x\,dx = \lim_{a\to-\infty}\int_{a}^{1} e^x\,dx = \lim_{a\to-\infty}[e^x]^1_{x=a} = \lim_{a\to-\infty}(e - e^a) = e - 0 = e,$$

and thus the integral converges to e. ∎

To evaluate $\int_{-\infty}^{\infty} f(x)\,dx$ we use the following definition.

Definition 9.3 If $f(x)$ is continuous on $(-\infty, \infty)$ and c is a real number, then

$$\int_{-\infty}^{\infty} f(x)\,dx = \int_{-\infty}^{c} f(x)\,dx + \int_{c}^{\infty} f(x)\,dx,$$

and $\int_{-\infty}^{\infty} f(x)\,dx$ converges if and only if both $\int_{-\infty}^{c} f(x)\,dx$ and $\int_{c}^{\infty} f(x)\,dx$ converge.

In general, we set $c = 0$ when we use Definition 9.3.

Example 9.5 Evaluate $\int_{-\infty}^{\infty} \operatorname{sech}x\,dx$.

See Figure 9.4 for a graph of $y = \operatorname{sech}x$.

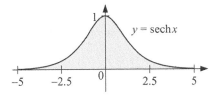

Figure 9.4. A graph of $y = \operatorname{sech}x$

Write $\int_{-\infty}^{\infty} \operatorname{sech}x\,dx = \int_{-\infty}^{0} \operatorname{sech} x\,dx + \int_{0}^{\infty} \operatorname{sech}x\,dx$, and first consider the indefinite integral $\int \operatorname{sech}x\,dx$ from (6.5):

$$\int \operatorname{sech}x\,dx = \arctan(\sinh x) + C.$$

Thus

$$\int_{0}^{\infty} \operatorname{sech}x\,dx = \lim_{b\to\infty}\int_{0}^{\infty} \operatorname{sech}x\,dx = \lim_{b\to\infty}[\arctan(\sinh x)]^b_{x=0}$$

$$= \lim_{b\to\infty}[\arctan(\sinh b) - 0] = \frac{\pi}{2}.$$

But since $\operatorname{sech}x$ is an even function we also have $\int_{-\infty}^{0} \operatorname{sech}x\,dx = \frac{\pi}{2}$, and hence $\int_{-\infty}^{\infty} \operatorname{sech}x\,dx$ converges to π. ∎

Example 9.6 Evaluate $\int_{-\infty}^{\infty} \frac{x}{x^2+1}\,dx$.

See Figure 9.5 for a graph of the integrand.

In Example 9.2 we saw that $\int_{0}^{\infty} x/(x^2 + 1)\,dx$ diverges, and hence so does $\int_{-\infty}^{\infty} x/(x^2 + 1)\,dx$. ∎

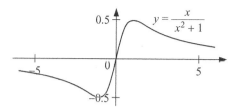

Figure 9.5. A graph of $y = x/(x^2 + 1)$

⊘ It is tempting to use a shortcut to evaluate an improper integral over the interval $(-\infty, \infty)$ by writing $\int_{-\infty}^{\infty} f(x)\,dx = \lim_{t\to\infty} \int_{-t}^{t} f(x)\,dx$. But this often leads to an incorrect answer, as using it with Example 9.6 illustrates:

$$\int_{-\infty}^{\infty} \frac{x}{x^2 + 1}\,dx = \lim_{t\to\infty} \int_{-t}^{t} \frac{x}{x^2 + 1}\,dx = \lim_{t\to\infty} \left[\frac{1}{2}\ln(x^2 + 1)\right]_{x=-t}^{t} = \lim_{t\to\infty}[0] = 0.$$

Thus the shortcut yields an incorrect answer (since the first $=$ sign should be $\neq$). In fact, it works only when $\int_{-\infty}^{\infty} f(x)\,dx$ is *already known* to converge, and hence it is *not* recommended.

Example 9.7 Evaluate $\int_{1}^{\infty} \frac{\ln x}{x^2}\,dx$.

See Figure 9.6 for a graph of $y = (\ln x)/x^2$.

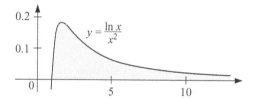

Figure 9.6. A graph of $y = (\ln x)/x^2$

We first evaluate the indefinite integral by parts, with $u = \ln x$ and $dv = (1/x^2)\,dx$:

$$\int \frac{\ln x}{x^2}\,dx = -\frac{\ln x}{x} + \int \frac{1}{x^2}\,dx = -\frac{\ln x}{x} - \frac{1}{x} + C = -\frac{\ln x + 1}{x} + C.$$

Then

$$\int_{1}^{\infty} \frac{\ln x}{x^2}\,dx = \lim_{b\to\infty} \left[-\frac{\ln x + 1}{x}\right]_{x=1}^{b} = 1 - \lim_{b\to\infty} \frac{\ln b + 1}{b}.$$

Both the numerator and denominator of $(\ln b + 1)/b$ become large as $b \to \infty$, and none of the familiar limit techniques from your PCE help much in evaluating this limit. In fact, it is not at all unusual for an improper integral whose antiderivative is found using integration by parts to yield a quotient in which both the numerator and denominator have infinite limits as $b \to \infty$ (or as $a \to -\infty$). So in the next section we present a limit technique called *L'Hôpital's*

Rule, which is useful in evaluating these (and other) limits, after which we shall return to Example 9.7. ■

9.1.1 Exercises

In Exercises 1 through 15, evaluate the integral.

1. $\displaystyle\int_0^\infty e^{-x/2}\,dx$ 2. $\displaystyle\int_0^\infty \frac{dx}{\sqrt{x+1}}$ 3. $\displaystyle\int_{-\infty}^2 \frac{dx}{(x-4)^2}$

4. $\displaystyle\int_{-\infty}^1 2^x\,dx$ 5. $\displaystyle\int_0^\infty \frac{x}{(x^2+1)^{3/2}}\,dx$ 6. $\displaystyle\int_1^\infty \frac{1}{x(x+1)}\,dx$

7. $\displaystyle\int_e^\infty \frac{dx}{x\ln x}$ 8. $\displaystyle\int_e^\infty \frac{dx}{x(\ln x)^2}$ 9. $\displaystyle\int_0^\infty \sin x\,dx$

10. $\displaystyle\int_0^\infty \frac{dx}{1+e^x}$ 11. $\displaystyle\int_0^\infty e^{-x}\sin x\,dx$ 12. $\displaystyle\int_1^\infty \operatorname{csch}x\,dx$

13. $\displaystyle\int_{-\infty}^\infty \frac{dx}{x^2+4}$ 14. $\displaystyle\int_{-\infty}^\infty xe^{-x^2}\,dx$ 15. $\displaystyle\int_{-\infty}^\infty e^{-|x|}\,dx$

16. Is it possible to assign a finite number to the area of the region in the first quadrant bounded by the graph of $f(x) = \tanh x$ and its horizontal asymptote $y = 1$? If so, evaluate it.

17. Is it possible to assign a finite number to the area of the region bounded by the graph of $y = 1/[1 + \cosh(2x)]$ and the x-axis? If so, evaluate it.

18. The *witch of Maria Agnesi* is the curve illustrated in Figure 9.7

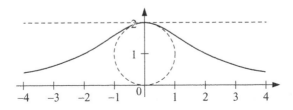

Figure 9.7. The witch of Maria Agnesi

This curve first appeared in the 1748 calculus text *Instituzioni analitiche ad uso della gioventù italiana*, written by Maria Gaetana Agnesi (1718–1799). The name is attributed to the confusion by a translator of the word *averisera* ("turning curve") with the word *aversiera* ("witch"). The witch can be defined as follows: let A be a point on the circle of radius 1 centered at $(0, 1)$ and draw the line from the origin through A intersecting the line $y = 2$ at N. The witch is the locus of points P in the plane for which the shaded triangles $\triangle ANP$ and $\triangle OAB$ are similar, as illustrated in Figure 9.8.

(a) Show that the equation of the witch is $y = 8/(x^2 + 4)$.

(b) Show that the area of the unbounded region between the witch and the x-axis is 4 times the area of the generating circle.

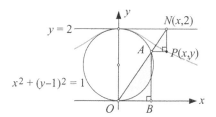

Figure 9.8. The definition of the witch of Agnesi

19. When the rectangular hyperbola $y = 1/x$ for $x \geq 1$ is revolved about the x-axis, the object formed is sometimes called *Gabriel's horn*. See Figure 9.9

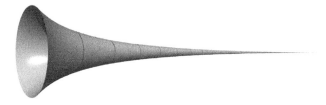

Figure 9.9. Gabriel's horn

Show that Gabriel's horn has finite volume but infinite surface area. (If this seems paradoxical, think about it one dimension down: are there regions in the plane with finite area but infinite perimeter?)

Let $f(x) = 1/x^p$ for p real, and let R be the region bounded by the x-axis and the graph of $y = f(x)$ for $x \geq 1$. Theorem 9.1 tells us that the area of R is finite if and only if $p > 1$. We now consider solids formed by rotating R about an axis or line.

20. For what values of p is the volume of the solid formed by rotating R about the x-axis finite? When the volume is finite, what is its volume?

21. For what values of p is the volume of the solid formed by rotating R about the line $x = 1$ finite? When the volume is finite, what is its value?

9.2 L'Hôpital's rule for quotients and products

How do you evaluate the limit of a quotient of two functions? With the limit quotient rule, of course—the limit of a quotient is the quotient of the limits (when the limit of the denominator is not zero) i.e.,

$$\lim \frac{f(x)}{g(x)} = \frac{\lim f(x)}{\lim g(x)}.$$

In this chapter "lim" stands for an ordinary limit ($\lim_{x \to a}$), a one-sided limit ($\lim_{x \to a^-}$ or $\lim_{x \to a^-}$), or a limit at infinity ($\lim_{x \to \infty}$ or $\lim_{x \to -\infty}$). But many important limits of quotients of functions in calculus cannot be evaluated with the limit quotient rule, for example $\lim_{x \to 0}(\sin x)/x$ and $\lim_{x \to (\pi/2)^-} \tan x / \sec x$, since the quotient rule is valid only when the limits of $f(x)$ and $g(x)$ *exist* and $\lim g(x) \neq 0$.

When the limits of the functions in the numerator and denominator of a quotient of two functions are both 0 or both infinite (either ∞ or $-\infty$) we say that the expression is *indeterminate*, or *an indeterminate form*, and use the symbols $0/0$ and ∞/∞ for the respective forms. For example, $\lim_{x\to 0}(\sin x)/x = 1$ and $\lim_{x\to 0}|x|/x$ does not exist, yet both limits have the form $0/0$. In this case the value of $f(x)/g(x)$ is not determined by the behavior of f and of g individually (as in limits where the quotient rule can be used), but by the relationship between the two functions.

⊘ Observe that $0/0$ and ∞/∞ are *symbols*, not fractions!

A very useful procedure for evaluating the limit of the quotient of two functions when the quotient is indeterminate is called L'Hôpital's rule:

Theorem 9.2 (L'Hôpital's rule) *Let f and g be differentiable function with $g'(x) \neq 0$ on an appropriate interval*. If $\lim f(x) = 0 = \lim g(x)$ or $\lim f(x) = \pm\infty = \lim g(x)$, and*

$$if \ \lim \frac{f'(x)}{g'(x)} = L, \ then \ \lim \frac{f(x)}{g(x)} = L.$$

In words: to evaluate the limit of a quotient of two functions that is indeterminate, you may replace the functions by their derivatives.

* The kind of interval where the derivatives of f and g must exist depends on the type of limit. For example, for an ordinary limit at a it's an open interval containing a (in this case $g(x) \neq 0$ must hold for all x in the interval except possibly a); for a limit from the left at a it's an open interval whose right endpoint is a; and for a limit at $+\infty$ it's an open interval unbounded on the right.

While we won't prove Theorem 9.2, we will give a geometric argument to show that it is plausible in the $0/0$ case with ordinary limits at a after a couple of examples.

Example 9.8 Evaluate $\lim_{x\to 0} \frac{\sin x}{x}$.

This is a $0/0$ indeterminate form, thus L'Hôpital's rule yields

$$\lim_{x\to 0} \frac{\sin x}{x} \overset{\text{L'H}}{=} \lim_{x\to 0} \frac{\cos x}{1} = 1. \quad \blacksquare$$

Above and in the examples to follow, we use $\overset{\text{L'H}}{=}$ as an "equals" sign to indicate where we have used L'Hôpital's rule.

⊘ When using L'Hôpital's rule, you replace the quotient of two functions by the quotient of their derivatives, you do *not* differentiate the quotient using the quotient rule for derivatives!

⊘ If the quotient of two functions is not indeterminate, using L'Hôpital's rule to find the limit will *almost always* give you an *incorrect* answer! For example: $\lim\limits_{x \to 2} x/(x-1) = 2$; but an inappropriate (and incorrect) use of L'Hôpital's rule yields: $\lim\limits_{x \to 2} x/(x-1) \overset{\text{L'H}}{=} \lim\limits_{x \to 2} 1/1 = 1$.

Back to Example 9.7 We can now complete our evaluation of $\int_1^\infty \dfrac{\ln x}{x^2}\, dx$:

$$\int_1^\infty \frac{\ln x}{x^2}\, dx = 1 - \lim_{b \to \infty} \frac{\ln b + 1}{b} \overset{\text{L'H}}{=} 1 - \lim_{b \to \infty} \frac{1/b}{1} = 1. \qquad \blacksquare$$

Here is an explanation of why L'Hôpital's rule works in the $0/0$ case with ordinary limits. In Figure 9.10(a) we see graphs of two differentiable functions $y = f(x)$ and $y = g(x)$ such that $f(a) = 0 = g(a)$. Thus $\lim_{x \to a} f(x)/g(x)$ is indeterminate. If we zoom in on the graph around the x-intercept $(a, 0)$, the graphs will appear more and more like straight lines, the lines being the two tangent lines $y = f'(a)(x - a)$ and $y = g'(a)(x - a)$, as shown in Figure 9.10(b).

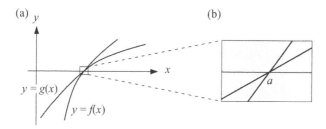

Figure 9.10. L'Hôpital's rule for the $0/0$ case

Hence, for x close to a, but not equal to a,

$$\frac{f(x)}{g(x)} \approx \frac{f'(a)(x - a)}{g'(a)(x - a)} = \frac{f'(a)}{g'(a)},$$

so that it is plausible that in the limit as $x \to a$ we have

$$\lim_{x \to a} \frac{f(x)}{g(x)} = \frac{f'(a)}{g'(a)} = \lim_{x \to a} \frac{f'(x)}{g'(x)}.$$

A proof of L'Hôpital's rule in this case can be constructed by observing that in a limit as $x \to a$ we have $x \neq a$, so that

$$\lim_{x \to a} \frac{f(x)}{g(x)} = \lim_{x \to a} \frac{f(x) - f(a)}{g(x) - g(a)} = \lim_{x \to a} \frac{\frac{f(x) - f(a)}{x - a}}{\frac{g(x) - g(a)}{x - a}} = \lim_{x \to a} \frac{f'(x)}{g'(x)},$$

providing the limits of the numerator and denominator in the next to last expression exist with $\lim_{x \to a} g'(x) \neq 0$.

Example 9.9 Evaluate $\int_0^\infty x e^{-x}\, dx$. See Figure 9.11.

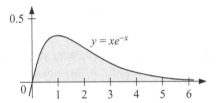

Figure 9.11. A graph of $y = xe^{-x}$

Following the three-step procedure from Section 9.1, we first find the antiderivative using integration by parts (we leave the details to you):

$$\int xe^{-x}\, dx = -(x+1)e^{-x} + C.$$

Although the antiderivative (with $C = 0$) is a product of two functions, we can write it as a quotient to evaluate the limit. Then we have

$$\int_0^\infty xe^{-x}\, dx = \lim_{b\to\infty}\left[-\frac{x+1}{e^x}\right]_{x=0}^b = \lim_{b\to\infty}\left[1 - \frac{b+1}{e^b}\right] \overset{\text{L'H}}{=} \lim_{b\to\infty}\left[1 - \frac{1}{e^b}\right] = 1.\ \blacksquare$$

In certain instances the application of L'Hôpital's rule yields another indeterminate form. In such cases a second or even third application of the rule may help in evaluating the limit.

Example 9.10 Evaluate $\lim_{x\to 0}\frac{x\sin x}{1-\cos x}$.

Since we have a $0/0$ form, we use L'Hôpital's rule:

$$\lim_{x\to 0}\frac{x\sin x}{1-\cos x} \overset{\text{L'H}}{=} \lim_{x\to 0}\frac{x\cos x + \sin x}{\sin x} \overset{\text{L'H}}{=} \lim_{x\to 0}\frac{-x\sin x + 2\cos x}{\cos x} = \frac{0+2}{1} = 2.\ \blacksquare$$

The first use of L'Hôpital's rule yielded a second indeterminate form, so we used the rule a second time, yielding a non-indeterminate form that we evaluated directly. It was only at this point did we learn that the two previous limits actually existed.

Example 9.11 Evaluate $\lim_{x\to\infty}\frac{\sqrt{x^2+1}}{x}$ and $\lim_{x\to-\infty}\frac{\sqrt{x^2+1}}{x}$.

Applying L'Hôpital's rule to the first limit yields

$$\lim_{x\to\infty}\frac{\sqrt{x^2+1}}{x} \overset{\text{L'H}}{=} \lim_{x\to\infty}\frac{x/\sqrt{x^2+1}}{1} = \lim_{x\to\infty}\frac{x}{\sqrt{x^2+1}} \overset{\text{L'H}}{=} \lim_{x\to\infty}\frac{1}{x/\sqrt{x^2+1}} = \lim_{x\to\infty}\frac{\sqrt{x^2+1}}{x},$$

so two applications of L'Hôpital's rule have brought us back to where we began! We look for another way to compute the limit. This limit can be evaluated by first simplifying the expression, and recalling that the limit of a continuous function is the function of the limit:

$$\lim_{x\to\infty}\frac{\sqrt{x^2+1}}{x} = \lim_{x\to\infty}\sqrt{\frac{x^2+1}{x^2}} = \lim_{x\to\infty}\sqrt{1 + \frac{1}{x^2}} = \sqrt{1 + \lim_{x\to\infty}\frac{1}{x^2}} = \sqrt{1+0} = 1.$$

What about the second limit? At first glance it appears this limit must also be 1. However, the limit at $-\infty$ is different:

$$\lim_{x\to-\infty}\frac{\sqrt{x^2+1}}{x} = \lim_{x\to-\infty}-\sqrt{\frac{x^2+1}{x^2}} = \cdots = -\sqrt{1+0} = -1,$$

since the simplification is different: $x = -\sqrt{x^2}$ when x is negative! ■

⊘ Although L'Hôpital's rule may be a new technique for you, it does not replace any of the techniques you learned in your PCE. Especially, it does not replace using algebra (carefully) to simplify expressions!

It is tempting to think that the converse of Theorem 9.2 also holds, that is, if $\lim f'(x)/g'(x)$ does not exist, then neither does $\lim f(x)/g(x)$. The next example shows that this is definitely *not* the case.

Example 9.12 Evaluate $\lim_{x \to 0} \frac{x^2 \sin(1/x)}{\sin x}$.

This is a 0/0 form, and L'Hôpital's rule yields

$$\lim_{x \to 0} \frac{x^2 \sin(1/x)}{\sin x} \stackrel{\text{L'H}}{=} \lim_{x \to 0} \frac{2x \sin(1/x) - \cos(1/x)}{\cos x}.$$

This limit does not exist since $\lim_{x \to 0} \cos(1/x)$ does not exist. However,

$$\lim_{x \to 0} \frac{x^2 \sin(1/x)}{\sin x} = \lim_{x \to 0} \frac{x}{\sin x} \cdot \lim_{x \to 0} \frac{\sin(1/x)}{1/x} = 1 \cdot 0 = 0. \quad ■$$

⊘ If, in Theorem 9.2, $\lim f'(x)/g'(x)$ does not exist, we can draw no conclusion about $\lim f(x)/g(x)$. For more examples of quotients where $\lim f'(x)/g'(x)$ does not exist yet $\lim f(x)/g(x)$ may or may not exist, see the article by R. J. Bumcrot mentioned in the Acknowledgments.

In addition to indeterminate quotients, there are *indeterminate products*. A limit of the form $\lim f(x)g(x)$ is indeterminate if $\lim f(x) = 0$ and $\lim g(x) = \pm\infty$, or if $\lim f(x) = \pm\infty$ and $\lim g(x) = 0$, and we denote this form by "$0 \cdot \infty$". Such products can always be written as indeterminate quotients. For example, if we have $\lim f(x) = 0$ and $\lim g(x) = \infty$, then

$$f(x)g(x) = \frac{f(x)}{1/g(x)} \text{ (0/0 form)} \quad \text{or} \quad f(x)g(x) = \frac{g(x)}{1/f(x)} \text{ (∞/∞ form)}.$$

We saw an indeterminate product $0 \cdot \infty$ in Example 9.9 and the procedure we followed there is the general procedure. We rewrite the expression as an indeterminate quotient and apply L'Hôpital's rule.

Example 9.13 Evaluate $\lim_{x \to 0^+} \sqrt{x} \ln x$. We can rewrite the product as a quotient in two ways:

$$\sqrt{x} \ln x = \frac{\ln x}{x^{-1/2}} \text{ or } \frac{\sqrt{x}}{1/\ln x}.$$

Remembering that the next step is to differentiate the numerator and denominator we choose the first form, since then the quotient of the derivatives will not contain a $\ln x$ term:

$$\lim_{x \to 0^+} \sqrt{x} \ln x = \lim_{x \to 0^+} \frac{\ln x}{x^{-1/2}} \stackrel{\text{L'H}}{=} \lim_{x \to 0^+} \frac{1/x}{(-1/2)x^{-3/2}} = -2 \lim_{x \to 0^+} \sqrt{x} = 0. \quad ■$$

9.2.1 Exercises

In Exercises 1 through 15, evaluate the limit if it exists.

1. $\lim_{x\to1}\dfrac{x^2+2x-3}{x^3-1}$
2. $\lim_{x\to0}\dfrac{\tan x}{x}$
3. $\lim_{x\to0}\dfrac{\sin^2 x}{\sin(x^2)}$
4. $\lim_{x\to0}\dfrac{\tan 3x}{\tan \pi x}$
5. $\lim_{x\to0}\dfrac{x-\sin x}{\tan x - x}$
6. $\lim_{x\to3^+}\dfrac{\sqrt{x^2-9}}{x-3}$
7. $\lim_{x\to\infty}\dfrac{(\ln x)^2}{x}$
8. $\lim_{x\to0^+}\dfrac{2^x-1}{x}$
9. $\lim_{x\to\infty}\dfrac{\sqrt{x^2-9}}{x-3}$
10. $\lim_{x\to\infty} x\sin\left(\dfrac{1}{x}\right)$
11. $\lim_{x\to0^+}\sin x\cdot\ln x$
12. $\lim_{x\to\infty} e^{-x}\text{sinhx}$
13. $\lim_{x\to\infty} x\ln\left(1+\dfrac{1}{x}\right)$
14. $\lim_{x\to\infty} x(a^{1/x}-1),\ a>0$
15. $\lim_{x\to0^+}\ln x\cdot\arctan x$

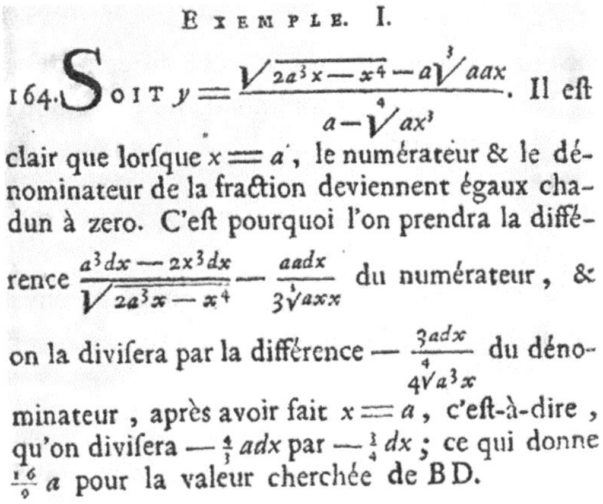

EXEMPLE. I.

164. SOIT $y = \dfrac{\sqrt{2a^3x - x^4} - a\sqrt[3]{aax}}{a - \sqrt[4]{ax^3}}$. Il eſt clair que lorſque $x = a$, le numérateur & le dénominateur de la fraction deviennent égaux chadun à zero. C'eſt pourquoi l'on prendra la différence $\dfrac{a^3dx - 2x^3dx}{\sqrt{2a^3x - x^4}} - \dfrac{aadx}{3\sqrt[3]{axx}}$ du numérateur, & on la diviſera par la différence $-\dfrac{3adx}{4\sqrt[4]{a^3x}}$ du dénominateur, après avoir fait $x = a$, c'eſt-à-dire, qu'on diviſera $-\frac{4}{3}adx$ par $-\frac{3}{4}dx$; ce qui donne $\frac{16}{9}a$ pour la valeur cherchée de BD.

Figure 9.12. L'Hôpital's example

16. Evaluate

$$\lim_{x\to a}\frac{\sqrt{2a^3x-x^4}-a\sqrt[3]{a^2x}}{a-\sqrt[4]{ax^3}}$$

for a positive constant a. This is L'Hôpital's example from his calculus book *Analyse des Infiniment Petits pour l'Intelligence des Lignes Courbes* (Analysis of the infinitely small to understand curves), published in 1696. See Figure 9.12

17. Consider a sector of the unit circle with angle θ ($0<\theta<\pi/2$) as shown in Figure 9.13. Let T_θ be the area of triangle ΔPQR and C_θ the area of the curved shape PQR. What is the limit of the ratio T_θ/C_θ as $\theta\to0^+$?

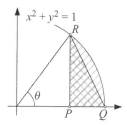

Figure 9.13. The circular sector in Exercise 17

In Exercises 18 through 23 determine if the improper integral is convergent or divergent; and if it is convergent, find its value.

18. $\displaystyle\int_1^\infty x^2 e^{-x}\,dx$ 19. $\displaystyle\int_1^\infty \frac{\ln x}{x}\,dx$ 20. $\displaystyle\int_1^\infty \frac{\ln x}{x^2}\,dx$

21. $\displaystyle\int_1^\infty \frac{dx}{x^2 + 5x + 6}$ 22. $\displaystyle\int_1^\infty \frac{dx}{x(x^2 + 1)}$ 23. $\displaystyle\int_1^\infty \frac{\arctan x}{x^2}\,dx$

The *Laplace transform* of a function is used to solve differential equations, and has applications in physics and engineering. If $f(t)$ is continuous for $t > 0$, the Laplace transform of f is the function F defined by the improper integral

$$F(s) = \int_0^\infty f(t)e^{-st}\,dt,$$

and the domain of F is the set of all s for which the integral converges. Find the Laplace transforms of the functions in Exercises 24 through 26.

24. $f(t) = at + b$ 25. $f(t) = e^{at}$ 26. $f(t) = \cos at$

27. Is it possible to assign a finite number to the volume of the solid obtained by revolving about the x-axis the region bounded by the positive x-axis and the graph of $y = x/(x^2 + 1)$ (see Figure 9.2)? If so, evaluate it.

28. It is easy to see that $\lim_{x\to\infty} e^{-\sin x}$ does not exist. But if we write this limit as

$$\lim_{x\to\infty} \frac{2x + \sin 2x}{(2x + \sin 2x)e^{\sin x}}$$

and apply L'Hôpital's rule and simplify, we obtain

$$\lim_{x\to\infty} \frac{4\cos x}{(2x + \sin 2x + 4\cos x)e^{\sin x}} = 0.$$

Can you explain the apparent contradiction? (Hint: be careful with cancellation when simplifying and recall the hypotheses of Theorem 9.2.)

9.3 Type II improper integrals

The second type of improper integral is one where the integrand has one or more infinite discontinuities in the interval of integration. We first encountered such an integral when we tried to use calculus to find the arc length of a quarter circle in Exercise 1 of Exploration 2.5.1 (in fact, a type II improper integral often arises in an arc length problem when the curve in question has a vertical tangent line). Let's repeat that exercise as the next example.

Example 9.14 Use calculus to find the circumference of a circle of radius r.

The upper half of the circle of radius r centered at the origin is the graph of the function $f(x) = \sqrt{r^2 - x^2}$. Since the circumference C is four times the arclength of the quarter-circle in the first quadrant, using the arc length formula (2.4) and $f'(x) = -x/\sqrt{r^2 - x^2}$ yields

$$C = 4 \int_0^r \frac{r}{\sqrt{r^2 - x^2}} \, dx.$$

The integrand approaches ∞ as x approaches r from the left, so we are unable to evaluate the integral as it has a discontinuity in the interval of integration. We will return to this example shortly. ∎

Since the integrand $r/\sqrt{r^2 - x^2}$ is continuous on $[0, r)$, we can approximate $\int_0^r r/\sqrt{r^2 - x^2} \, dx$ with integrals like

$$\int_0^{r-0.1} \frac{r}{\sqrt{r^2 - x^2}} \, dx, \qquad \int_0^{r-0.01} \frac{r}{\sqrt{r^2 - x^2}} \, dx, \qquad \int_0^{r-0.001} \frac{r}{\sqrt{r^2 - x^2}} \, dx,$$

and so on, which motivates the following definition.

Definition 9.4 If $f(x)$ is continuous on $[a, b)$, then

$$\int_a^b f(x) \, dx = \lim_{t \to b^-} \int_a^t f(x) \, dx$$

if the limit exists as a real number L, in which case we say that $\int_a^b f(x) \, dx$ *converges to L*. If the limit does not exist, we say that $\int_a^b f(x) \, dx$ *diverges*.

Back to Example 9.14 We have

$$C = 4 \int_0^r \frac{r}{\sqrt{r^2 - x^2}} \, dx = 4r \lim_{t \to r^-} \int_0^t \frac{1}{\sqrt{r^2 - x^2}} \, dx.$$

We now use the same three step process we used for type I improper integrals: evaluate the indefinite integral; use the fundamental theorem of calculus, and take the limit (here as $t \to r^-$). For the indefinite integral we use the trigonometric substitution $\theta = \arcsin(x/r)$:

$$\int \frac{1}{\sqrt{r^2 - x^2}} \, dx = \int d\theta = \theta + C = \arcsin\left(\frac{x}{r}\right) + C.$$

Thus

$$\lim_{t \to r^-} \int_0^t \frac{1}{\sqrt{r^2 - x^2}} \, dx = \lim_{t \to r^-} \left[\arcsin\left(\frac{x}{r}\right) \right]_{x=0}^t$$

$$= \lim_{t \to r^-} \left[\arcsin\left(\frac{t}{r}\right) \right] = \arcsin 1 = \frac{\pi}{2}$$

and so

$$C = 4 \int_0^r \frac{r}{\sqrt{r^2 - x^2}} \, dx = 4r \lim_{t \to r^-} \int_0^t \frac{1}{\sqrt{r^2 - x^2}} \, dx = 4r \frac{\pi}{2} = 2\pi r. \quad ∎$$

Example 9.15 Evaluate $\int_{-1}^{1} \frac{1}{\sqrt{1+x}}\, dx$. See Figure 9.14 for a graph of $y = 1/\sqrt{1+x}$.

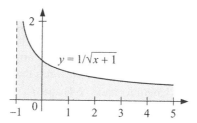

$y = 1/\sqrt{x+1}$

Figure 9.14. A graph of $y = 1/\sqrt{1+x}$

Since the domain of the integrand is all $x > -1$, the integral is improper: the integrand approaches ∞ as x approaches -1. But we can approximate $\int_{-1}^{1} \frac{1}{\sqrt{1+x}}\, dx$ with integrals like

$$\int_{-0.9}^{1} \frac{1}{\sqrt{1+x}}\, dx, \quad \int_{-0.99}^{1} \frac{1}{\sqrt{1+x}}\, dx, \quad \int_{-0.999}^{1} \frac{1}{\sqrt{1+x}}\, dx,$$

which motivates the following definition.

Definition 9.5 If $f(x)$ is continuous on $(a, b]$, then

$$\int_{a}^{b} f(x)\, dx = \lim_{t \to a^+} \int_{t}^{b} f(x)\, dx$$

if the limit exists as a real number L, in which case we say that $\int_{a}^{b} f(x)\, dx$ *converges to L.* If the limit does not exist, we say that $\int_{a}^{b} f(x)\, dx$ *diverges.*

Thus in this example, we have

$$\int_{-1}^{1} \frac{1}{\sqrt{1+x}}\, dx = \lim_{t \to -1^+} \int_{t}^{1} \frac{1}{\sqrt{1+x}}\, dx = \lim_{t \to -1^+} \left[2\sqrt{1+x} \right]_{x=t}^{1}$$

$$= \lim_{t \to -1^+} \left[2\sqrt{2} - 2\sqrt{1+t} \right] = 2\sqrt{2},$$

and thus $\int_{-1}^{1} \frac{1}{\sqrt{1+x}}\, dx$ converges to $2\sqrt{2}$. ∎

Example 9.16 Evaluate $\int_{-1}^{3} \frac{1}{\sqrt{|x^2-1|}}\, dx$. See Figure 9.15 for a graph of the integrand.

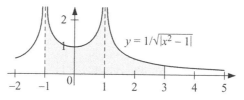

$y = 1/\sqrt{|x^2 - 1|}$

Figure 9.15. A graph of $y = 1/\sqrt{|x^2 - 1|}$

The procedure we have presented is for type II improper integrals where there is just one infinite discontinuity, occurring at either the right or left endpoint of the interval of integration. So if we have discontinuities at both endpoints, or at interior points, we first partition the interval of integration so that each integral has a single endpoint discontinuity. Since our integrand has discontinuities at -1 and $+1$, we have

$$\int_{-1}^{3} \frac{1}{\sqrt{|x^2 - 1|}}\, dx = \int_{-1}^{0} \frac{1}{\sqrt{1 - x^2}}\, dx + \int_{0}^{1} \frac{1}{\sqrt{1 - x^2}}\, dx + \int_{1}^{3} \frac{1}{\sqrt{x^2 - 1}}\, dx$$

or

$$\int_{-1}^{3} \frac{1}{\sqrt{|x^2 - 1|}}\, dx = 2\int_{0}^{1} \frac{1}{\sqrt{1 - x^2}}\, dx + \int_{1}^{3} \frac{1}{\sqrt{x^2 - 1}}\, dx$$

by the symmetry of the integrand on $(-1, 1)$ as seen in Figure 9.15. The first integral on the right of the $=$ sign is improper at the right endpoint, and the second is improper at the left endpoint. We leave the completion of the example as an exercise. ∎

Here is a companion to Theorem 9.1.

Theorem 9.3 *Let $b > 0$. Then the improper integral $\int_0^b 1/x^p\, dx$ converges for $p < 1$ and diverges for $p \geq 1$.*

Proof. We first note that the integrand is continuous on $(0, b]$ for all p, and, as with Theorem 9.1, it suffices to prove the theorem for $b = 1$. Now consider the case $p = 1$:

$$\int_{0}^{1} \frac{1}{x}\, dx = \lim_{x \to 0^+} \int_{t}^{1} \frac{1}{x}\, dx = \lim_{t \to 0^+} \ln x \Big|_{x=t}^{1} = \lim_{t \to 0^+} [\ln 1 - \ln t] = \infty,$$

and hence $\int_0^1 \frac{1}{x^p}\, dx$ diverges for $p = 1$. When $p \neq 1$ we have

$$\int_{0}^{1} \frac{1}{x^p}\, dx = \lim_{t \to 0^+} \int_{t}^{1} x^{-p}\, dx = \lim_{t \to 0^+} \frac{x^{1-p}}{1 - p} \Big|_{x=t}^{1}$$

$$= \lim_{t \to 0^+} \left[\frac{1}{1 - p} - \frac{t^{1-p}}{1 - p} \right] = \begin{cases} \frac{1}{1-p}, & p < 1, \\ \infty, & p > 1, \end{cases}$$

so that $\int_0^1 \frac{1}{x^p}\, dx$ converges (and so does $\int_0^b \frac{1}{x^p}\, dx$) for $p < 1$ and diverges for $p \geq 1$. □

Example 9.17 Find the area of the region in the fourth quadrant between the graph of $y = \ln x$ and its vertical asymptote. See Figure 9.16.

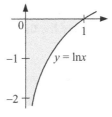

Figure 9.16. A graph of the region in Example 9.17

An integral that represents the area A of this region is $A = -\int_0^1 \ln x\, dx$ (do you see why there is a negative sign?), which is improper at its left endpoint. Since the indefinite integral (obtained using integration by parts) is $\int \ln x\, dx = x \ln x - x + C$, we have

$$A = -\lim_{t \to 0^+} \int_t^1 \ln x\, dx = -\lim_{t \to 0^+} [x \ln x - x]_{x=t}^1$$

$$= 1 + \lim_{t \to 0^+} \frac{\ln t}{1/t} \overset{\text{L'H}}{=} 1 + \lim_{t \to 0^+} \frac{1/t}{-1/t^2} = 1 + 0 = 1.$$

Thus A converges to 1. ■

Example 9.18 Evaluate $\int_1^4 \frac{1}{(x-2)^2}\, dx$.
   Since the integrand is undefined at $x = 2$ we write

$$\int_1^4 \frac{1}{(x-2)^2}\, dx = \int_1^2 \frac{1}{(x-2)^2}\, dx + \int_2^4 \frac{1}{(x-2)^2}\, dx.$$

Then

$$\int_1^2 \frac{1}{(x-2)^2}\, dx = \lim_{t \to 2^-} \int_1^t \frac{1}{(x-2)^2}\, dx = \lim_{t \to 2^-} \left[\frac{-1}{x-2}\right]_{x=1}^t = \lim_{t \to 2^-} \left[\frac{-1}{t-2} + \frac{1}{-1}\right] = \infty,$$

so that $\int_1^2 \frac{1}{(x-2)^2}\, dx$ diverges, and hence so does $\int_1^4 \frac{1}{(x-2)^2}\, dx$. ■

⊘ If we had not noticed the discontinuity at $x = 2$ and used the fundamental theorem of calculus as if the integrand were continuous on $[1, 4]$, we would have (incorrectly) obtained

$$\int_1^4 \frac{1}{(x-2)^2}\, dx = \left[\frac{-1}{x-2}\right]_{x=1}^4 = \left(\frac{-1}{2} - \frac{-1}{-1}\right) = -\frac{3}{2}.$$

⊘ Before using the fundamental theorem of calculus to evaluate any definite integral $\int_a^b f(x)\, dx$, be sure to check that the integrand $f(x)$ is continuous on $[a, b]$! Graphing the integrand over the interval $[a, b]$ may help you spot any discontinuities in the function.

Example 9.19 In Figure 9.17 we see the graphs of one arch of the cosine curve, and the upper half of the ellipse $x^2 + y^2/2 = 1$. Which curve has the greater arc length?

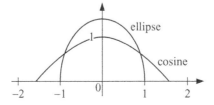

Figure 9.17. Graphs of $y = \cos x$ and $x^2 + y^2/2 = 1$

Using (2.4) the two arc lengths are

$$L_{\text{cosine}} = 2 \int_0^{\pi/2} \sqrt{1 + \sin^2 x} \, dx \quad \text{and} \quad L_{\text{ellipse}} = 2 \int_0^1 \sqrt{\frac{1 + x^2}{1 - x^2}} \, dx.$$

The integral for L_{ellipse} is a type II improper integral. To compare the two integrals, let's try the substitution $x = \sin\theta$ in the integral for the arc length of the ellipse. Then we have

$$L_{\text{ellipse}} = 2 \int_0^{\pi/2} \sqrt{\frac{1 + \sin^2\theta}{1 - \sin^2\theta}} \cos\theta \, d\theta = 2 \int_0^{\pi/2} \sqrt{1 + \sin^2\theta} \, d\theta = L_{\text{cosine}}.$$

Consequently, the two curves have the same arc length! However, to evaluate either integral we must resort to a numeric method such as the trapezoidal rule. ■

⊘ A substitution may remove an endpoint discontinuity in a convergent type II improper integral.

Our next example in this section is an integral that is both type I and type II, that is, the interval of integration has infinite length and the integrand has an infinite discontinuity in the interval. We once again partition the interval into two or more segments so that the resulting integrals are type I or type II.

Example 9.20 Evaluate $\int_0^\infty \frac{e^{-\sqrt{x}}}{\sqrt{x}} \, dx$.

The interval of integration has infinite length and the integrand is undefined at 0, so we will break the interval at any positive number, such as 1, to yield a type I and a type II improper integral:

$$\int_0^\infty \frac{e^{-\sqrt{x}}}{\sqrt{x}} \, dx = \int_0^1 \frac{e^{-\sqrt{x}}}{\sqrt{x}} \, dx + \int_1^\infty \frac{e^{-\sqrt{x}}}{\sqrt{x}} \, dx.$$

In each of the integrals to the right of the equal sign, the substitution $u = \sqrt{x}$ will simplify matters, yielding

$$\int_0^\infty \frac{e^{-\sqrt{x}}}{\sqrt{x}} \, dx = 2 \int_0^1 e^{-u} \, du + 2 \int_1^\infty e^{-u} \, du.$$

Once again a substitution has removed the endpoint discontinuity in a convergent type II improper integral, as happened in the preceding example. Since an antiderivative of e^{-u} is $-e^{-u}$, we have

$$\int_0^\infty \frac{e^{-\sqrt{x}}}{\sqrt{x}} \, dx = 2 \left[-e^{-u} \right]_{u=0}^1 + 2 \lim_{b \to \infty} \int_1^b e^{-u} \, du = 2 \left(1 - \frac{1}{e} \right) + 2 \lim_{b \to \infty} \left[-e^{-u} \right]_{u=1}^b$$

$$= 2 \left(1 - \frac{1}{e} \right) + 2 \lim_{b \to \infty} \left(\frac{1}{e} - \frac{1}{e^b} \right) = 2. \quad ■$$

In certain instances it is necessary to know if an improper integral converges or diverges prior to using the procedure in Definition 9.1; and in such cases the comparison test in the following theorem is useful.

Theorem 9.4 (Comparison test for improper integrals) *If f and g are continuous and $0 \le f(x) \le g(x)$ for all $x \ge a$, then $0 \le \int_a^\infty f(x)\,dx \le \int_a^\infty g(x)\,dx$. Hence if $\int_a^\infty g(x)\,dx$ converges, so does $\int_a^\infty f(x)\,dx$, and if $\int_a^\infty f(x)\,dx$ diverges, so does $\int_a^\infty g(x)\,dx$.*

A plausibility argument for Theorem 9.4 considers the areas of the regions under the graphs of $y = f(x)$ and $y = g(x)$ over the interval $[a, \infty)$. Similar versions of the theorem cover the intervals of integration $(-\infty, b]$ and $(-\infty, \infty)$ and type II improper integrals.

Example 9.21 Does $\int_0^\infty e^{-x^2/2}\,dx$ converge or diverge?
   Since

$$\int_0^\infty e^{-x^2/2}\,dx = \int_0^1 e^{-x^2/2}\,dx + \int_1^\infty e^{-x^2/2}\,dx$$

and $\int_0^1 e^{-x^2/2}\,dx$ is a proper integral, it suffices to consider $\int_1^\infty e^{-x^2/2}\,dx$. For $x \ge 1$, $-x^2 \le -x$ which implies that $0 \le e^{-x^2/2} \le e^{-x/2}$. Since $\int_1^\infty e^{-x/2}\,dx$ converges, so does $\int_1^\infty e^{-x^2/2}\,dx$ (and consequently so does $\int_0^\infty e^{-x^2/2}\,dx$). ∎

9.3.1 Exercises

In Exercises 1 through 12 determine if the improper integral is convergent or divergent; and if it is convergent, find its value.

1. $\displaystyle\int_1^5 \frac{dx}{\sqrt{x-1}}$ 2. $\displaystyle\int_0^8 \frac{dx}{x^{2/3}}$ 3. $\displaystyle\int_0^3 \frac{dx}{\sqrt{9-x^2}}$

4. $\displaystyle\int_0^{\pi/2} \tan x\,dx$ 5. $\displaystyle\int_0^1 x \ln x\,dx$ 6. $\displaystyle\int_0^{\pi/2} \sec\theta\,d\theta$

7. $\displaystyle\int_{-1}^8 x^{-1/3}\,dx$ 8. $\displaystyle\int_0^2 \frac{dx}{(x-1)^{2/3}}$ 9. $\displaystyle\int_{-2}^2 \frac{dx}{\sqrt{4-x^2}}$

10. $\displaystyle\int_2^\infty \frac{dx}{x^2-4}$ 11. $\displaystyle\int_0^\infty \frac{dx}{\sqrt{x}(1+x)}$ 12. $\displaystyle\int_0^\infty \frac{dx}{x^{2/3}+x^{4/3}}$

13. Complete the evaluation of the improper integral in Example 9.16.

14. When the curve $y = \ln x$ in Example 9.17 is revolved about the y-axis for x in $(0, 1]$, it generates an object known as the *funnel surface*. Find the volume and surface area of the funnel.

Let $f(x) = 1/x^p$ for p real, and let R be the region bounded by the y-axis, the x-axis, and the graph of $y = f(x)$ for $0 < x \le 1$. Theorem 9.3 tells us that the area of R is finite if and only if $p < 1$. We now consider solids formed by rotating R about an axis in Exercises 15 and 16.

15. For what values of p is the volume of the solid formed by rotating R about the x-axis finite? When the volume is finite, what is its value?

16. For what values of p is the volume of the solid formed by rotating R about the y-axis finite? When the volume is finite, what is its value?

17. Let $f(x)$ and $g(x)$ be continuous on $[a, \infty)$.

(a) Prove that if both $\int_a^\infty f(x)\,dx$ and $\int_a^\infty g(x)\,dx$ converge, then $\int_a^\infty (f(x)+g(x))\,dx$ converges.

(b) If both $\int_a^\infty f(x)\,dx$ and $\int_a^\infty g(x)\,dx$ diverge, must $\int_a^\infty (f(x)+g(x))\,dx$ diverge as well?

18. Let $f(x)$ be continuous on $[a, \infty)$. Prove that if $\int_a^\infty |f(x)|\,dx$ converges, then so does $\int_a^\infty f(x)\,dx$. (Hints: (a) show that $0 \le f(x)+|f(x)| \le 2|f(x)|$, (b) show that $\int_a^\infty (f(x)+|f(x)|)\,dx$ converges, (c) write $f(x) = (f(x)+|f(x)|) - |f(x)|$ to conclude that $\int_a^\infty f(x)\,dx$ converges.)

Show that the improper integrals in Exercises 19 and 20 converge.

19. $\displaystyle\int_1^\infty \frac{\sin x}{x^2}\,dx$ 20. $\displaystyle\int_1^\infty \frac{\sin x}{x}\,dx$ (Hint: integrate by parts first).

21. The graph of $y = \operatorname{sech}^{-1} x - \sqrt{1-x^2}$ for x in $(0, 1]$ is called a *tractrix*, as illustrated in Figure 9.18(a). The y-axis is a vertical asymptote of the tractrix. When the tractrix is rotated about the y-axis, it forms the top half of a solid called a *pseudosphere*, as illustrated in Figure 9.18(b).

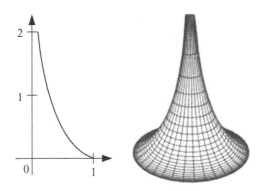

Figure 9.18. A tractrix and the top half of a psuedosphere

(a) Find the area of the region between the graph of the tractrix and the positive x- and y-axes, and the arc length of the tractrix between $x = x_0$ (for x_0 in $(0, 1)$) and $x = 1$.

(b) Find the volume and lateral surface area of the top half of the pseudosphere.

9.4 L'Hôpital's rule for differences and powers

If $\lim f(x) = +\infty$ and $\lim g(x) = +\infty$, what can you say about $\lim [f(x) - g(x)]$? In general, not much, since the size of $f(x) - g(x)$ depends on the relative magnitudes of $f(x)$ and $g(x)$. Thus another indeterminate form is "$\infty - \infty$", representing $\lim [f(x) - g(x)]$ when $\lim f(x) = +\infty$ and $\lim g(x) = +\infty$.

To evaluate this limit using L'Hôpital's rule, we again must first convert the difference $f(x) - g(x)$ into a quotient. The requisite algebra to do this depends on the functions $f(x)$ and $g(x)$. We illustrate with several examples.

Example 9.22 Evaluate $\lim_{x \to 0} \left(\frac{1}{x} - \frac{1}{\sin x} \right)$.

Using a common denominator for the two fractions yields an indeterminate quotient, and hence

$$\lim_{x \to 0} \left(\frac{1}{x} - \frac{1}{\sin x} \right) = \lim_{x \to 0} \frac{\sin x - x}{x \sin x} \overset{\text{L'H}}{=} \lim_{x \to 0} \frac{\cos x - 1}{x \cos x + \sin x} \overset{\text{L'H}}{=} \lim_{x \to 0} \frac{- \sin x}{x \sin x + 2 \cos x}$$

$$= \frac{0}{0 + 2} = 0. \blacksquare$$

Example 9.23 Evaluate $\lim_{x \to \infty} (\sqrt{x^2 + x} - x)$.

Since the expression is the difference of two square roots (here $x = \sqrt{x^2}$), we multiply and divide by the conjugate $\sqrt{x^2 + x} + x$:

$$\lim_{x \to \infty} \left(\sqrt{x^2 + x} - x \right) = \lim_{x \to \infty} \left(\sqrt{x^2 + x} - x \right) \cdot \frac{\sqrt{x^2 + x} + x}{\sqrt{x^2 + x} + x} = \lim_{x \to \infty} \frac{x}{\sqrt{x^2 + x} + x}.$$

Since the quotient is now a ∞/∞ form, it is tempting to apply L'Hôpital's rule. But the presence of the term $\sqrt{x^2 + x}$ in the denominator will lead to some rather ugly derivatives. Hence we fall back on a procedure from your PCE—multiplying the numerator and denominator by a negative power of x (here $1/x$) to reduce the magnitude of each:

$$\lim_{x \to \infty} \left(\sqrt{x^2 + x} - x \right) = \lim_{x \to \infty} \frac{x}{\sqrt{x^2 + x} + x} \cdot \frac{1/x}{1/x} = \lim_{x \to \infty} \frac{1}{\sqrt{1 + (1/x)} + 1} = \frac{1}{1 + 1} = \frac{1}{2}. \blacksquare$$

Example 9.24 Evaluate $\int_1^{\infty} \sqrt{x^2 + x} - x \, dx$.

In Example 9.23 we showed that $\lim_{x \to \infty} (\sqrt{x^2 + x} - x) = 1/2$, and hence for some $a > 1$ we have $\sqrt{x^2 + x} - x > 1/4$ for $x > a$. But $\int_a^{\infty} (1/4) \, dx$ diverges, and so by Theorem 9.4 so does $\int_a^{\infty} \sqrt{x^2 + x} - x \, dx$ (and so does $\int_1^{\infty} \sqrt{x^2 + x} - x \, dx$). $\blacksquare$

Our final indeterminate forms are limits of exponential expressions of the form $f(x)^{g(x)}$. The three forms are "0^0" meaning $\lim f(x) = 0$ and $\lim g(x) = 0$, "1^{∞}" meaning $\lim f(x) = 1$ and $\lim g(x) = \pm\infty$, and "∞^0" meaning $\lim f(x) = +\infty$ and $\lim g(x) = 0$. Our next example shows that the three forms really are indeterminate.

Example 9.25 Evaluate (a) $\lim_{x \to 0^+} x^{\alpha / \ln x}$, (b) $\lim_{x \to 1} x^{\alpha / \ln x}$, and (c) $\lim_{x \to \infty} x^{\alpha / \ln x}$, where α is an arbitrary nonzero real number.

The limits are examples of the forms 0^0, 1^{∞}, and ∞^0. If we let $h(x) = x^{\alpha / \ln x}$, then the domain of h is $x > 0$, $x \neq 1$. But for any x in this domain, $h(x)$ is positive and $\ln[h(x)] = (\alpha / \ln x) \ln x = \alpha$. Hence $h(x) = e^{\alpha}$ for all $x > 0$, $x \neq 1$, and so each of the limits is e^{α} for an arbitrary α, that is, each is indeterminate. $\blacksquare$

To evaluate the limit of an indeterminate exponential expression using L'Hôpital's rule, again we must first convert it to a quotient. We proceed as follows for all three forms. Letting $y = f(x)^{g(x)}$ we first evaluate the logarithm of each side:

$$\ln y = g(x) \cdot \ln f(x).$$

The product will be indeterminate (do you see why?), so we convert it to an indeterminate quotient and apply L'Hôpital's rule to evaluate $\lim \ln y = L$. The final step (and one that is easy to forget) is to exponentiate:

$$\lim f(x)^{g(x)} = \lim y = \lim e^{\ln y} = e^{\lim \ln y} = e^L$$

(the step $\lim e^{\ln y} = e^{\lim \ln y}$ is justified by the continuity of the exponential function). We illustrate with several examples.

Example 9.26 Evaluate $\lim_{x \to \infty} x^{1/x}$.

This is a "∞^0" indeterminate form. Letting $y = x^{1/x}$ yields $\ln y = \frac{1}{x} \ln x$, so that

$$\lim_{x \to \infty} \ln y = \lim_{x \to \infty} \frac{\ln x}{x} \overset{\text{L'H}}{=} \lim_{x \to \infty} \frac{1/x}{1} = 0$$

and

$$\lim_{x \to \infty} x^{1/x} = \lim_{x \to \infty} y = \lim_{x \to 0} e^{\ln y} = e^0 = 1. \ \blacksquare$$

Example 9.27 Evaluate $\lim_{x \to 0}(1 + ax)^{b/x}$ where a and b are positive constants.

This is a "1^∞" indeterminate form. If we set $y = (1 + ax)^{b/x}$ we can take the natural logarithm of each side when $|x| < |1/a|$ to yield $\ln y = \frac{b}{x} \ln(1 + ax)$. Thus

$$\lim_{x \to 0} \ln y = \lim_{x \to 0} \frac{b \ln(1 + ax)}{x} \overset{\text{L'H}}{=} \lim_{x \to 0} \frac{ab/(1 + ax)}{1} = ab,$$

and thus

$$\lim_{x \to 0}(1 + ax)^{b/x} = \lim_{x \to 0} y = \lim_{x \to 0} e^{\ln y} = e^{\lim_{x \to 0} \ln y} = e^{ab}. \ \blacksquare$$

Example 9.28 Evaluate (a) $\lim_{x \to 0^+} x^x$, (b) $\lim_{x \to 0^+} x^{x^x}$, and (c) $\lim_{x \to 0^+} x^{x^{x^x}}$. In a tower of exponents, the expression is evaluated from top to bottom, so that in (b) we have $x^{x^x} = x^{(x^x)}$ and in (c) $x^{x^{x^x}} = x^{(x^{(x^x)})}$.

(a) This is a "0^0" indeterminate form. Let $y = x^x$ so that $\ln y = x \ln x$. Then we have

$$\lim_{x \to 0^+} \ln y = \lim_{x \to 0^+} x \ln x = \lim_{x \to 0^+} \frac{\ln x}{1/x} \overset{\text{L'H}}{=} \lim_{x \to 0^+} \frac{1/x}{-1/x^2} = - \lim_{x \to 0^+} x = 0,$$

so that

$$\lim_{x \to 0^+} x^x = \lim_{x \to 0^+} e^{x \ln x} = e^{\lim_{x \to 0^+} x \ln x} = e^0 = 1.$$

(b) This limit is *not* indeterminate, since the limit of x^x is 1, and "0^1" is not an indeterminate form. Hence $\lim_{x \to 0^+} x^{x^x} = 0^1 = 0$.

(c) This is another "0^0" indeterminate form. Proceeding as in (a), we $y = x^{x^{x^x}}$ so that $\ln y = x^{x^x} \ln x = \frac{\ln x}{x^{(-x^x)}}$. Noting that the derivative of $x^{(-x^x)}$ is likely to be rather complicated, we try something else. Since x^x approaches 1 as x approaches 0 from the right, there is an interval $(0, \varepsilon)$ for some ε between 0 and 1 where $x^x > 1/2$. Hence for x in this interval we have $x^{x^x} < x^{1/2}$ and $\ln x < 0$, so that for x in $(0, \varepsilon)$,

$$\sqrt{x} \ln x < x^{x^x} \ln x < 0.$$

Then in the limit as x approaches 0 from the right,

$$0 = \lim_{x \to 0^+} \sqrt{x} \ln x \le \lim_{x \to 0^+} x^{x^x} \ln x \le 0$$

(see Example 9.12 for the leftmost limit). So by the squeeze theorem $\lim_{x \to 0^+} x^{x^x} \ln x = 0$. Thus

$$\lim_{x \to 0^+} x^{x^{x^x}} = \lim_{x \to 0^+} e^{x^{x^x} \ln x} = e^{\lim_{x \to 0^+} x^{x^x} \ln x} = e^0 = 1. \quad \blacksquare$$

9.4.1 Exercises

In Exercises 1 through 12, evaluate the limit if it exists.

1. $\lim_{x \to 1^+} \left(\dfrac{1}{\ln x} - \dfrac{1}{x - 1} \right)$

2. $\lim_{x \to 2} \left(\dfrac{x}{x^2 - 4} - \dfrac{1}{x^2 - 2x} \right)$

3. $\lim_{x \to (\pi/2)^-} (\sec x - \tan x)$

4. $\lim_{x \to \infty} (\cosh x - \sinh x)$

5. $\lim_{x \to \infty} \left(x - \sqrt{x^2 - x} \right)$

6. $\lim_{x \to \infty} \left(x 2^{1/x} - x \right)$

7. $\lim_{x \to \infty} \left(1 + \dfrac{1}{x} \right)^x$

8. $\lim_{x \to 0^+} \left(1 + \dfrac{1}{x} \right)^x$

9. $\lim_{x \to 0^+} (\sin x)^{\tan x}$

10. $\lim_{x \to 0^+} (x + e^x)^{1/x}$

11. $\lim_{x \to \infty} (x + e^x)^{1/x}$

12. $\lim_{x \to 0} \left(\dfrac{\sin x}{x} \right)^{1/x}$

13. If you deposit P dollars into a savings account that pays interest at the rate $100r\%$ compounded n times a year, after t years the amount A in the account will be

$$A = P \left(1 + \frac{r}{n} \right)^{nt}.$$

Use L'Hôpital's rule to show that as the number of compoundings per year increases without bound, the limiting formula for the amount in the account is $A = Pe^{rt}$.

14. Evaluate $\lim_{x \to 0^+} |x|^{1/x}$ and $\lim_{x \to 0^-} |x|^{1/x}$.

In the next two exercises, you will show that "$0^{+\infty}$" and "$0^{-\infty}$" are not indeterminate forms. In each exercise let $f(x) \ge 0$ and $\lim f(x) = 0$, where each occurrence of "$\lim$" is the same ordinary, one-sided, or infinite limit.

15. If $\lim g(x) = \infty$, show that $\lim f(x)^{g(x)} = 0$.

16. If $\lim g(x) = -\infty$, show that $\lim f(x)^{g(x)} = \infty$.

9.5 Continuous probability models

A *histogram* is a visual representation of a data set wherein the frequency of data with a specified value is represented as the area of a rectangle. In Figure 9.19 we have a histogram illustrating the age distribution of college students in Canada in 2007 (drawn for ages 17 to 39 inclusive—ages 40 and higher comprised 10% of the student population, and we haven't drawn that portion of the histogram).

For example, the proportion of college students in Canada in 2007 who were teenagers was 33% (.33 = .06 + .13 + .14) corresponding to the sum of the areas of the first three rectangles. Another way to state this is to say that the relative frequency of teenagers among Canadian college students in 2007 was 33%. Statements such as this can be rephrased in terms of probability:

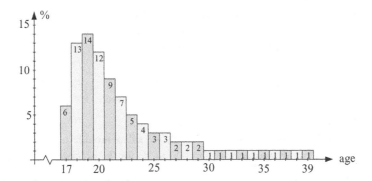

Figure 9.19. A histogram of the ages of college students in Canada in 2007

The *probability* that a randomly selected college student in Canada in 2007 was a teenager was 0.33.

The histogram in Figure 9.19 resembles the diagram for a Riemann sum for a function (not drawn in the figure) used to approximate the area under the curve. Such a function must be nonnegative, and the total area under the graph must be 1 (corresponding to a sum of 100% for the rectangles in the entire histogram). Such a function is called a *probability density*.

Definition 9.6 A *probability density function* (or *pdf*) $p(x)$ is a function with domain $(-\infty, \infty)$ such that

(i) $p(x) \geq 0$ for all x,

(ii) $\int_{-\infty}^{\infty} p(x)\,dx = 1$.

A typical pdf has a graph like the one illustrated in Figure 9.20.

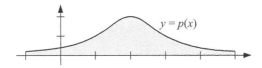

Figure 9.20. The graph of a typical probability density function

One application of a probability density is to build a *probability model*, a mathematical representation of some random phenomenon. Different random phenomena require different models, so it has proven useful to have a variety of probability densities from which to choose. In this section we will illustrate the use of improper integrals in constructing probability densities.

If $f(x)$ is a nonnegative function for which the improper integral $\int_{-\infty}^{\infty} f(x)\,dx$ converges to the positive number A, then $p(x) = f(x)/A$ is a pdf—both parts of Definition 9.6 will be satisfied by $p(x)$. In the next few examples we will construct some pdfs that have wide application in probability models.

Example 9.29 Consider the function

$$p(x) = \begin{cases} 0, & x < 0 \\ Ke^{-2x}, & x \geq 0. \end{cases}$$

Find a value for K so that $p(x)$ is a probability density function.

The constant K must satisfy $1 = \int_{-\infty}^{\infty} p(x)\,dx = K \int_0^{\infty} e^{-2x}\,dx$, that is, K is the reciprocal of the area A under the graph of $y = e^{-2x}$ over the interval $[0, \infty)$. But

$$A = \int_0^{\infty} e^{-2x}\,dx = \lim_{b \to \infty} \int_0^b e^{-2x}\,dx = \lim_{b \to \infty} \left[-\frac{1}{2}e^{-2x} \right]_{x=0}^b = \lim_{b \to \infty} \left[\frac{1}{2} - \frac{1}{2}e^{-2b} \right] = \frac{1}{2},$$

and thus $K = 1/A = 2$.

To make our probability models more useful, we often include an unspecified constant (called a *parameter*) in the expression for $p(x)$. Doing so for the preceding example yields

Example 9.30 *The exponential probability density.* For a positive real number λ, the function

$$p(x) = \begin{cases} 0, & x < 0, \\ Ke^{-\lambda x}, & x \geq 0, \end{cases}$$

is a probability density function if $K = \lambda$.

The calculus to show that $K = \lambda$ is analogous to that in the preceding example:

$$A = \int_0^{\infty} e^{-\lambda x}\,dx = \lim_{b \to \infty} \int_0^b e^{-\lambda x}\,dx = \lim_{b \to \infty} \left[-\frac{1}{\lambda}e^{-\lambda x} \right]_{x=0}^b = \lim_{b \to \infty} \left[\frac{1}{\lambda} - \frac{1}{\lambda}e^{-\lambda b} \right] = \frac{1}{\lambda},$$

and thus $K = 1/A = \lambda$. In Figure 9.21 we see graphs of $y = \lambda e^{-\lambda x}$ for $\lambda = 1/2$, 1, and 2.

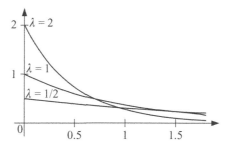

Figure 9.21. Graphs of three exponential density functions

Exponential models are used for random phenomena such as time intervals between events such as incoming phone calls and cars arriving at an intersection, for the lifetimes of electronic devices, and for the time until decay of a radioactive isotope.

The parameter λ controls the shape of the density, and has important statistical implications that you can learn about in a course in mathematical statistics. ■

Example 9.31 *The Erlang probability density.* The Erlang distribution (named for the Danish mathematician Agner Krarup Erlang (1878–1929)) is a generalization of the exponential distribution. For a nonnegative integer n and positive real number λ, consider the function

$$f(x) = \begin{cases} 0, & x < 0, \\ Kx^n e^{-\lambda x}, & x \geq 0. \end{cases}$$

As in the preceding example we seek a value of K so that $f(x)$ is a probability density function. So we must find the area under the graph of $y = x^n e^{-\lambda x}$ over the interval $[0, \infty)$. First we evaluate the indefinite integral using integration by parts:

$$\int x^n e^{-\lambda x}\, dx = -\frac{x^n}{\lambda} e^{-\lambda x} + \frac{n}{\lambda} \int x^{n-1} e^{-\lambda x}\, dx.$$

Thus

$$\int_0^\infty x^n e^{-\lambda x}\, dx = \lim_{b \to \infty} \left[-\frac{x^n}{\lambda} e^{-\lambda x} \right]_{x=0}^b + \frac{n}{\lambda} \int_0^\infty x^{n-1} e^{-\lambda x}\, dx = 0 + \frac{n}{\lambda} \int_0^\infty x^{n-1} e^{-\lambda x}\, dx.$$

Repeating integration by parts another $n - 1$ times yields

$$\int_0^\infty x^n e^{-\lambda x}\, dx = \frac{n!}{\lambda^n} \int_0^\infty e^{-\lambda x}\, dx = \frac{n!}{\lambda^n} \cdot \frac{1}{\lambda} = \frac{n!}{\lambda^{n+1}},$$

and thus $K = \lambda^{n+1}/n!$. The parameters n and λ control the shape of the Erlang pdf, and in Figure 9.22 we see graphs of $y = \lambda^{n+1} x^n e^{-\lambda x}/n!$ for $\lambda = 2$ and several values of n. ∎

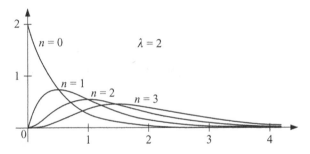

Figure 9.22. Graphs of four Erlang density functions

Example 9.32 *The normal probability density.* One of the most important probability densities is the standard normal, the canonical bell-shaped curve. In this example we first find the constant K so that the area under the curve is 1, and then later introduce location and shape parameters.

The standard normal pdf is traditionally denoted by the Greek letter ϕ and the variable is usually denoted by z, so that we have

$$\phi(z) = K e^{-z^2/2} \text{ for all } z.$$

To find K we must find the area A under the graph fo $y = e^{-x^2/2}$. This function does not have a simple antiderivative, so we must employ a different technique. Note

$$A = \int_{-\infty}^\infty e^{-z^2/2}\, dz = 2 \int_0^\infty e^{-z^2/2}\, dz$$

(the integrand is an even function, and the improper integral $\int_0^\infty e^{-z^2/2}\, dx$ converges to $A/2$ by Example 9.21), so

$$A^2/4 = \int_0^\infty e^{-z^2/2}\, dz \cdot \int_0^\infty e^{-y^2/2}\, dy.$$

Now substitute $z = xy$, $dz = y\,dx$ in the first integral, and insert this number (recall the integral converges) into the second integral:

$$A^2/4 = \int_0^\infty e^{-(xy)^2/2} y\,dx \cdot \int_0^\infty e^{-y^2/2}\,dy = \int_0^\infty \left(\int_0^\infty e^{-(xy)^2/2} y\,dx \right) e^{-y^2/2}\,dy.$$

Simplification yields

$$A^2/4 = \int_0^\infty \left(\int_0^\infty y e^{-(x^2+1)y^2/2}\,dx \right) dy = \int_0^\infty \left(\int_0^\infty y e^{-(x^2+1)y^2/2}\,dy \right) dx. \qquad (9.1)$$

Do you see what occurred in the final step? We exchanged dx and dy in the inner and outer integrals, a step that needs some justification. In Section 1.3 we encountered iterated integrals over regions that were rectangles of the form $[a, b] \times [c, d]$, i.e., both dimensions were finite. A formal proof that this exchange is permissible in iterated type I improper integrals requires the study of multiple integrals, which you will encounter in a course in multivariate calculus. Consequently, we will proceed assuming that (1.12) holds for unbounded rectangles. In Figure 9.23 we see a graph of the integrand in (9.1) over the rectangle $[0, 3] \times [0, 3]$.

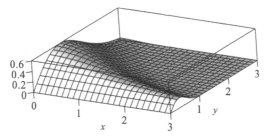

Figure 9.23. A graph of $h(x, y) = y e^{-(x^2+1)y^2/2}$ over $[0, 3] \times [0, 3]$

We now return to our computation of $A^2/4$. If we make the substitution $u = (x^2 + 1)y^2/2$, $du = (x^2 + 1)y\,dy$ (so that $y\,dy = [1/(x^2 + 1)]\,du$) in the inner integral, we have

$$\int_0^\infty y e^{-(x^2+1)y^2/2}\,dy = \frac{1}{x^2 + 1} \int_0^\infty e^{-u}\,du = \frac{1}{x^2 + 1} \cdot 1 = \frac{1}{x^2 + 1}.$$

Hence

$$A^2/4 = \int_0^\infty \left(\frac{1}{x^2 + 1} \right) dx = \lim_{b \to \infty} (\arctan b) = \frac{\pi}{2}, \qquad (9.2)$$

thus $A = \sqrt{2\pi}$ and $\phi(z) = \frac{1}{\sqrt{2\pi}} e^{-z^2/2}$ is the standard normal pdf.

If we let $z = (x - \mu)/\sigma$ for a real number μ and a positive real number σ, we shift the graph of $\phi(z)$ horizontally μ units (to the right if μ is positive, to the left if μ is negative), and expand or contract (depending whether σ is larger or smaller than 1) the graph in the horizontal direction by a factor of σ. For an expansion or contraction the area is also multiplied by σ, so we must divide ϕ by σ to maintain area 1 under the graph. Thus we have a two-parameter family of normal probability density functions given by

$$f(x) = \frac{1}{\sigma\sqrt{2\pi}} e^{-\frac{1}{2}\left(\frac{x-\mu}{\sigma}\right)^2} \quad \text{for all } x.$$

The standard abbreviation for the normal probability model with the pdf displayed above with parameters μ and σ is $N(\mu, \sigma)$. Thus the standard normal with pdf ϕ is denoted $N(0, 1)$. In Figure 9.24 we have graphs of the pdfs for $N(0, 1)$, $N(-1, 2)$, $N(1, 1/2)$, and $N(-1, 1/3)$. In spite of its rather complicated form, the normal probability model is one of the most important in the theory and application of probability and statistics. Probability models based on normal densities are used in many instances where the variable of interest is a sum or average of many factors. For example, the weight of an adult male human is a result of many factors such as heredity, diet, environment, etc., and consequently a normal probability model is often appropriate to describe weight. Similar variables abound in biology, medicine, psychology, sociology, finance, etc. ∎

9.5.1   Exercises

For each of the functions p in Exercises 1 through 4, find a constant K so that p is a probability density function (in each a is a positive constant):

1. $p(x) = K/(x^2 + a^2)$ for all x (the *Cauchy* density).

2. $p(x) = K\operatorname{sech}(x/a)$ for all x (the *hyperbolic secant* density).

3. $p(x) = K\operatorname{sech}^2(x/a)$ for all x (the *logistic* density).

4. $p(x) = \begin{cases} Kx^{-(a+1)}, & x \geq 1 \\ 0, & x < 1 \end{cases}$ (the *Pareto* density).

The *mean* (or *expected value*) μ associated with a pdf $p(x)$ is defined analogously to the center of mass $\bar{x}$ of a rod with mass density $p(x)$, as in Section 5.2.2 (see (5.4)). Thus

$$\mu = \frac{\displaystyle\int_{-\infty}^{\infty} xp(x)\,dx}{\displaystyle\int_{-\infty}^{\infty} p(x)\,dx} = \frac{\displaystyle\int_{-\infty}^{\infty} xp(x)\,dx}{1} = \int_{-\infty}^{\infty} xp(x)\,dx$$

if the integral converges, otherwise μ is undefined. Find the mean μ associated with the probability densities in Exercises 5 through 9.

5. The exponential density in Example 9.30.

6. The Erlang density in Example 9.31.

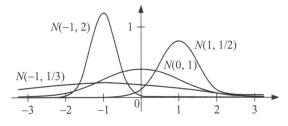

Figure 9.24. Graphs of four normal density functions

7. The standard normal density in Example 9.32.

8. The Cauchy density in Exercise 1.

9. The Pareto density in Exercise 4.

The *gamma function* $\Gamma(\alpha)$ is defined for real numbers $\alpha > 0$ and given by the improper integral

$$\Gamma(\alpha) = \int_0^\infty x^{\alpha-1} e^{-x} \, dx.$$

10. Show that (a) $\Gamma(\alpha + 1) = \alpha \Gamma(\alpha)$ and (b) for a positive integer n, $\Gamma(n) = (n-1)!$. (Hint: in (a) use integration by parts or the lemma in Exploration 1.4.3.) Hence the gamma function generalizes factorials to positive real number arguments.

11. Show that $\Gamma(1/2) = \sqrt{\pi}$. (Hint: try the substitution $x = t^2/2$. Does the result look familiar?)

The *gamma probability density* is a two-parameter function given by

$$f(x) = \begin{cases} 0, & x < 0 \\ K x^{\alpha-1} e^{-x/\beta}, & x \geq 0 \end{cases}$$

for $\alpha > 0$ and $\beta > 0$. Special cases of the gamma density include the exponential density in Example 9.30 ($\alpha = 1$ and $\beta = 1/\lambda$) and the Erlang density in Example 9.31 ($\alpha = n+1$ and $\beta = 1/\lambda$).

12. Evaluate K in terms of α and β for the gamma density.

13. Find the mean μ associated with the gamma density.

The *moment generating function $M(t)$* associated with the pdf $p(x)$ is defined by the improper integral

$$M(t) = \int_{-\infty}^\infty e^{tx} p(x) \, dx$$

if the integral converges. It is used in probability and statistics to evaluate *moments* associated with $p(x)$, such as the mean and variance. Find the moment generating function associated with the densities in Exercises 14 through 16.

14. The exponential density in Example 9.30.

15. The gamma density preceding Exercise 12.

16. The standard normal density in Example 9.32.

17. One use for the moment generating function is in finding the mean μ associated with the pdf. In a probability and statistics course you will show that $\mu = M'(0)$. Verify this for the densities in Exercises 14 through 16.

9.6 Explorations

9.6.1 A useful improper integral

Consider the following two questions about type I improper integrals:

Question 1. If a function $f(x)$ is continuous and $\int_0^\infty f(x)\,dx$ converges, must $\lim_{x\to\infty} f(x) = 0$?

Question 2. If a function $f(x)$ is continuous for all real x but unbounded as $x \to \infty$, must the integral $\int_0^\infty f(x)\,dx$ diverge?

Think about these questions for a few minutes.

Okay, that's long enough. It seems like the answer to each must be *yes*, right? But that's not correct, both answers are *no*!

In this exploration, you will learn more about one of the Fresnel integrals introduced in Exercise 11 in Section 8.3.4:

$$\int_0^\infty \sin(x^2)\,dx. \tag{9.3}$$

(There is also a Fresnel integral with integrand $\cos(x^2)$.) The integrand $f(x) = \sin(x^2)$ is graphed in Figure 9.25.

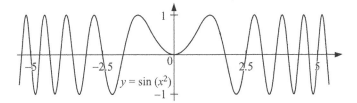

Figure 9.25. A graph of $y = \sin(x^2)$

Exercise 1. Show that the integral in (9.3) converges. Here are several hints:

1. You need only show that $\int_1^\infty \sin(x^2)\,dx$ converges.
2. Make the substitution $t = x^2$, then integrate by parts.
3. Show that the integral from the integration by parts converges.

When you study complex analysis, you will be able to show that $\int_0^\infty \sin(x^2)\,dx$ converges to $\sqrt{\pi/8}$.

Exercise 2. Now use (9.3) to answer Question 1.

Exercise 3. Consider the integral $\int_0^\infty x \sin(x^4)\,dx$. Show that it also converges. (Hint: a simple substitution will do.)

Exercise 4. Use the integral in Exercise 3 to answer Question 2.

9.6.2 The *n*th root of *n* factorial

When you study infinite series in Chapters 10 and 11, it will be advantageous to know that the behavior of the *n*th root of $n!$ as $n \to \infty$. We begin with a numerical investigation.

Exercise 1. Evaluate $\sqrt[n]{n!}$ for some values of n, say 10, 100, 1000, 10000, and 100000. Do you see a pattern? If not, look at $n/\sqrt[n]{n!}$.

You should have obtained values similar to those in Table 9.1.

Table 9.1. The nth root of n factorial

n	10	100	1000	10000	100000
$\sqrt[n]{n!}$	4.5287	37.9927	369.4916	3680.8272	36790.3999
$\dfrac{n}{\sqrt[n]{n!}}$	2.2081	2.6321	2.7064	2.7168	2.7181

While $\sqrt[n]{n!} \to +\infty$ may not be surprising, it appears that $\frac{n}{\sqrt[n]{n!}} \to e$, which may be. In this exploration you will prove that

$$\lim_{n \to \infty} \frac{\sqrt[n]{n!}}{n} = \frac{1}{e}, \tag{9.4}$$

which establishes the behavior of both $\sqrt[n]{n!}$ and $\frac{n}{\sqrt[n]{n!}}$.

Exercise 2. Why does (9.4) imply $\lim_{n \to \infty} \sqrt[n]{n!} = +\infty$?

Exercise 3. Show that $\ln\left(\frac{\sqrt[n]{n!}}{n}\right) = \sum_{k=1}^{n} \ln(k/n) \cdot (1/n)$.

Exercise 4. Show that $\sum_{k=1}^{n} \ln(k/n) \cdot (1/n)$ is a Riemann sum for the improper integral $\int_0^1 \ln x \, dx$. (Hint: let $x_k^* = x_k = k/n$.)

Exercise 5. Invoke Example 9.17 to conclude $\lim_{n \to \infty} \ln\left(\frac{\sqrt[n]{n!}}{n}\right) = -1$, which establishes (9.4).

9.6.3 Numerical integration of improper integrals

In Exploration 7.3.3 you saw how substitutions could used to improve the performance of Simpson's rule for proper integrals whose integrands had unbounded derivatives. In this exploration you will discover that similar substitutions can be used with improper integrals. In the exercises S_n represents a Simpson's rule approximation with n subintervals, as given in (7.8) on page 161. We begin with an exercise with a type II integral.

Exercise 1. The integral $\int_1^3 (x^2 - 1)^{-1/3} \, dx$ is improper since its integrand is undefined at $x = 1$. Evaluate S_{20} for this integral. Then find a substitution so that the integrand and its derivatives are bounded on the interval of integration, and evaluate its S_{20}.

For the given integral $S_{20} \approx 1.57271519$, which (as we shall see) is not correct even to the first decimal place. If you set $u = (x^2 - 1)^{1/3}$ then the integral becomes

$$(3/2) \int_0^{\sqrt[3]{2}} u(u^3 + 1)^{-1/2} \, du,$$

which is proper. Then $S_{20} \approx 1.70644534$, which is correct to six decimal places.

For a type I improper integral, all the numerical methods in Chapter 7 fail since the interval of integration has infinite length. To adapt the methods to a type I improper integral for which we are unable to use the fundamental theorem we seek a substitution that transforms an interval of integration with infinite length to one with finite length.

Exercise 2. Consider the integral $\int_1^\infty dx/\sqrt{x^3+1}$. Since the function $1/\sqrt{x^3+1}$ has no closed form antiderivative, we are unable to use the methods in this chapter to evaluate the integral. But a substitution of the form $x = 1/u^n$ for a positive integer n will change the interval of integration from x in $[1, \infty)$ to u in $(0, 1]$. Show that the substitution yields

$$\int_1^\infty \frac{dx}{\sqrt{x^3+1}} = n \int_0^1 \frac{u^{(n/2)-1}}{\sqrt{u^{3n}+1}}\, du.$$

What values of n yield a proper u-integral?

Exercise 3. The u-integral in Exercise 2 is proper for $n \geq 2$, so use S_{20} to approximate its value when $n = 2$.

Exercise 4. In the previous exercise, you should obtain $S_{20} \approx 1.8947609974$, which is accurate to eight decimal places. But suppose the improper integral we wish to evaluate is $\int_0^\infty dx/\sqrt{x^3+1}$. Will the substitution $x = 1/u^n$ work on it?

You answered "no" to the question in the previous exercise since the substitution transforms the interval of integration from x in $[0, \infty)$ to u in $(0, \infty)$, and so the resulting u-integral is also improper.

Exercise 5. Does writing $\int_0^\infty dx/\sqrt{x^3+1} = \int_0^1 dx/\sqrt{x^3+1} + \int_1^\infty dx/\sqrt{x^3+1}$ help?

Of course it does, since the first integral to the right of the equals sign in the exercise is proper, for this integral we have $S_{20} \approx 0.9096044032$, and hence

$$\int_0^\infty \frac{dx}{\sqrt{x^3+1}} \approx 1.8947609974 + 0.9096044032 = 2.8043654006.$$

9.7 Acknowledgments

1. Example 9.12 and Exercise 17 in Section 9.2.1 is from R. C. Buck, *Advanced Calculus* (3rd edition), McGraw-Hill Book Company, New York, 1978.
2. For more examples of quotients where $\lim f'(x)/g'(x)$ does not exist yet $\lim f(x)/g(x)$ may or may not exist, see R. J. Bumcrot, Subtleties in L'Hôpital's rule, *The College Mathematics Journal*, **15** (1985), pp. 51–52.
3. The argument in Example 9.28(c) is adapted from J. N. Ash, The limit of $x^{x^{\cdot^{\cdot^{\cdot^x}}}}$ as x tends to zero, *Mathematics Magazine*, **69** (1996), pp. 207–209.
4. The data on the age distribution of college students in Canada is from "Trends in the Age Composition of College and University Students and Graduates" on the Statistics Canada website at www.statcan.gc.ca/pub/81-004-x/2010005/article/11386-eng.htm
5. Exploration 9.6.1 is adapted from counterexamples 5.12 and 5.15 in S. Klymchuk, *Counterexamples in Calculus*, Mathematical Association of America, Washington, 2010.
6. Exploration 9.6.2 is adapted from C. C. Mumma II, $N!$ and the root test, *The American Mathematical Monthly*, **93** (1986), p. 561.
7. Exploration 9.6.3 is adapted from C. W. Avery and F. P. Soler, Applications of transformations to numerical integration, *The College Mathematics Journal*, **19** (1988), pp. 166–168.

10

Infinite Series (Part I)

Even as the finite encloses an infinite series
And in the unlimited limits appear,
So the soul of immensity dwells in minutia
And in narrowest limits no limits inhere
What joy to discern the minute in infinity!

Jacob Bernoulli, *Ars Conjectandi* (1713)

The three primary mathematical objects encountered in a single-variable calculus course are *derivatives*, *integrals*, and *infinite series*. The common concept underlying each is the *limit*. The derivative is a limit (of a difference quotient), the definite integral is a limit (of a Riemann sum), and an infinite series is a limit (of a sequence of partial sums). You are well aware of the applications of derivatives and integrals. Infinite series have a great many applications throughout mathematics, and you will encounter them in nearly every course in mathematics and the physical, biological, and social sciences for which a calculus course is a prerequisite.

This chapter and the next will introduce you to a variety of types of infinite series and to their applications. In this chapter we deal primarily with series of constants, and in the next chapter we study power series and their use in approximating certain functions by polynomials.

10.1 An introduction to series and sequences

We begin with a couple of simple examples.

Example 10.1 Suppose we have a square of paper with area 1 and cut it in half vertically, as shown in Figure 10.1, creating two rectangles each with area 1/2. Then we cut the right-hand rectangle in half, to create two squares each with area 1/4. Then we cut the square in the upper right-hand corner in half vertically, creating two rectangles each with area 1/8. If we continue this process indefinitely we will have cut the square into infinitely many pieces whose total area is the same as the area of the original square, so that

$$\frac{1}{2} + \frac{1}{4} + \frac{1}{8} + \frac{1}{16} + \cdots = 1. \quad \blacksquare$$

245

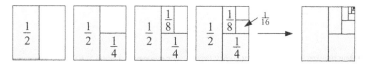

Figure 10.1. Dissecting a square with area 1

Example 10.2 If you evaluate the fraction 4/9 on a simple calculator, you will see something like Figure 10.2:

Figure 10.2. Dividing 4 by 9 on a calculator

The calculator shows you part of the repeating decimal for 4/9:

$$.44444444\cdots = \frac{4}{9}.$$

Each 4 in the decimal represents a fraction whose denominator is a power of 10, so that we have

$$\frac{4}{10} + \frac{4}{100} + \frac{4}{1000} + \frac{4}{10000} + \cdots = \frac{4}{9}. \quad \blacksquare$$

The results in Examples 10.1 and 10.2 are examples of a *geometric series* (called geometric since its terms are elements of a geometric progression), an expression of the form

$$a + ar + ar^2 + ar^3 + \cdots + ar^n + \cdots. \tag{10.1}$$

In Example 10.1 we have $a = 1/2$ and $r = 1/2$, while in Example 10.2 we have $a = 4/10$ and $r = 1/10$.

Example 10.3 The expression

$$2 - \frac{2}{3} + \frac{2}{9} - \frac{2}{27} + \cdots + 2\left(-\frac{1}{3}\right)^n + \cdots$$

is a geometric series, with $a = 2$ and $r = -1/3$. $\blacksquare$

In general an *infinite series* is an expression of the form

$$a_1 + a_2 + a_3 + \cdots + a_k + \cdots \quad \text{or} \quad a_0 + a_1 + a_2 + \cdots + a_k + \cdots \tag{10.2}$$

(for a geometric series we have $a_k = ar^k$). When the expression represents a real number, we say that the series has a *sum*. For example, the series in Example 10.1 has the sum 1 (or sums to 1), and the series in Example 10.2 sums to 4/9. We will be more precise about what we mean by an infinite series in general and geometric series in particular later in this chapter. We will also define what we mean by "the series has a sum."

⊘ It is good practice when using (10.2) for a series to present enough terms before the ellipsis (the "$\cdots$") so that the pattern is clear. For example $1 + \frac{1}{2} + \cdots$ is ambiguous, since the terms replaced by the ellipsis may be reciprocals of the positive integers, reciprocals of powers of 2, or even reciprocals of the prime numbers. When in doubt, give a formula for the general term a_n if possible.

In this chapter and the next we will use the sigma notation that you encountered in your PCE for Riemann sums and write (10.2) as

$$\sum_{k=1}^{\infty} a_k = a_1 + a_2 + a_3 + \cdots + a_k + \cdots$$

and

$$\sum_{k=0}^{\infty} a_k = a_0 + a_1 + a_2 + \cdots + a_k + \cdots .$$

Example 10.4 (The harmonic series) If we set $a_k = 1/k$ for all $k \geq 1$ we have the *harmonic series*

$$\sum_{k=1}^{\infty} \frac{1}{k} = 1 + \frac{1}{2} + \frac{1}{3} + \frac{1}{4} + \cdots + \frac{1}{k} + \cdots . \tag{10.3}$$

Does this series have a sum, as did the series in Examples 10.1 and 10.2? This is an important example, and we shall return to it later. ∎

Why the name harmonic?
The word comes from the Greek αρμονία("joint, agreement, concord"), since when things are joined together nicely, harmony results. In their study of music, the Greeks learned that when the lengths of strings were in certain ratios such as 1/2, 1/3, and 1/4 and played simultaneously, the sounds fit together well, resulting in pleasant music. As music and arithmetic were often studied together, the word was also applied to a sequence of numbers whose reciprocals form an arithmetic progression.

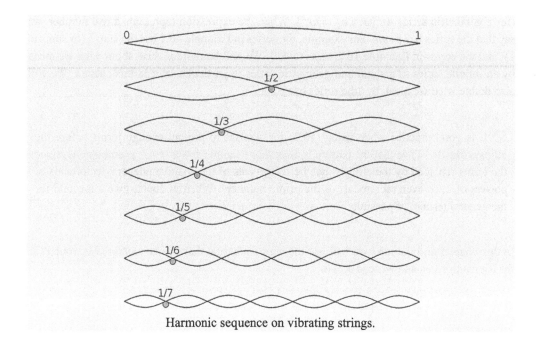

Harmonic sequence on vibrating strings.

The question we asked in Example 10.4 is the first basic question in the study of infinite series: *when does a series represent a real number*, or *when does the series have a sum*? Clearly some don't. For example, the series $1 + 1 + 1 + 1 + \cdots$ does not, since the sum of finitely many terms increases without bound as we add more and more 1s. This observation is the key to answering the question. As we add up the terms in the series, we generate a collection of *partial sums* of the series, and this collection is an example of a type of function called a *sequence*:

Definition 10.1 A *sequence* is a function whose domain is a set $\{n, n + 1, n + 2, n + 3, \ldots\}$ of consecutive nonnegative integers. In most instances n is 1 or 0. Sequences are frequently given by listing the elements of the range, e.g., for the sequence function $f(n) = x_n$ we write $\{x_n\}_{n=1}^{\infty} = \{x_1, x_2, x_3, \ldots\} = \{f(1), f(2), f(3), \ldots\}$.

For the infinite series (10.2) the elements of the sequence of partial sums $\{S_n\}_{n=1}^{\infty} = \{S_1, S_2, S_3, \ldots\}$ look like

$$S_1 = a_1 \qquad\qquad\qquad\qquad S_1 = a_1$$
$$S_2 = a_1 + a_2 \qquad\qquad\qquad S_2 = S_1 + a_2$$
$$S_3 = a_1 + a_2 + a_3 \qquad \text{or} \qquad S_3 = S_2 + a_2$$
$$\vdots \qquad\qquad\qquad\qquad\qquad \vdots$$
$$S_n = a_1 + a_2 + \cdots + a_n, \text{ etc.} \qquad S_n = S_{n-1} + a_n, \text{ etc.}$$

In the right-hand side of the above display we see the efficient way to compute the sequence of partial sums: successively add a new term to the preceding partial sum.

The sequence $\{S_n\}_{n=k}^\infty$ of partial sums of a series is defined in terms of the sequence $\{a_n\}_{n=k}^\infty$ of terms of the series. Each of the two sequences *define* the series: $\{a_n\}_{n=k}^\infty$ tells us what numbers are being added together; $\{S_n\}_{n=k}^\infty$ tells us the order in which the numbers are added.

We will often write $\{a_n\}$ and $\{S_n\}$ in place of $\{a_n\}_{n=k}^\infty$ and $\{S_n\}_{n=k}^\infty$.

Back to Example 10.1 From Figure 10.1 it is easy to compute the sequence $\{S_n\}$ of partial sums for this series—each term is 1 minus the area of the unshaded square or rectangle, so that $S_n = (1/2) + (1/2)^2 + (1/2)^3 + \cdots + (1/2)^n = 1 - (1/2)^n$. ∎

Definition 10.2 A sequence $\{x_n\}$ has a limit L if we can make x_n as close to L as we wish by taking n sufficiently large, that is, for any number $\varepsilon > 0$ we can find an integer N such that whenever $n > N$ we have $|x_n - L| < \varepsilon$. We then write $\lim_{n\to\infty} x_n = L$ and say that $\{x_n\}$ *converges* to L. If $\lim_{n\to\infty} x_n$ does not exist, we say that $\{x_n\}$ *diverges*.

We will often write $\lim x_n = L$ (or simply $x_n \to L$) in place of $\lim_{n\to\infty} x_n = L$ since the limit as $n \to \infty$ is the only limit of interest for a sequence.

Applying Definition 10.3 to the sequence of partial sums of a series yields the following definition.

Definition 10.3 Let $\sum_{k=1}^\infty a_k$ be a series whose sequence $\{S_n\}$ of partial sums is given by $S_n = \sum_{k=1}^n a_k$. Then $\sum_{k=1}^\infty a_k$ *converges to S* if $\lim S_n = S$, and we call S the *sum* of the series; $\sum_{k=1}^\infty a_k$ *diverges* if $\lim S_n$ does not exist, and the series does not have a sum. When $\sum_{k=1}^\infty a_k$ converges to S, we often write $\sum_{k=1}^\infty a_k = S$.

⊘ Be careful not to confuse a series, say $1/2 + 1/4 + 1/8 + \cdots$, with its sequence $\{1/2, 1/4, 1/8, \ldots\}$ of terms. Both converge, but the *sum* of the series is 1, while the *limit* of the sequence of terms is 0.

The numbering of the terms in a series (and in the sequence of partial sums) need not start with 1. For example, $\sum_{n=1}^\infty (1/n)$, $\sum_{n=0}^\infty [1/(n+1)]$, and $\sum_{n=7}^\infty [1/(n-6)]$ all represent the harmonic series in Example 10.4. In addition, for any positive integer k, the first k terms of a series can be replaced by other terms (or deleted) without affecting the convergence or divergence of the series (although the sum of a convergent series will change). We often say that the convergence or divergence of a series occurs in the *tail* of the series (the terms beyond the kth term for any value of k).

Returning to sequences, the terms in many (but certainly not all) sequences can be expressed as $x_n = f(n)$ where f is a function whose domain is the interval $[1, \infty)$. In this case we can use techniques from your PCE to evaluate $\lim x_n$ as $\lim_{x\to\infty} f(x)$.

Example 10.5 Let r be a real number, and consider the sequence $\{r^n\}_{n=1}^\infty$. For what values of r does $\{r^n\}$ converge? Clearly the sequence converges for $r = 0$ and for $r = 1$, and diverges for $r = -1$. For r positive but not equal to 1 we consider the function $f(x) = r^x$. From your PCE you know that when $r > 1$ we have $\lim_{x\to\infty} r^x = +\infty$, and for $0 < r < 1$ we have $\lim_{x\to\infty} r^x = 0$.

Hence $\{r^n\}$ converges for $0 < r < 1$ and diverges for $r > 1$. For r negative, $r^n = |r|^n$ when n is even and $r^n = -|r|^n$ when n is odd. So if $r < -1$, we have $|r| > 1$ so that the terms in $\{r^n\}$ with n even increase without bound and the terms with n odd decrease without bound, and hence $\lim r^n$ does not exist. For $-1 < r < 0$ we can write $-|r|^n \le r^n \le |r|^n$ where $0 < |r| < 1$ and apply the squeeze theorem to conclude that $\lim r^n = 0$. In summary, we have shown that $\{r^n\}$ converges if and only if $-1 < r \le 1$, and that

$$\lim r^n = \begin{cases} 0 & \text{if } |r| < 1, \\ 1 & \text{if } r = 1. \end{cases} \quad \blacksquare$$

The squeeze theorem can also be used to prove that if $|x_n| \to 0$, then $x_n \to 0$. See Exercise 5 at the end of this section.

In Theorem 10.1 we present some important limit properties of sequences. The proofs are similar to the corresponding results for functions, and are omitted.

Theorem 10.1 (Arithmetic properties of sequences) *Let $\{x_n\}$ and $\{y_n\}$ be convergent sequences, and c any constant. Then:*

1. $\lim c = c$ *and* $\lim cx_n = c \lim x_n$,

2. $\lim (x_n + y_n) = \lim x_n + \lim y_n$ *and* $\lim (x_n - y_n) = \lim x_n - \lim y_n$,

3. $\lim x_n y_n = \lim x_n \cdot \lim y_n$,

4. $\lim \dfrac{x_n}{y_n} = \dfrac{\lim x_n}{\lim y_n}$ *if every $y_n \ne 0$ and $\lim y_n \ne 0$,*

5. $\lim x_n^c = (\lim x_n)^c$ *if $c > 0$ and $x_n > 0$.*

In the next theorem we present an important relationship between the convergence of the series $\sum_{k=1}^{\infty} a_k$ (i.e., the limit of the sequence $\{S_n\}_{n=1}^{\infty}$ where $S_n = \sum_{k=1}^{n} a_k$) and the limit of the sequence $\{a_n\}_{n=1}^{\infty}$.

Theorem 10.2 *If $\sum_{k=1}^{\infty} a_k$ converges, then $\lim a_n = 0$.*

Proof. Suppose $\sum_{k=1}^{\infty} a_k$ converges to the number S. Then $\lim S_n = S$. Similarly $\lim S_{n-1} = S$. But $a_n = S_n - S_{n-1}$ so that $\lim a_n = \lim (S_n - S_{n-1}) = S - S = 0$. $\qquad \square$

What can we conclude about $\sum_{k=1}^{\infty} a_k$ if $\lim a_n \ne 0$? Theorem 10.2 tells us that clearly the series must then diverge. Thus we have a procedure (traditionally called a *test*) for showing that a series diverges:

Corollary 10.3 (The divergence test) *If $\lim a_n \ne 0$ then $\sum_{k=1}^{\infty} a_k$ diverges.*

Example 10.6 Does the series $\frac{1}{1} + \frac{2}{3} + \frac{3}{5} + \frac{4}{7} + \frac{5}{9} + \cdots$ converge or diverge? Do you see the pattern for the terms? Since the numerator of a_n is n and the denominator is one less than twice the numerator, we have

$$a_n = \frac{n}{2n-1} = \frac{1}{2 - (1/n)} \to \frac{1}{2} \neq 0$$

and so the series diverges. ∎

What can we conclude about $\sum_{k=1}^{\infty} a_k$ if $\lim a_n = 0$? In Examples 10.1 and 10.2 we saw two series for which $\lim a_n = 0$ and the series was convergent. But there are divergent series for which $\lim a_n = 0$. An important one is the harmonic series from Example 10.4.

⊘ *Be Careful!* If $\lim a_n = 0$, then Corollary 10.3 doesn't apply, and we say that the *divergence test fails*, and hence the series may converge or diverge. Or as Jacob Bernoulli wrote in his classic work *Ars Conjectandi*, "the sum of an infinite series whose final term vanishes perhaps is infinite, perhaps finite."

Back to Example 10.4 To prove that the harmonic series $1 + \frac{1}{2} + \frac{1}{3} + \cdots + \frac{1}{n} + \cdots$ diverges, we interpret each term $1/k$ of the series as the area of a rectangle with base 1 and height $1/k$, and compare the nth partial sum $H_n = 1 + \frac{1}{2} + \frac{1}{3} + \cdots + \frac{1}{n}$ to the area under the graph of $y = 1/x$ over the interval $[1, n+1]$. See Figure 10.3 (the number above each shaded rectangle is its area).

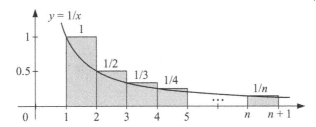

Figure 10.3. A partial sum of the harmonic series

Thus

$$H_n = \sum_{k=1}^{n} \frac{1}{k} > \int_{1}^{n+1} \frac{1}{x}\, dx = [\ln x]_{1}^{n+1} = \ln(n+1).$$

Since $\lim[\ln(n+1)] = \infty$, we have $\lim H_n = \infty$ and hence the harmonic series diverges. But it diverges very slowly. With the aid of a calculator or computer, it is easy to compute some values of H_n as shown in Table 10.1.

Table 10.1. Partial sums of the harmonic series

n	10	100	1,000	10,000	100,000	1,000,000
H_n	2.9290	5.1874	7.4855	9.7876	12.0901	14.3927

The partial sums H_n of the harmonic series are called *harmonic numbers*. For some properties of the harmonic numbers, see Exercises 10, 11, and 12. ∎

In Section 10.4 we will continue with this idea of representing the terms of a series with positive terms by areas of rectangles and comparing the series to an integral.

The limit properties for sequences presented earlier can be applied to the sequence of partial sums of a series to prove the following theorem (see Exercise 14):

Theorem 10.4 (Arithmetic properties of series) *Let $\sum_{n=1}^{\infty} a_n$ and $\sum_{n=1}^{\infty} b_n$ be series and c a nonzero real number.*

1. *If $\sum_{n=1}^{\infty} a_n$ converges to S, then $\sum_{n=1}^{\infty} ca_n$ converges to cS.*

2. *If $\sum_{n=1}^{\infty} a_n$ diverges, so does $\sum_{n=1}^{\infty} ca_n$.*

3. *If $\sum_{n=1}^{\infty} a_n$ converges to S and $\sum_{n=1}^{\infty} b_n$ converges to T. Then $\sum_{n=1}^{\infty}(a_n + b_n)$ converges to S + T.*

4. *If $\sum_{n=1}^{\infty} a_n$ converges and $\sum_{n=1}^{\infty} b_n$ diverges, then $\sum_{n=1}^{\infty}(a_n + b_n)$ diverges.*

Example 10.7 Do the following series converge or diverge?

(a) $\dfrac{1}{100} + \dfrac{1}{200} + \dfrac{1}{300} + \cdots + \dfrac{1}{100n} + \cdots$
(b) $\displaystyle\sum_{n=1}^{\infty} \dfrac{10^n - 2^{n+2}}{20^n}$
(c) $\displaystyle\sum_{n=1}^{\infty} \dfrac{2^n + n}{n \cdot 2^n}$

In (a), the terms of the series are $1/100$ times the corresponding terms of the harmonic series, and since $\sum_{n=1}^{\infty} \frac{1}{n}$ diverges, so does $\sum_{n=1}^{\infty} \frac{1}{100} \cdot \frac{1}{n}$.

In (b), the term simplifies:

$$\frac{10^n - 2^{n+2}}{20^n} = \frac{1}{2^n} - \frac{4}{10^n}.$$

From Examples 10.1 and 10.2, we know

$$\sum_{n=1}^{\infty} \frac{1}{2^n} = 1 \quad \text{and} \quad \sum_{n=1}^{\infty} \frac{4}{10^n} = \frac{4}{9},$$

and so

$$\sum_{n=1}^{\infty} \frac{10^n - 2^{n+2}}{20^n} \quad \text{converges to } 1 - \frac{4}{9} = \frac{5}{9}.$$

In (c), the term simplifies:

$$\frac{2^n + n}{n \cdot 2^n} = \frac{1}{n} + \frac{1}{2^n};$$

and since $\sum_{n=1}^{\infty} \frac{1}{n}$ diverges and $\sum_{n=1}^{\infty} \frac{1}{2^n}$ converges, $\sum_{n=1}^{\infty} \frac{2^n+n}{n \cdot 2^n}$ diverges. ∎

Reductio ad absurdum **proofs that the harmonic series diverges**
Reductio ad absurdum is Latin for "reduction to the absurd," and refers to a form of proof where a statement is proven to be false by following its implications to an absurd conclusion. In mathematics these proofs often take a form called proof by contradiction. To show that the harmonic series diverges, we follow the implications of the statement "The harmonic series converges." There are many such proofs; here is a simple one. If the harmonic series converges to S, then the sum of the even numbered terms $\frac{1}{2} + \frac{1}{4} + \frac{1}{6} + \cdots$ clearly converge to $\frac{1}{2}S$ since each term is one half the corresponding term in the harmonic series. So the sum of the odd numbered terms $1 + \frac{1}{3} + \frac{1}{5} + \cdots$ must also converge to $\frac{1}{2}S$. But this is absurd, since $1 > \frac{1}{2}, \frac{1}{3} > \frac{1}{4}, \frac{1}{5} > \frac{1}{6}$, etc. Thus the harmonic series must diverge.

10.1.1 Exercises

1. In Figure 10.1 we illustrated the geometric series with $a = r = 1/2$ by cutting a square of paper with area 1 into smaller squares and rectangles. If the rectangle in Figure 10.4(a) has area 1, it illustrates the geometric series with $a = r = 1/3$, i.e., $1/3 + 1/9 + 1/27 + \cdots = 1/2$. What geometric series and their sums are illustrated in Figures 10.4 (b) and (c) if in each case the original square of paper has area 1?

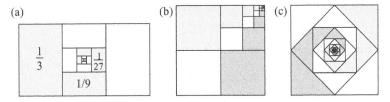

Figure 10.4. More geometric disections

2. A *golden rectangle* (a $1 \times \varphi$ rectangle where $\varphi = (1 + \sqrt{5})/2$ is the golden ratio encountered in Section 2.1) has the property that if we cut off a square from it (as in Figure 10.5 (b)), the new rectangle is similar to the original. The process can be continued indefinitely (see Figure 10.5 (c)). Conclude that

$$1 + \frac{1}{\varphi^2} + \frac{1}{\varphi^4} + \frac{1}{\varphi^6} + \cdots = \varphi.$$

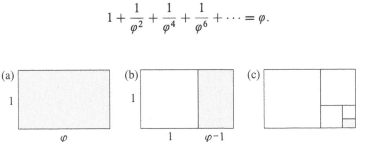

Figure 10.5. Golden rectangles

Determine whether the series in Exercises 3–8 converge or diverge. If the series converges, find its sum.

3. $\displaystyle\sum_{n=1}^{\infty} \frac{1}{\sqrt[n]{2}}$ 4. $\dfrac{1}{100} + \dfrac{1}{101} + \dfrac{1}{102} + \cdots$ 5. $\dfrac{1}{4} + \dfrac{1}{8} + \dfrac{1}{16} + \cdots$

6. $5 + \dfrac{5}{2} + \dfrac{5}{4} + \dfrac{5}{8} + \cdots$ 7. $\displaystyle\sum_{n=1}^{\infty} \left(\frac{n}{n+1} \right)^n$ 8. $1 + \dfrac{3}{4} + \dfrac{5}{9} + \dfrac{7}{16} + \dfrac{9}{25} + \cdots$

9. One of the prettiest fractals is *Sierpiński's triangle*, also known as the Sierpiński sieve or gasket. It was first described by the Polish mathematician Waclaw Sierpiński (1882–1969) in 1915. To construct one, begin with an equilateral triangle and delete the central one-quarter, then delete the central one-quarter of the remaining smaller equilateral triangles, and continue. The first four steps are shown in Figure 10.6. The Sierpiński triangle is the limit after infinitely many steps.

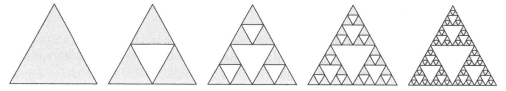

Figure 10.6. Sierpiński's triangle

Show that the Sierpiński triangle has zero area but a boundary of infinite length. (Hint: the area and boundary sequences are geometric.)

10. (a) Show that the nth harmonic number H_n satisfies

$$\ln(n + 1) < H_n < 1 + \ln n.$$

(Hint: integrate each term of the inequality $\frac{1}{k+1} \le \frac{1}{x} \le \frac{1}{k}$ from k to $(k + 1)$ then add the results for $k = 1, 2, \ldots, n - 1$.)
(b) Find bounds on the sum $\frac{1}{101} + \frac{1}{102} + \frac{1}{103} + \cdots + \frac{1}{500}$. (Hint: how does the sum compare to $\int_{100}^{500} (1/x)\, dx$ and to $\int_{101}^{501} (1/x)\, dx$?)

11. If H_k denotes the kth harmonic number, show that $\sum_{k=1}^{n-1} H_k = nH_n - n$. (Hint: in the $n \times n$ grid of numbers in Figure 10.7, summing the numbers by rows yields nH_n. Now separately sum the numbers in the white and gray regions.)

1	$\frac{1}{2}$	$\frac{1}{3}$	$\cdots$		$\frac{1}{n-1}$	$\frac{1}{n}$
1	$\frac{1}{2}$	$\frac{1}{3}$	$\cdots$		$\frac{1}{n-1}$	$\frac{1}{n}$
1	$\frac{1}{2}$	$\frac{1}{3}$	$\cdots$		$\frac{1}{n-1}$	$\frac{1}{n}$
$\vdots$	$\vdots$	$\vdots$			$\vdots$	$\vdots$
1	$\frac{1}{2}$	$\frac{1}{3}$	$\cdots$		$\frac{1}{n-1}$	$\frac{1}{n}$

Figure 10.7. Summing harmonic numbers

12. (a) Show that the nth harmonic number H_n has the following integral representation: for any positive integer n,

$$H_n = \int_0^1 \frac{1 - t^n}{1 - t}\, dt.$$

(b) The integral in (a) allows us to compute H_n for some fractional values of n between 0 and 1. Show that

$$H_{1/4} = 4 - 3\ln 2 - \frac{\pi}{2}, \quad H_{1/2} = 2 - 2\ln 2, \quad \text{and} \quad H_{3/4} = \frac{4}{3} - 3\ln 2 + \frac{\pi}{2}.$$

(Hint: use the substitution $u = t^{1/4}$ followed by partial fractions for $H_{1/4}$, and similarly for $H_{1/2}$ and $H_{3/4}$.) The numbers $H_x = \int_0^1 \frac{1-t^x}{1-t}\, dt$ are known as the *fractional harmonic numbers*, and will reappear in Exploration D.3.

13. Prove that if $\lim |x_n| = 0$, then $\lim x_n = 0$. (Hint: $-|x_n| \le x_n \le |x_n|$.)

14. Prove Theorem 10.4.

15. Suppose $\sum_{n=1}^{\infty} a_n$ and $\sum_{n=1}^{\infty} b_n$ both diverge. Does $\sum_{n=1}^{\infty} (a_n + b_n)$ also diverge?

16. Here is another *reductio ad absurdum* proof that the harmonic series diverges. It begins as all such proofs do—assume that the harmonic series converges with sum S. Then:
   (a) Show that for any $n > 1$, $\frac{1}{n-1} + \frac{1}{n} + \frac{1}{n+1} > \frac{3}{n}$.
   (b) Use the result in (a) to obtain the contradiction

$$S = 1 + \left(\frac{1}{2} + \frac{1}{3} + \frac{1}{4}\right) + \left(\frac{1}{5} + \frac{1}{6} + \frac{1}{7}\right) + \left(\frac{1}{8} + \frac{1}{9} + \frac{1}{10}\right) + \cdots \ge 1 + S.$$

10.2 Geometric and telescoping series: evaluating partial sums

One way to determine whether or not a series converges is to evaluate the limit of its sequence of partial sums. Two types of series for which this can be done are the geometric series and the telescoping series.

10.2.1 Geometric series

The nth partial sum S_n of the geometric series (10.1) with first term a and common ratio r is

$$S_n = a + ar + ar^2 + \cdots + ar^n.$$

To find the limit of the sequence $\{S_n\}$ we first need to find a closed form for S_n, i.e., an expression for S_n without the ellipsis. This is easy when $a = 0$ (then $S_n = 0$) and when $r = 1$ (then $S_n = (n+1)a$) so let's assume $a \ne 0$ and $r \ne 1$. To find a closed form expression for S_n we look at two ways to compute S_{n+1}. One way is to add the next term ar^{n+1} to S_n, i.e., $S_{n+1} = S_n + ar^{n+1}$, while a second way is to multiply S_n by r and add a: $S_{n+1} = a + rS_n$.

These two ways to compute S_{n+1} yield the same number, hence

$$S_n + ar^{n+1} = a + rS_n, \quad \text{or} \quad (1-r)S_n = a(1 - r^{n+1}),$$

and thus

$$S_n = \begin{cases} a\dfrac{1 - r^{n+1}}{1 - r}, & r \neq 1, \\ (n+1)a, & r = 1. \end{cases} \tag{10.4}$$

When $a \neq 0$ and $r = 1$ the sequence $\{S_n\}$ diverges (do you see why?). When $a \neq 0$ and $r \neq 1$ we use the result from Example 10.5 to conclude that $\{S_n\}$ converges to $a/(r-1)$ if and only if $|r| < 1$. Hence we have proven

Theorem 10.5 (Geometric series convergence) *The geometric series*

$$\sum_{k=0}^{\infty} ar^k = a + ar + ar^2 + \cdots + ar^n + \cdots$$

(with $a \neq 0$) converges for $|r| < 1$ and diverges for $|r| \geq 1$. When $|r| < 1$ its sum is

$$a + ar + ar^2 + \cdots + ar^n + \cdots = \frac{a}{1-r}. \tag{10.5}$$

Back to Example 10.3 Since $|r| = |-1/3| < 1$, the series converges, and

$$\sum_{n=0}^{\infty} 2\left(-\frac{1}{3}\right)^n = 2 - \frac{2}{3} + \frac{2}{9} - \frac{2}{27} + \cdots = \frac{2}{1 - (1/3)} = \frac{3}{2}. \quad \blacksquare$$

The sum $a/(1-r)$ of a geometric series with $a > 0$ and $0 < r < 1$ can be illustrated geometrically with rectangles as shown in Figure 10.8. The rectangle labeled "a" has area a since its height is $a/(1-r)$ and its base is $1 - r$, the rectangle labeled "ar" has area ar since its height is $a/(1-r)$ and its base is $r - r^2 = r(1-r)$, and so on, and the entire gray rectangle has area $a/(1-r)$, thus illustrating (10.5).

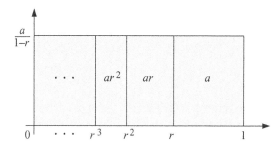

Figure 10.8. Summing geometric series geometrically

Furthermore, the nth partial sum S_n is represented by the area of a rectangle with height $a/(1-r)$ and base $1 - r^{n+1}$ so that

$$S_n = \frac{a}{1-r} \cdot (1 - r^{n+1}) = a\frac{1 - r^{n+1}}{1 - r}.$$

Example 10.8 Consider the geometric series with first term 1 and common ratio x:

$$1 + x + x^2 + \cdots + x^{n-1} + \cdots .$$

The series defines a *function f with domain* $(-1, 1)$ (since the series converges if and only if $|x| < 1$), and for any x in the interval $(-1, 1)$,

$$f(x) = 1 + x + x^2 + \cdots + x^{n-1} + \cdots = \frac{1}{1-x}.$$

The function in this example is called a *power series* (since its terms involve powers of x) and its domain is often called its *interval of convergence* (we will see later why we use the word "interval" here). Its partial sums are polynomials. With your graphing calculator you can graph $y = 1/(1-x)$ and several of the partial sum polynomials ($y = 1 + x$, $y = 1 + x + x^2$, $y = 1 + x + x^2 + x^3$, etc.) and observe the convergence to $y = 1/(1-x)$ for $|x| < 1$ and the divergence for $|x| \geq 1$.

Other power series can be obtained using other powers of x for both r and a:

$$\frac{1}{1+x} = 1 - x + x^2 - x^3 + \cdots , \quad (a = 1, r = -x)$$

$$\frac{1}{1+x^2} = 1 - x^2 + x^4 - x^6 + \cdots , \quad (a = 1, r = -x^2)$$

$$\frac{x}{1-x^3} = x + x^4 + x^7 + x^{10} + \cdots , \quad (a = x, r = x^3) \quad \text{etc.}$$

The interval of convergence for each is $(-1, 1)$. ∎

Example 10.9 Can you find a power series whose sum is $2/(3 - x)$? We rephrase the question as: Can you find an a and r so that $a/(1 - r) = 2/(3 - x)$? Yes, multiplying numerator and denominator by $1/3$ yields

$$\frac{2}{3-x} = \frac{2/3}{1-(x/3)}$$

so that $a = 2/3$ and $r = x/3$. Hence if $|x/3| < 1$, or $|x| < 3$, we have

$$\sum_{n=0}^{\infty} \frac{2}{3} \left(\frac{x}{3}\right)^n = \frac{2}{3} + \frac{2x}{3^2} + \frac{2x^2}{3^3} + \frac{2x^3}{3^4} + \cdots = \frac{2/3}{1-(x/3)} = \frac{2}{3-x}.$$

The interval of convergence is $(-3, 3)$. ∎

Example 10.10 For x in $(0, 1)$ construct a square with sides $1 + x + x^2 + \cdots = 1/(1-x)$ as shown in Figure 10.9, and use the terms of the series to partition the square into squares and rectangles.

Computing the area of the entire square in two different ways yields

$$1 + 2x + 3x^2 + 4x^3 + \cdots = \frac{1}{(1-x)^2},$$

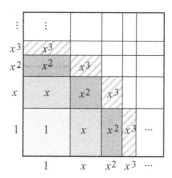

Figure 10.9. A geometric series partition of a square

which we are tempted to write as

$$\frac{d}{dx}(1 + x + x^2 + x^3 + \cdots) = \frac{d}{dx}\frac{1}{1 - x}.$$

This suggests that we may be able to differentiate each term in a convergent power series, and the series of derivatives will converge to the derivative of the sum of the original series. The study of power series is a major focus of the next chapter, where we will explore the use of differential and integral calculus with power series. ∎

10.2.2 Telescoping series

If in the series $\sum_{n=1}^{\infty} a_n$ you can replace each term a_n by a *difference of simpler terms*, some of the terms may cancel in the evaluation of the partial sum S_n. We illustrate with several examples.

Example 10.11 Consider the series

$$\sum_{n=1}^{\infty} \frac{1}{n(n + 1)} = \frac{1}{1 \cdot 2} + \frac{1}{2 \cdot 3} + \frac{1}{3 \cdot 4} + \cdots + \frac{1}{n(n + 1)} + \cdots,$$

where each term is the reciprocal of the product of a pair of consecutive integers. The divergence test fails since $\lim[1/(n(n + 1))] = 0$. But since the denominator of each term is a product of distinct linear factors, we can use the partial fractions technique from Chapter 3 to write

$$\frac{1}{n(n + 1)} = \frac{1}{n} - \frac{1}{n + 1}$$

and hence

$$S_n = \left(1 - \frac{1}{2}\right) + \left(\frac{1}{2} - \frac{1}{3}\right) + \left(\frac{1}{3} - \frac{1}{4}\right) + \cdots + \left(\frac{1}{n} - \frac{1}{n + 1}\right).$$

When we remove the parentheses in S_n and simplify, the terms with denominators $2, 3, \ldots, n$ cancel, leaving us with

$$S_n = 1 - \frac{1}{n + 1}.$$

Since $\lim S_n = 1$, the series converges to 1.

We can illustrate this result with the graphs of $y = x^{n-1}$ for x in $[0, 1]$ and $n = 1, 2, 3, \ldots$ shown in Figure 10.10 for $n = 1, 2, \ldots, 10$. The areas between the graphs of $y = x^{n-1}$ and

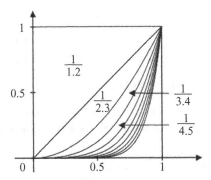

Figure 10.10. The telescoping series $\sum_{n=1}^{\infty} 1/(n(n+1))$

$y = x^n$ for $n = 1, 2, 3, \ldots$ are the terms in the series, since

$$\int_0^1 \left(x^{n-1} - x^n\right) dx = \left[\frac{x^n}{n} - \frac{x^{n+1}}{n+1}\right]_{x=0}^1 = \frac{1}{n} - \frac{1}{n+1} = \frac{1}{n(n+1)}. \ \blacksquare$$

Example 10.12 Consider the series

$$\sum_{n=1}^{\infty} \ln\left(1 + \frac{1}{n}\right).$$

Again the divergence test fails, since $\lim(\ln[1 + 1/n]) = \ln[1 + \lim(1/n)] = \ln(1 + 0) = 0$. So we consider the partial sum S_n. Since $\ln(1 + 1/n) = \ln[(n+1)/n] = \ln(n+1) - \ln n$, we have

$$S_n = [\ln 2 - \ln 1] + [\ln 3 - \ln 2] + [\ln 4 - \ln 3] + \cdots + [\ln n - \ln(n-1)] + [\ln(n+1) - \ln n]$$
$$= \ln(n+1) - \ln 1 = \ln(n+1).$$

Since $\lim S_n = \lim \ln(n+1) = \infty$, the series diverges. $\blacksquare$

Example 10.13 Consider the series

$$\sum_{n=1}^{\infty} \frac{1}{n(n+1)(n+2)} = \frac{1}{1 \cdot 2 \cdot 3} + \frac{1}{2 \cdot 3 \cdot 4} + \frac{1}{3 \cdot 4 \cdot 5} + \cdots.$$

Again the divergence test fails. Using partial fractions we can write

$$\frac{1}{n(n+1)(n+2)} = \frac{1}{2}\left(\frac{1}{n} - \frac{2}{n+1} + \frac{1}{n+2}\right)$$

so that

$$S_n = \frac{1}{2}\left[\left(1 - \frac{2}{2} + \frac{1}{3}\right) + \left(\frac{1}{2} - \frac{2}{3} + \frac{1}{4}\right) + \left(\frac{1}{3} - \frac{2}{4} + \frac{1}{5}\right) + \left(\frac{1}{4} - \frac{2}{5} + \frac{1}{6}\right) + \cdots \right.$$
$$\left. + \left(\frac{1}{n-2} - \frac{2}{n-1} + \frac{1}{n}\right) + \left(\frac{1}{n-1} - \frac{2}{n} + \frac{1}{n+1}\right) + \left(\frac{1}{n} - \frac{2}{n+1} + \frac{1}{n+2}\right)\right].$$

The terms with denominators $3, 4, \ldots, n$ cancel, leaving

$$S_n = \frac{1}{2}\left[1 - \frac{2}{2} + \frac{1}{2} + \frac{1}{n+1} - \frac{2}{n+1} + \frac{1}{n+2}\right] = \frac{1}{2}\left[\frac{1}{2} - \frac{1}{n+1} + \frac{1}{n+2}\right].$$

Hence $\lim S_n = \frac{1}{4}$ and the series converges to $\frac{1}{4}$. ■

Series like the ones in Examples 10.11, 10.12, and 10.13 are called *telescoping* since their partial sums collapse much like the old-fashioned pirate's spyglass, as shown in Figure 10.11.

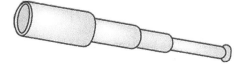

Figure 10.11. An old-fashioned spyglass

10.2.3 Exercises

In Exercises 1 through 11 determine if the series converges or diverges. If it converges, find its sum.

1. $1 - \dfrac{1}{2} + \dfrac{1}{4} - \dfrac{1}{8} + \cdots$

2. $\dfrac{1}{1 \cdot 3} + \dfrac{1}{3 \cdot 5} + \dfrac{1}{5 \cdot 7} + \cdots$

3. $\displaystyle\sum_{n=1}^{\infty} \dfrac{(-1)^{n-1}}{3^{n+1}}$

4. $0.123123123\ldots$

5. $\displaystyle\sum_{n=0}^{\infty} e^{-n}$

6. $0.999999\ldots$

7. $\displaystyle\sum_{n=1}^{\infty} \dfrac{3^n + 5^n}{15^n}$

8. $\displaystyle\sum_{n=0}^{\infty} \dfrac{(-1)^n}{2^{n/2}}$

9. $\displaystyle\sum_{n=2}^{\infty} \dfrac{1}{n^2 - 1}$

10. $\displaystyle\sum_{n=1}^{\infty} \dfrac{1}{\sqrt{n+1} + \sqrt{n}}$

11. $\displaystyle\sum_{n=1}^{\infty} \dfrac{2n+1}{(n^2+n)^2}$ (Hint: $2n + 1 = (n+1)^2 - n^2$.)

12. The *triangular numbers* $\{t_n\}_{n=1}^{\infty} = \{1, 3, 6, 10, \ldots\}$ enumerate the number of balls in triangular arrays, as shown in Figure 10.12. Thus for each $n \geq 1$,

$$t_n = 1 + 2 + 3 + \cdots + n.$$

Figure 10.12. The first four triangular numbers

(a) Show that for each $n \geq 1$, $t_n = n(n+1)/2$. (Hint: arrange $2t_n$ balls in an n-by-$(n+1)$ rectangular array.)

(b) Show that the series of reciprocals of the triangular numbers converges to 2, i.e.,

$$\frac{1}{1} + \frac{1}{3} + \frac{1}{6} + \cdots + \frac{1}{t_n} + \cdots = 2.$$

13. The sequence of *Fibonacci numbers* $\{F_n\}_{n=1}^{\infty} = \{1, 1, 2, 3, 5, 8, 13, \ldots\}$ is defined by $F_1 = F_2 = 1$ and $F_n = F_{n-1} + F_{n-2}$ for $n \geq 3$, that is, the first two terms are 1 and thereafter each term is the sum of the preceding two. For $n \geq 5$, $F_n \geq n$ so that $F_n \to \infty$. Find the sums of the series

$$\frac{1}{1 \cdot 2} + \frac{1}{2 \cdot 3} + \frac{2}{3 \cdot 5} + \frac{3}{5 \cdot 8} + \frac{5}{8 \cdot 13} + \cdots + \frac{F_n}{F_{n+1} F_{n+2}} + \cdots$$

and

$$\frac{1}{1 \cdot 2} + \frac{2}{1 \cdot 3} + \frac{3}{2 \cdot 5} + \frac{5}{3 \cdot 8} + \frac{8}{5 \cdot 13} + \cdots + \frac{F_{n+1}}{F_n F_{n+2}} + \cdots.$$

14. Show that $\sum_{n=0}^{\infty} \arctan\left(1/(n^2 + n + 1)\right) = \pi/2$.
(Hint: first show that $\arctan\left(1/(n^2 + n + 1)\right) = \arctan(1/n) - \arctan(1/(n+1))$.)

15. Show that $\dfrac{1}{2!} + \dfrac{2}{3!} + \dfrac{3}{4!} + \cdots + \dfrac{n}{(n+1)!} + \cdots = 1$ (recall that $n! = 1 \cdot 2 \cdot 3 \cdots n$).
(Hint: first show that $\dfrac{n}{(n+1)!} = \dfrac{1}{n!} - \dfrac{1}{(n+1)!}$.)

16. Show that $\displaystyle\sum_{n=1}^{\infty} \sin \frac{3}{2^{n+1}} \sin \frac{1}{2^{n+1}} = \frac{1}{2}(1 - \cos 1)$.
(Hint: $2 \sin \alpha \sin \beta = \cos(\alpha - \beta) - \cos(\alpha + \beta)$.)

10.3 Monotone sequences

In the previous section we were able to determine if a geometric or telescoping series converges because we were able to find a formula for the nth partial sum S_n, and evaluate the limit. But in general it is difficult or impossible to find a formula for S_n. In this section we learn how to determine if certain sequences of partial sums converge without having a formula for S_n. To discuss how this can be done, we need some definitions.

Definition 10.4 A sequence $\{x_n\}$ of real numbers is *increasing* if $x_n \leq x_{n+1}$ for all n, and *decreasing* if $x_n \geq x_{n+1}$ for all n. A *monotone* (or *monotonic*) sequence is a sequence that is either increasing or decreasing.

Example 10.14 The sequences $\{1, 1, 2, 2, 3, 3, \ldots\}$, $\{1/2, 2/3, 3/4, \ldots\}$, and $\{2, 2, 2, 2, \ldots\}$ are increasing, and the sequences $\{3, 2, 1, 0, -1, -2, \ldots\}$ and $\{1/2, 1/3, 1/8, 1/16, \ldots\}$ are decreasing. All five are monotone. The sequences $\{1, -1, 1, -1, 1, \ldots\}$ and $\{1, -2, 3, -4, 5, \ldots\}$ are neither increasing nor decreasing, and thus not monotone. ∎

The words increasing, decreasing, and monotone should sound familiar to you, since they were used to describe functions in your PCE, for example, a function f is increasing if $f(a) \leq f(b)$ whenever $a < b$. Our use of these words is consistent with their use for functions since sequences are functions (see Definition 10.1).

⊘ Some mathematicians (including authors of other calculus textbooks) call increasing sequences *nondecreasing* and reserve the word increasing for sequences such that $x_n < x_{n+1}$ and call decreasing sequences *nonincreasing* reserving the word decreasing for sequences such that $x_n > x_{n+1}$. We will use the words *strictly increasing* for sequences such that $x_n < x_{n+1}$, and the words *strictly decreasing* for sequences such that $x_n > x_{n+1}$.

Definition 10.5 A sequence $\{x_n\}$ of real numbers is *bounded above* if there exists a real number M such that $x_n \leq M$ for all n and is *bounded below* if there exists a real number m such that $x_n \geq m$ for all n. A sequence of real numbers is *bounded* if it is bounded above and bounded below.

Example 10.15 The sequence $\{3, 2, 1, 0, -1, -2, \dots\}$ is bounded above but not below, the sequence $\{1, 1, 2, 2, 3, 3, \dots\}$ is bounded below but not above, the sequences $\{1/2, 2/3, 3/4, \dots\}$, $\{2, 2, 2, 2, \dots\}$, and $\{1, -1, 1, -1, 1, \dots\}$ are bounded (i.e., bounded above and below), and the sequence $\{1, -2, 3, -4, 5, \dots\}$ is unbounded above and below. ■

All increasing sequences are bounded below (by the first term of the sequence) and all decreasing sequences are bounded above (again by the first term of the sequence). So to show that an increasing sequence is bounded, we need only show that it is bounded above, and to show that a decreasing sequence is bounded, we need only show that it is bounded below.

An important property of sequences that are both monotone and bounded is succinctly expressed in the following theorem.

Theorem 10.6 *If the sequence $\{x_n\}$ is bounded and monotone, then $\{x_n\}$ converges. Specifically, increasing sequences bounded above converge, and decreasing sequences bounded below converge.*

Before discussing the proof of Theorem 10.6, we ask: What do these results have to do with series? Suppose all the terms $\{a_k\}$ of the series $\sum a_k$ are positive. Then its sequence $\{S_n\}$ of partial sums is (strictly) increasing, and so if we can show that the sequence $\{S_n\}$ has an upper bound, then by Theorem 10.6 it, and the series, must converge.

Example 10.16 Show that the series $\sum_{k=1}^{\infty} \frac{1}{k^2} = 1 + \frac{1}{4} + \frac{1}{9} + \cdots + \frac{1}{n^2} + \cdots$ converges. Since this is a positive term series, its sequence $\{S_n\} = \{\sum_{k=1}^{n} 1/k^2\}$ of partial sums is increasing. So to show that the series converges, we need only show that $\{S_n\}$ is bounded above. Figure 10.13 shows that squares with areas $1, 1/4, 1/9$, and so on easily fit into a 1-by-2 rectangle: we write 2 as the sum of a geometric series with $a = 1$ and $r = 1/2$ and stack the squares with areas $1/(2^k)^2$ through $1/(2^{k+1} - 1)^2$ in a column above $1/2^k$. So, for any positive integer n, $S_n < 2$. Hence the sequence of partial sums (and thus the series) converges to a number between 1 and 2. ■

Proof of Theorem 10.6. We prove the theorem for increasing sequences; the proof for decreasing sequences is similar. The proof of this theorem requires a result from advanced calculus known as the *completeness property* of the real numbers: Any nonempty set of real numbers that has an upper bound has a least upper bound (and similarly any set of real numbers that is bounded

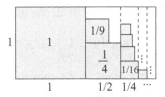

Figure 10.13. The series $\sum_{k=1}^{\infty} 1/k^2$ converges

below has a greatest lower bound). So if M is any upper bound of the set (or sequence) $\{x_n\}$, then there exists a number L such that L is also an upper bound and $L \le M$. Now suppose that $\{x_n\}$ is an increasing sequence bounded above, and that L is its least upper bound. We show that $\{x_n\}$ converges to L. Since L is the least upper bound of $\{x_n\}$, for any number $\varepsilon > 0$, $L - \varepsilon$ is not an upper bound of $\{x_n\}$, so that for some integer N, $L - \varepsilon < x_N$. But $\{x_n\}$ is increasing, so that when $n > N$, we have $L - \varepsilon < x_N \le x_n$. But the least upper bound L is an upper bound, so $x_n \le L < L + \varepsilon$. Putting the last two inequalities together we have $L - \varepsilon < x_n < L + \varepsilon$, or $|x_n - L| < \varepsilon$ whenever $n > N$. Thus $\{x_n\}$ converges to L. □

We also have a companion result:

Theorem 10.7 *If $\{x_n\}$ is convergent and monotone, then $\{x_n\}$ is bounded.*

Proof. Again we prove the theorem for an increasing sequence $\{x_n\}$. Since x_1 is a lower bound, we need only prove $\{x_n\}$ has an upper bound. Assume $\{x_n\}$ converges to L, so that for any $\varepsilon > 0$ there exists an N such that when $n > N$ we have $|x_n - L| < \varepsilon$, or $L - \varepsilon < x_n < L + \varepsilon$. So $x_n < L + \varepsilon$ for all $n > N$. But $\{x_n\}$ is increasing, so $x_n < L + \varepsilon$ holds for all $n \le N$ as well, and thus $L + \varepsilon$ is an upper bound for $\{x_n\}$. □

Combining Theorems 10.6 and 10.7 yields

Theorem 10.8 (Monotone sequence theorem) *A monotone sequence is convergent if and only if it is bounded.*

In the next two sections we will use the monotone sequence theorem to establish tests for the convergence of certain positive term series—a series of the form $\sum a_k$ where $a_k > 0$. The sequence of partial sums of such a series is monotone (increasing), so that the series converges if and only if its sequence of partial sums has an upper bound.

10.3.1 Exercises

In Exercises 1 through 6, use Theorem 10.6 to show that the sequence converges.

1. $\left\{ \cos\left(\dfrac{1}{n}\right) \right\}$ 2. $\left\{ \dfrac{n}{2^n} \right\}$ 3. $\left\{ \dfrac{2^n}{n!} \right\}$

4. $\left\{ \dfrac{2^n}{1 + 2^n} \right\}$ 5. $\left\{ \dfrac{1 \cdot 3 \cdot 5 \cdots (2n-1)}{2 \cdot 4 \cdot 6 \cdots (2n)} \right\}$ 6. $\left\{ \dfrac{(n!)^2}{(2n)!} \right\}$

7. You may recall the following limit from your PCE, one of the most important in calculus:

$$\lim_{n \to \infty} \left(1 + \frac{1}{n}\right)^n = e.$$

However, you may not have seen a proof. A rigorous proof begins with showing that the limit exists, as you will now prove by showing that the sequence $\{s_n\}$ where $s_n = (1 + 1/n)^n$ converges.

(a) Show that for any $n \geq 2$ and $0 \leq a < b$,

$$b^{n+1} - a^{n+1} < (b - a)(n + 1)b^n.$$

(Hint: compare the area under the graph of $y = (n + 1)x^n$ over the interval $[a, b]$ to the area of the circumscribing rectangle with base $b - a$ and height $(n + 1)b^n$.)

(b) Show that the inequality in (a) is equivalent to

$$b^n [(n + 1)a - nb] < a^{n+1}. \tag{10.6}$$

(c) Show that $s_n < s_{n+1}$. (Hint: let $a = 1 + \frac{1}{n+1}$ and $b = 1 + \frac{1}{n}$ in (10.6) and show that the term in brackets reduces to 1.)

(d) Show that $s_{2n} < 4$. (Hint: let $a = 1$ and $b = 1 + \frac{1}{2n}$ in (10.6).)

(e) Why do the results in (c) and (d) imply that $\{s_n\}$ converges?

(f) Now show $\lim_{n \to \infty} s_n = e$. (Hint: L'Hôpital's rule!)

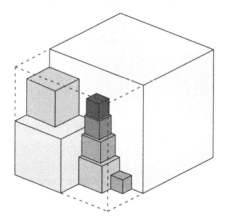

Figure 10.14. The series $\sum_{k=1}^{\infty} 1/k^3$ converges

8. Modify Example 10.16 to show that the series

$$\sum_{k=1}^{\infty} \frac{1}{k^3} = 1 + \frac{1}{8} + \frac{1}{27} + \cdots + \frac{1}{n^3} + \cdots$$

converges. (Hint: see Figure 10.14.)

10.4 The integral test: series and improper integrals

In Section 10.1 we showed that the harmonic series diverges by comparing its partial sums to integrals over intervals of finite length. In this section we extend that idea, and exploit the remarkably close relationship between certain positive term infinite series and type I improper integrals. To set the stage, consider a function f that is continuous, positive, and decreasing on the interval $[1, \infty)$. Now let $a_k = f(k)$ for $k = 1, 2, 3, \ldots$. How do the improper integral $\int_1^\infty f(x)\,dx$ and the infinite series $\sum_{k=1}^\infty a_k$ compare? Does the convergence or divergence of one tell us anything about the convergence or divergence of the other?

Example 10.17 If $f(x) = 1/x$, we have the divergent improper integral $\int_1^\infty \frac{1}{x}\,dx$ (recall Theorem 8.1) and the divergent harmonic series $\sum_{k=1}^\infty \frac{1}{k}$. Both the integral and series diverge. ∎

Example 10.18 If $f(x) = 1/[x(x + 1)]$, we have the convergent improper integral $\int_1^\infty \frac{1}{x(x+1)}\,dx$ (converging to $\ln 2$, see Exercise 6 in Section 9.1.1) and the convergent telescoping series $\sum_{k=1}^\infty \frac{1}{k(k+1)}$ (with sum 1, see Example 10.11 in Section 10.2.2). Although both converge, they converge to different numbers. ∎

In these two examples the improper integral and the infinite series have the same behavior–either both converge or both diverge. As we shall show, this is a consequence of the similarity in the definition of convergence for the two:

$$\int_1^\infty f(x)\,dx \text{ converges if and only if } \lim_{b\to\infty} \int_1^b f(x)\,dx \text{ exists (Definition 9.1)}$$

$$\sum_{k=1}^\infty a_k \text{ converges if and only if } \lim_{n\to\infty} \sum_{k=1}^n a_k \text{ exists (Definition 10.3)}.$$

This leads us to

Theorem 10.9 (The integral test) *Let f be a function that is continuous, positive, and decreasing on the interval $[1, \infty)$, and set $a_n = f(n)$ for $n = 1, 2, 3, \ldots$. Then the series $\sum_{n=1}^\infty a_n$ and the integral $\int_1^\infty f(x)\,dx$ either both converge or both diverge.*

The theorem is called the integral test since we will use the techniques from Chapter 8 to determine the convergence or divergence of an infinite series by testing whether or not an improper integral converges. In the statement of the theorem, the interval $[1, \infty)$ can be replaced by $[k, \infty)$, and the series and integral by $\sum_{n=k}^\infty a_k$ and $\int_k^\infty f(x)\,dx$ for any positive integer k. Before proving the theorem, here is an example of its use.

Back to Example 10.16 For another proof that the series $\sum_{k=1}^\infty \frac{1}{k^2}$ converges, we use the integral test. Since $a_n = 1/n^2$, we let $f(x) = 1/x^2$. Since f is continuous, positive, and decreasing on $[1, \infty)$, we have

$$\int_1^\infty \frac{1}{x^2} = \lim_{b\to\infty} \int_1^b \frac{1}{x^2}\,dx = \lim_{b\to\infty} \left[-\frac{1}{x}\right]_{x=1}^b$$

$$= \lim_{b\to\infty} \left[-\frac{1}{b} + 1\right] = 0 + 1 = 1,$$

and since the integral converges, so does the series (but clearly not to 1, the value of the improper integral). We shall return to this example later in this section to estimate the actual sum of the series. ■

Proof of Theorem 10.9. In this proof we use the monotone sequence theorem with two sequences: the sequence of partial sums of the series, and the sequences $\{\int_1^n f(x)\,dx\}$ and $\{\int_1^{n+1} f(x)\,dx\}$ employed in evaluating the improper integral $\int_1^\infty f(x)\,dx$. We begin with a visual comparison of the nth partial sum $S_n = \sum_{k=1}^n a_k$ of the series and the integral $\int_1^{n+1} f(x)\,dx$ in Figure 10.15.

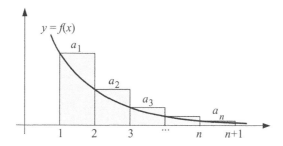

Figure 10.15. A comparison of partial sums to an integral

Since $f(x)$ is positive and decreasing on $[1, \infty)$, we have for each integer $n \geq 1$

$$\int_1^{n+1} f(x)\,dx \leq S_n.$$

If $\sum_{n=1}^\infty a_n$ converges (to S say), then S is an upper bound for the sequence $\{\int_1^{n+1} f(x)\,dx\}$ and hence $\int_1^\infty f(x)\,dx$ converges. If $\int_1^\infty f(x)\,dx$ diverges, then the sequence $\{\int_1^{n+1} f(x)\,dx\}$ is unbounded, hence so is $\{S_n\}$ and $\sum_{n=1}^\infty a_n$ diverges.

Now we arrange the rectangles whose areas are the terms of the series a little differently, as is shown in Figure 10.16.

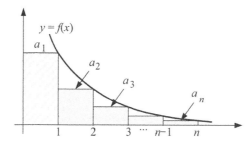

Figure 10.16. A second comparison of partial sums to an integral

Thus, for $n \geq 2$ we have

$$\sum_{k=2}^n a_k = S_n - a_1 \leq \int_1^n f(x)\,dx$$

or

$$S_n \le a_1 + \int_1^n f(x)\,dx.$$

If $\int_1^\infty f(x)\,dx$ converges, (to A say) then $\int_1^n f(x)\,dx \le \int_1^\infty f(x)\,dx$ so that $S_n \le a_1 + A$, and thus $\sum_{n=1}^\infty a_n$ converges. If $\sum_{n=1}^\infty a_n$ diverges, then the sequence $\{S_n\}$ is unbounded, so that the sequence $\{\int_1^n f(x)\,dx\}$ is unbounded and thus $\int_1^\infty f(x)\,dx$ diverges. Consequently, the series $\sum_{n=1}^\infty a_n$ and the integral $\int_1^\infty f(x)\,dx$ either both converge or both diverge. $\square$

Example 10.19 Does the series $\sum_{n=2}^\infty \frac{\ln n}{n^2}$ converge or diverge? The function $f(x) = \ln x/x^2$ is continuous, positive, and decreasing on $[1, \infty)$ (which you should verify), so we proceed with the integral test. Employing integration by parts and L'Hôpital's rule yields

$$\int_1^\infty \frac{\ln x}{x^2}\,dx = \lim_{b\to\infty} \int_1^b \frac{\ln x}{x^2}\,dx = \lim_{b\to\infty}\left[-\frac{\ln x}{x}\Big|_{x=1}^b + \int_1^b \frac{1}{x^2}\,dx\right]$$

$$= \lim_{b\to\infty}\left[-\frac{1+\ln x}{x}\right]_{x=1}^b = 1 - \lim_{b\to\infty}\frac{1+\ln b}{b} \overset{\text{L'H}}{=} 1 - \lim_{b\to\infty}\frac{1/b}{1} = 1 - 0 = 1$$

and so the series converges. ∎

Example 10.20 (The p-series) The series $1 + \frac{1}{2^p} + \frac{1}{3^p} + \cdots + \frac{1}{n^p} + \cdots$ is known as a *p-series*. When $p = 1$ we have the divergent harmonic series, and when $p = 2$ we have the convergent series in Example 10.16. For what values of p does a p-series converge?

When $p = 0$, the series is $1 + 1 + 1 + \cdots$ which clearly diverges. When $p < 0$ we have $\lim_{n\to\infty} 1/n^p = \infty$ and so the series diverges. When $p > 0$, the function $f(x) = x^{-p}$ is continuous, positive, and decreasing on $[1, \infty)$ so we can use the integral test. In Theorem 8.1, we learned that $\int_1^\infty (1/x^p)\,dx$ converges for $p > 1$ and diverges for $p \le 1$. Hence

> The p-series $\displaystyle\sum_{n=1}^\infty \frac{1}{n^p}$ converges for $p > 1$ and diverges for $p \le 1$. ∎

The Riemann zeta function

What is the domain of the function $\zeta(x) = \sum_{n=1}^\infty n^{-x}$? Since this series is the p-series with p replaced by x, the domain is the interval $(1, \infty)$. Replacing the real variable x by a complex variable $z = x + iy$ (where x and y are real and $i = \sqrt{-1}$) enlarges the domain and yields the function $\zeta(z) = \sum_{n=1}^\infty n^{-z}$ known as the *Riemann zeta function* (ζ is the lower case Greek letter zeta). This very important function plays a critical role in many fields, including physics, analytic number theory, and probability and statistics. One of the most important unsolved problems in mathematics is to find all complex solutions to the equation $\zeta(z) = 0$. A conjecture known as the *Riemann hypothesis* states that if $z = x + iy$ is a solution to $\zeta(z) = 0$ with $y \ne 0$ then $x = 1/2$.

10.4.1 Estimating the sum of a convergent series

Suppose we have been able to show that the series $\sum_{k=1}^{\infty} a_k$ converges using the integral test. We can approximate the sum S of the series by any one of the partial sums S_n since the sequence $\{S_n\}$ converges to S. But since the terms of the series are positive, each S_n is an underestimate of S. To measure how much S_n underestimates S, we consider the remainder R_n given by

$$R_n = S - S_n = a_{n+1} + a_{n+2} + a_{n+3} + \cdots,$$

another infinite series that converges by the integral test. In fact, illustrations similar to 10.15 and 10.16 (see Exercise 2) readily show that

$$\int_{n+1}^{\infty} f(x)\,dx \leq R_n \leq \int_{n}^{\infty} f(x)\,dx. \tag{10.7}$$

Since $S = S_n + R_n$ we now add S_n to each term in the above double inequality to obtain the following interval which is guaranteed to contain S:

$$\boxed{S_n + \int_{n+1}^{\infty} f(x)\,dx \leq S \leq S_n + \int_{n}^{\infty} f(x)\,dx.} \tag{10.8}$$

Back to Example 10.16 Let's use (10.8) and $S_{20} = 1.59616332\ldots$ to construct an interval that contains the sum S of the series $\sum_{n=1}^{\infty} 1/n^2$. Since $\int_{n}^{\infty} \left(1/x^2\right) dx = 1/n$ (you should check this!), we add $1/21$ and $1/20$ to S_{20} to obtain

$$1.64378229 \cdots \leq S \leq 1.64616324.\ldots$$

Hence we know the first two decimal places of S, i.e., $S \approx 1.64$. ■

When the function f in the integral test is concave up in addition to being continuous, positive, and decreasing (as is often the case), we can improve the interval estimate for S in (10.8). In Chapter 7 we learned that for the definite integral of a function that is concave up on a finite interval, the trapezoidal rule overestimates the value of the integral, while the midpoint rule underestimates it. It is easy to show that the same is true for convergent type I improper integrals.

Using the midpoint rule to approximate the value of $\int_{n+1/2}^{\infty} f(x)\,dx$ with interval widths $\Delta x = 1$ yields $\int_{k-1/2}^{k+1/2} f(x)\,dx \geq f(k)$ for $k \geq n+1$ and

$$\int_{n+1/2}^{\infty} f(x)\,dx \geq \sum_{k=n+1}^{\infty} f(k) = \sum_{k=n+1}^{\infty} a_k = R_n$$

(recalling that $f(k) = a_k$) so that $R_n \leq \int_{n+1/2}^{\infty} f(x)\,dx$.

Similarly, using the trapezoidal rule to approximate the value of $\int_{n+1}^{\infty} f(x)\,dx$ with interval widths $\Delta x = 1$ yields $\int_{k}^{k+1} f(x)\,dx \leq \frac{1}{2}\left[f(k) + f(k+1)\right]$ for $k \geq n+1$ and

$$\int_{n+1}^{\infty} f(x)\,dx \leq \sum_{k=n+1}^{\infty} f(k) - \frac{1}{2}f(n+1) = R_n - \frac{1}{2}a_{n+1}$$

so that $\int_{n+1}^{\infty} f(x)\,dx + \frac{1}{2}a_{n+1} \le R_n$. Thus

$$\int_{n+1}^{\infty} f(x)\,dx + \frac{1}{2}a_{n+1} \le R_n \le \int_{n+1/2}^{\infty} f(x)\,dx.$$

So when the function f in the integral test is concave up in addition to being continuous, positive, and decreasing, (10.8) becomes

$$S_n + \int_{n+1}^{\infty} f(x)\,dx + \frac{1}{2}a_{n+1} \le S \le S_n + \int_{n+1/2}^{\infty} f(x)\,dx. \qquad (10.9)$$

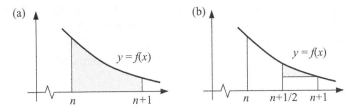

Figure 10.17. Comparing the widths of the intervals in (10.8) and (10.9)

The width of the interval for S in (10.8) is $\int_n^{n+1} f(x)\,dx$, while the width of the interval for S in (10.9) is $\int_{n+1/2}^{n+1} f(x)\,dx - \frac{1}{2}f(n+1)$. The improvement in the precision of the estimate is illustrated in Figure 10.17, where the area of the shaded region in (a) is numerically equal to the width of the interval for S in (10.8), and the area of the shaded region in (b) is numerically equal to the width of the interval for S in (10.9).

Back to Example 10.16 Since $f(x) = 1/x^2$ is concave up, we can construct an interval for S from (10.9). Adding $1/21 + 1/(2 \cdot 21^2)$ to S_{20} for the lower bound and $1/(20.5)$ to S_{20} for the upper bound yields

$$1.64491607\cdots \le S \le 1.64494373\ldots.$$

Hence we actually know the first *four* decimal places of S, i.e., $S \approx 1.6449$. For the exact (and somewhat surprising) value of S, see Exploration 11.8.2 in the next chapter. ∎

10.4.2 Exercises

In Exercises 1 through 9 determine if the series converges or diverges.

1. $\displaystyle\sum_{n=1}^{\infty} \frac{1}{n^2 + 1}$

2. $\displaystyle\sum_{n=1}^{\infty} \frac{n}{e^n}$

3. $\displaystyle\sum_{n=2}^{\infty} \frac{1}{n \ln n}$

4. $\displaystyle\sum_{n=1}^{\infty} \frac{1}{n\sqrt{n^2 + 1}}$

5. $\displaystyle\sum_{n=1}^{\infty} \frac{e^{1/n}}{n^2}$

6. $\displaystyle\sum_{n=2}^{\infty} \frac{1}{n (\ln n)^3}$

7. $\displaystyle\sum_{n=1}^{\infty} \frac{1}{\sqrt{n}(n + 1)}$

8. $\displaystyle\sum_{n=3}^{\infty} \frac{1}{n(\ln n)(\ln \ln n)}$

9. $\displaystyle\sum_{n=1}^{\infty} \frac{1}{n^3 + n}$

10. Draw figures similar to Figure 10.15 and Figure 10.16 to illustrate the two inequalities in (10.7).

11. The sum $\zeta(3)$ of the p-series with $p = 3$ is known as *Apéry's constant*, after Roger Apéry (1916–1994), who in 1979 proved that $\zeta(3)$ is irrational. Use the 20th partial sum of the series and (10.9) to find $\zeta(3)$ rounded to five decimal places.

12. Show that (10.9) and the 10th partial sum suffice to estimate the sum $\zeta(4)$ of the p-series with $p = 4$ to five decimal places. Since $\zeta(4)$ is known to equal $\pi^4/90$, show that the average of the bounds in (10.9) gives $\zeta(4)$ rounded to six places.

13. In Exercise 19 of Section 9.1.1 you learned about Gabriel's horn, the object obtained by revolving $y = f(x) = 1/x$ for $x \geq 1$ about the x-axis. If you replace the function f by $f(x) = 1/n$ for $n \leq x < n + 1, n = 1, 2, 3, \ldots$ you obtain *Gabriel's wedding cake*, the first five layers of which are shown in the Figure 10.18 below. Show that Gabriel's is a cake you can eat but you cannot frost, that is, like his horn it has finite volume but infinite surface area.

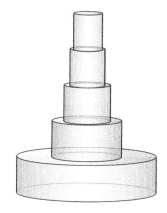

Figure 10.18. Gabriel's wedding cake

10.5 Comparison tests for positive term series

In the preceding two sections we have learned about the convergence or divergence of a large collection of series, including geometric, telescoping, and p-series, and series for which the integral test is applicable. In this section we develop some powerful tests for determining the convergence or divergence of some positive term series whose terms can be compared, via inequalities or limits, to those in the series mentioned above. These tests are naturally called *comparison tests*.

Theorem 10.10 (The direct comparison test) *Let $\sum_{n=1}^{\infty} a_n$ and $\sum_{n=1}^{\infty} b_n$ be series such that $0 \leq a_n \leq b_n$ for all n.*

1. *If $\sum_{n=1}^{\infty} b_n$ converges, then so does $\sum_{n=1}^{\infty} a_n$, and*
2. *if $\sum_{n=1}^{\infty} a_n$ diverges, then so does $\sum_{n=1}^{\infty} b_n$.*

In words, we have: If the series with larger terms converges, so does the series with smaller terms; if the series with smaller terms diverges, so does the series with larger terms. As in earlier tests, the phrase "for all n" can be replaced by "for all $n \geq k$" for any positive integer k.

Example 10.21 Do the series

$$\frac{1}{2} + \frac{1}{4} + \frac{1}{6} + \cdots + \frac{1}{2n} + \cdots \quad \text{and} \quad 1 + \frac{1}{3} + \frac{1}{5} + \cdots + \frac{1}{2n-1} + \cdots$$

converge or diverge? These series are the even denominator and odd denominator portions of the divergent harmonic series, so we suspect that at least one diverges. Clearly $\sum_{n=1}^{\infty}[1/(2n)]$ diverges since it is $\sum_{n=1}^{\infty}(1/2)(1/n)$ (recall Theorem 10.4). But the two series can be compared via

$$0 < \frac{1}{2n} < \frac{1}{2n-1},$$

and since $\sum_{n=1}^{\infty}[1/(2n)]$ diverges, so does $\sum_{n=1}^{\infty}[1/(2n-1)]$. ∎

Proof of Theorem 10.10. Let $\{A_n\}$ and $\{B_n\}$ be the sequences of partial sums for $\sum_{n=1}^{\infty} a_n$ and $\sum_{n=1}^{\infty} b_n$, resepectively. Then $0 \leq a_n \leq b_n$ implies that $\{A_n\}$ and $\{B_n\}$ are increasing sequences and that $A_n \leq B_n$. We now use the monotone sequence theorem 10.8. For (a) the convergence of $\sum_{n=1}^{\infty} b_n$ implies that $\{B_n\}$ is bounded, hence $\{A_n\}$ is also bounded, so that $\sum_{n=1}^{\infty} a_n$ converges. For (b), the divergence of $\sum_{n=1}^{\infty} a_n$ implies that $\{A_n\}$ is unbounded, hence $\{B_n\}$ is also unbounded, and $\sum_{n=1}^{\infty} b_n$ diverges. □

To make effective use of the direct comparison test, you must use your intuition. If you suspect that a given series diverges, you need to show that its terms are larger than the corresponding terms of another series known to diverge (as in Example 10.21); if you suspect that a given series converges, you must show that its terms are smaller than the corresponding terms of another series known to converge. Developing this intuition will take some practice.

Example 10.22 Consider the series

$$\sum_{n=1}^{\infty} \frac{n!}{n^n} = 1 + \frac{1 \cdot 2}{2 \cdot 2} + \frac{1 \cdot 2 \cdot 3}{3 \cdot 3 \cdot 3} + \cdots .$$

If we suspect it converges, we want to compare it to a convergent series with larger terms, and if we suspect it diverges, we want to compare it to a divergent series with smaller terms. The denominator n^n seems to grow *much* faster that the numerator $n!$, and so we suspect convergence. Indeed for $n \geq 2$ we have

$$\frac{n!}{n^n} = \frac{1}{n} \cdot \frac{2}{n} \cdot \frac{3}{n} \cdots \frac{n-1}{n} \cdot \frac{n}{n} \leq \frac{1}{n} \cdot \frac{2}{n} \cdot 1 \cdots 1 \cdot 1 = \frac{2}{n^2}.$$

and since $\sum_{n=1}^{\infty}(2/n^2)$ converges (why?), so does $\sum_{n=1}^{\infty}(n!/n^n)$. ∎

One difficulty you may encounter in trying to apply the direct comparison test is the algebra required to establish the inequality $a_n \leq b_n$ between the corresponding terms of the series. The following test replaces establishing the inequality by the evaluation of a limit.

Theorem 10.11 (The limit comparison test) *Let $\sum_{n=1}^{\infty} a_n$ and $\sum_{n=1}^{\infty} b_n$ be two positive term series, and suppose $\lim (a_n/b_n) = c > 0$. Then the two series either both converge or both diverge.*

⊘ In the limit comparison test, "$\lim (a_n/b_n) = c > 0$" means that the limit *exists* and is *finite* and *positive*.

Example 10.23 Consider the series

$$\sum_{n=1}^{\infty} \frac{1}{2^n - 1} = 1 + \frac{1}{3} + \frac{1}{7} + \cdots .$$

Direct comparison with the geometric series with $a = r = 1/2$ fails since $1/(2^n - 1) > 1/2^n$. But for large n, $2^n - 1$ is approximately 2^n, so the series may behave like the geometric series with $a = r = 1/2$. In the limit comparison test, we have

$$\lim \frac{1/(2^n - 1)}{1/2^n} = \lim \frac{2^n}{2^n - 1} = \lim \frac{1}{1 - 2^{-n}} = 1 > 0,$$

and hence the series $\sum_{n=1}^{\infty} (1/(2^n - 1))$ converges since the geometric series with $r = 1/2$ converges. ∎

Proof of Theorem 10.11. Since the limit of the sequence $\{a_n/b_n\}$ is the positive number c, there exists an integer N such that for $n > N$ all the terms a_n/b_n lie in the interval $(c/2, 3c/2)$, that is, $c/2 < a_n/b_n < 3c/2$. Multiplying by b_n yields the double inequality

$$(c/2)b_n < a_n < (3c/2)b_n.$$

If $\sum_{n=1}^{\infty} a_n$ converges, then so does $\sum_{n=1}^{\infty} (c/2)b_n$ by the direct comparison test, and hence $\sum_{n=1}^{\infty} b_n$ converges. If $\sum_{n=1}^{\infty} a_n$ diverges, then so does $\sum_{n=1}^{\infty} (3c/2)b_n$ by the direct comparison test, and hence $\sum_{n=1}^{\infty} b_n$ diverges. □

Example 10.24 Does the series $\sum_{n=1}^{\infty} [\sin(1/n)]/n$ converge or diverge?
Since $1/n$ lies in the interval $(0, 1]$, the series has positive terms. Furthermore, $\sin(1/n) < 1$, so that $[\sin(1/n)]/n < 1/n$, which is inconclusive since the harmonic series diverges. But for x near 0, $\sin x$ is approximately x, so $\sin(1/n)$ is approximately $1/n$. So we suspect that $[\sin(1/n)]/n$ is approximately $1/n^2$ for large n. Using the limit comparison test, we have

$$\lim \frac{[\sin(1/n)]/n}{1/n^2} = \lim \frac{\sin(1/n)}{1/n} = 1 \text{ (do you see why?)}.$$

Since $\sum_{n=1}^{\infty} 1/n^2$ converges, so does $\sum_{n=1}^{\infty} [\sin(1/n)]/n$. ∎

☺ The series most commonly used for comparisons in the two comparison tests are geometric series and p-series.

10.5.1 Exercises

In Exercises 1 through 12 determine if the series converges or diverges.

1. $\displaystyle\sum_{n=1}^{\infty} \frac{1}{1+n+n^2}$

2. $\displaystyle\sum_{n=1}^{\infty} \frac{1+n+n^2}{\sqrt{n^3+n^4+n^5}}$

3. $\displaystyle\sum_{n=1}^{\infty} \arctan\left(\frac{1}{n}\right)$

4. $\frac{1}{3} + \frac{1}{3}\cdot\frac{2}{5} + \frac{1}{3}\cdot\frac{2}{5}\cdot\frac{3}{7} + \cdots$

5. $\displaystyle\sum_{n=0}^{\infty}(1-\sin n)e^{-n}$

6. $\displaystyle\sum_{n=2}^{\infty} \frac{1}{\sqrt[3]{n^3+n}}$

7. $\displaystyle\sum_{n=1}^{\infty} \frac{\sqrt{n+1}-\sqrt{n}}{n+1}$

8. $\frac{1}{2} + \frac{2}{5} + \frac{3}{10} + \frac{4}{17} + \frac{5}{26} + \cdots$

9. $\displaystyle\sum_{n=0}^{\infty} \frac{2^n+3^n}{3^n+4^n}$

10. $\displaystyle\sum_{n=1}^{\infty} \sin\left(1/n\right)$

11. $\displaystyle\sum_{n=1}^{\infty} \tan\left(1/n\right)$

12. $\displaystyle\sum_{n=1}^{\infty}[1-\cos(1/n)]$

13. The direct comparison test can be used to give another proof that the harmonic series diverges. To do so, show that $x \geq \ln(1+x)$ for $x > -1$, let $x = 1/n$, and recall Example 10.10.

14. A *generalized harmonic series* is a series of the form $\sum_{n=k}^{\infty} 1/(a+dn)$ where $a \geq 0$ and $d > 0$. For example, the series in Examples 10.7(a) and 10.21 are generalized harmonic series. Prove that every generalized harmonic series diverges.

15. Let $\sum_{k=1}^{\infty} a_n$ and $\sum_{n=1}^{\infty} b_n$ be convergent series of positive terms.

    (a) Show that $\sum_{n=1}^{\infty} a_n b_n$ converges.
    (b) Show that $\sum_{n=1}^{\infty} \sqrt{a_n b_n}$ converges. (Hint: $\left(\sqrt{a_n} - \sqrt{b_n}\right)^2 \geq 0$.)
    (c) Show that $\sum_{n=1}^{\infty} \sqrt{a_n^2 + b_n^2}$ converges. (Hint: $a_n^2 + b_n^2 \leq (a_n + b_n)^2$.)

16. In Example 10.20 we noted that the "p" in a p-series is a *constant*. If it's not, the series may diverge for $p > 1$. To verify this, show that the p-series with $p = 1 + (1/n)$, i.e., $\sum_{n=1}^{\infty} 1/n^{1+(1/n)}$ diverges. (Hint: show that $n^{1/n} < 2$ and use the direct comparison test.)

10.6 The alternating series test

We now consider series with both positive and negative terms. The simplest cases of such a series are *alternating series*, series where the terms alternate between positive and negative. Let $\{b_n\}$ be an infinite sequence of positive numbers. Then

$$b_1 - b_2 + b_3 - b_4 + \cdots + (-1)^{n+1}b_n + \cdots = \sum_{n=1}^{\infty}(-1)^{n+1}b_n$$

and

$$-b_1 + b_2 - b_3 + b_4 + \cdots + (-1)^n b_n + \cdots = \sum_{n=1}^{\infty}(-1)^n b_n$$

are alternating series. We've seen alternating series in earlier sections. For example, any geometric series with $r < 0$ is an alternating series.

Example 10.25 The geometric series with first term 1 and common ratio $-1/2$ is an alternating series converging to $2/3$ by Theorem 10.5. In Figure 10.19 we use an isosceles trapezoid with area 1 to illustrate the first five partial sums and their limit. ∎

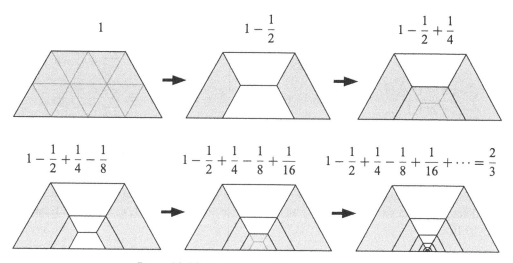

Figure 10.19. Partial sums of an alternating series

⊘ The terms of an alternating series must *strictly alternate* in sign; it is not sufficient for the series merely to possess both positive and negative terms. For example, the series $1 + (1/2) - (1/3) + (1/4) + (1/5) - (1/6) + \cdots$ is not an alternating series. The sequence $\{a_n\}$ of terms is an alternating series must be either $a_n = (-1)^{n+1}b_n$ or $a_n = (-1)^n b_n$ where $\{b_n\}$ is a sequence of positive numbers.

It is a rather simple matter to determine if an alternating series converges.

Theorem 10.12 (The alternating series test) *An alternating series $\sum_{n=1}^{\infty}(-1)^{n+1}b_n$ (or $\sum_{n=1}^{\infty}(-1)^n b_n$) with $b_n > 0$ for all n converges if*

$$\text{(i) } \lim b_n = 0 \text{ and (ii) } b_{n+1} \leq b_n \text{ for all } n.$$

In other words, an alternating series converges if the terms of the sequence $\{b_n\}$ *decrease to zero.* Again, the phrase "for all n" in (ii) above can be replaced by "for all $n \geq k$" for any positive integer k.

Example 10.26 (The alternating harmonic series) An important alternating series is the alternating harmonic series

$$1 - \frac{1}{2} + \frac{1}{3} - \frac{1}{4} + \cdots + \frac{(-1)^{n+1}}{n} + \cdots = \sum_{n=1}^{\infty} \frac{(-1)^{n+1}}{n}. \tag{10.10}$$

Here $b_n = 1/n$. Since $\lim(1/n) = 0$ and $1/(n+1) \le 1/n$ the alternating harmonic series converges. We will find its sum later in this section. ∎

Proof of Theorem 10.12. We prove the theorem for $\sum_{n=1}^{\infty}(-1)^{n+1}b_n$. The proof for $\sum_{n=1}^{\infty}(-1)^n b_n$ is similar. Assume that $b_n \to 0$ and $b_{n+1} \le b_n$ for all n. The idea of the proof is exhibited in Figure 10.20, where we compute the partial sums by successively adding and subtracting the elements of the sequence $\{b_n\}$ and then plot the partial sums on an axis.

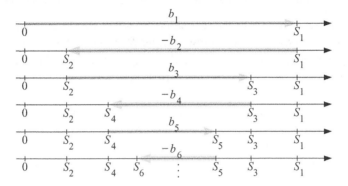

Figure 10.20. Sequence of partial sums of an alternating series

We begin by examining two subsequences of partial sums—the sequence $\{S_{2n}\}$ of even numbered partial sums, and the sequence $\{S_{2n+1}\}$ of odd numbered partial sums. Since

$$S_{2n+2} = S_{2n} + (b_{2n+1} - b_{2n+2}) \ge S_{2n},$$

$\{S_{2n}\}$ is increasing, and since

$$S_{2n} = b_1 - (b_2 - b_3) - (b_4 - b_5) - \cdots - (b_{2n-2} - b_{2n-1}) - b_{2n} \le b_1$$

the sequence $\{S_{2n}\}$ is bounded above by b_1. Thus by Theorem 10.6, $\{S_{2n}\}$ converges, say to S, i.e., $\lim S_{2n} = S$.

For the sequence $\{S_{2n+1}\}$ we have $S_{2n+1} = S_{2n} + b_{2n+1}$, and hence

$$\lim S_{2n+1} = \lim S_{2n} + \lim b_{2n+1} = S + 0 = S.$$

Since both the sequence of even numbered partial sums and the sequence of odd numbered partial sums converge to S, the entire sequence $\{S_n\}$ of partial sums converges to S, and hence the series $\sum_{n=1}^{\infty} b_n$ converges.   □

Here are some hints for applying the alternating series test to an alternating series.

Step 1. Examine $\lim b_n$ first. If the limit is not 0, the series diverges, since for an alternating series, evaluating this limit is equivalent to the divergence test. If the limit is 0, go on to Step 2.

Step 2. There are a variety of ways to show that the inequality $b_{n+1} \le b_n$ holds. Some are:

(a) show that $b_{n+1} \le b_n$ (i.e., do it directly),
(b) show that $b_{n+1} - b_n \le 0$,
(c) show that $b_{n+1}/b_n \le 1$, and
(d) if $b_n = f(n)$ for a differentiable function f, show that $f'(x) \le 0$ for $x \ge 1$.

Example 10.27 Does the series

$$\sum_{n=1}^{\infty} (-1)^{n+1} \frac{n}{n^2 + 1}$$

converge? Here $b_n = n/(n^2 + 1)$, and $\lim b_n = 0$. Is $b_{n+1} \le b_n$? Of the methods suggested, perhaps the simplest for this example is the last. If we let $f(x) = x/(x^2 + 1)$, then $f'(x) = (1 - x^2)/(x^2 + 1)^2$ so that $f'(x) \le 0$ for all $x \ge 1$. Thus the series converges. ■

If an alternating series converges by the alternating series test, we can approximate its sum S by the partial sums S_N and S_{N+1} for any positive integer N. The sum S is an upper bound for the sequence $\{S_{2n}\}$ and similarly a lower bound for the sequence $\{S_{2n+1}\}$. Consequently, S will lie between any pair S_N and S_{N+1} of consecutive partial sums, which yields an interval guaranteed to contain S.

Since the distance between S_N and S_{N+1} is b_{N+1} and the distance $|S - S_N|$ between S and S_N is smaller, we have $|S - S_N| < b_{N+1}$ for any positive integer N. In other words, the error in approximating S by the partial sum S_N is always less that the first term b_{N+1} in $\{b_n\}$ not used in the computation of S_N.

⊘ The results in the preceding two paragraphs hold *only* for alternating series that satisfy properties (i) and (ii) in the alternating series test in Theorem 10.12.

Back to Example 10.26 If A denotes the sum of the alternating harmonic series and

$$A_n = 1 - \frac{1}{2} + \frac{1}{3} - \frac{1}{4} + \cdots + \frac{(-1)^{n+1}}{n}$$

its nth partial sum, then

$$|A_n - A| < \frac{1}{n + 1}. \tag{10.11}$$

Using A_n to approximate A will require about $n = 10^k$ terms to obtain k decimal places of accuracy. Deriving (10.11) in a different fashion reveals the value of A:

$$A_n = 1 - \frac{1}{2} + \frac{1}{3} - \frac{1}{4} + \cdots + \frac{(-1)^{n+1}}{n} = \int_0^1 (1 - x + x^2 - x^3 + \cdots + (-x)^{n-1})\, dx$$

$$= \int_0^1 \frac{1 - (-x)^n}{1 + x}\, dx \text{ (from (10.5))}$$

$$= \int_0^1 \frac{1}{1 + x}\, dx - (-1)^n \int_0^1 \frac{x^n}{1 + x}\, dx$$

$$= \ln 2 - (-1)^n \int_0^1 \frac{x^n}{1 + x}\, dx.$$

Hence

$$|A_n - \ln 2| = \int_0^1 \frac{x^n}{1 + x}\, dx < \int_0^1 x^n\, dx = \frac{1}{n + 1}. \tag{10.12}$$

Since (10.11) and (10.12) both hold for all n, $A = \ln 2$. ■

Example 10.28 Consider the series

$$\frac{1}{2!} - \frac{1}{3!} + \frac{1}{4!} - \cdots = \sum_{n=2}^{\infty} \frac{(-1)^n}{n!}.$$

Since $b_n = \frac{1}{n!}$, we have $\lim b_n = \lim 1/n! = 0$ and $b_{n+1}/b_n = 1/(n+1) \le 1$, and so the series converges, say to S. Since S lies between $S_8 = 0.367879188\ldots$ and $S_9 = 0.367879464\ldots$, we know the first six decimal places of S: $S \approx 0.367879$. In Section 11.5 we will find the exact value of S. ∎

10.6.1 Exercises

In Exercises 1 through 9 determine if the series converges or diverges.

1. $\displaystyle\sum_{n=1}^{\infty}(-1)^{n+1}n^{-1/2}$

2. $\displaystyle\frac{\ln 2}{2} - \frac{\ln 3}{3} + \frac{\ln 4}{4} - \frac{\ln 5}{5} + \cdots$

3. $\displaystyle\sum_{n=1}^{\infty} \frac{(-1)^n}{\ln n}$

4. $\displaystyle\frac{1}{2} - \frac{2}{5} + \frac{3}{10} - \frac{4}{17} + \frac{5}{26} - \cdots$

5. $\displaystyle\sum_{n=1}^{\infty}(-1)^{n+1}\frac{n}{2n+1}$

6. $\displaystyle\sum_{n=0}^{\infty} \frac{\cos n\pi}{n+1}$

7. $\displaystyle\sum_{n=1}^{\infty}(-1)^{n+1}\arctan\left(\frac{1}{n}\right)$

8. $\displaystyle\sum_{n=0}^{\infty} \frac{\cos n\pi}{\cosh n}$

9. $\displaystyle\sum_{n=1}^{\infty}(-1/n)^n$

In Exercises 10 through 12, show that the series converges and approximate its sum correct to three decimal places.

10. $\displaystyle\sum_{n=0}^{\infty} \frac{(-1)^n}{(2n+1)!}$

11. $\displaystyle\sum_{n=1}^{\infty} \frac{(-1)^{n+1}}{n \cdot 10^n}$

12. $\displaystyle\sum_{n=0}^{\infty} \frac{(-1)^n}{(2n+1)2^n}$

13. *Catalan's constant G*, named for the French-Belgian mathematician Eugène Charles Catalan (1814–1894), finds applications in combinatorial analysis and can be defined by the infinite series

$$G = \sum_{n=0}^{\infty} \frac{(-1)^n}{(2n+1)^2}.$$

Find the numeric value of G rounded to three decimal places. It is not known if G is rational or irrational.

14. Show that the series

$$1 - \frac{1}{3} + \frac{1}{5} - \cdots + \frac{(-1)^{n-1}}{2n-1} + \cdots$$

converges to $\pi/4$. (Hint: the solution parallels Example 10.26.)

15. In Exercise 12 in Section 10.2.3, you found the sum of the reciprocals of the triangular numbers. Show that the sum of the reciprocals of odd-numbered triangular numbers is

$$1 + \frac{1}{6} + \frac{1}{15} + \cdots + \frac{1}{t_{2n-1}} + \cdots = 2\ln 2.$$

(Hint: first show that $\frac{1}{t_{2n-1}} = 2\left(\frac{1}{2n-1} - \frac{1}{2n}\right)$.) What is the sum of the reciprocals of the even-numbered triangular numbers, $\frac{1}{3} + \frac{1}{10} + \frac{1}{21} + \cdots + \frac{1}{t_{2n}} + \cdots$?

16. Show that

$$\sum_{n=0}^{\infty} \frac{1}{(4n+1)(4n+3)} = \frac{\pi}{8}.$$

(Hint: use partial fractions and the result of Exercise 14.)

17. (a) Show that the series

$$\sum_{k=1}^{\infty} \frac{(-1)^{k+1}}{2k(2k+1)(2k+2)} = \frac{1}{2\cdot 3\cdot 4} - \frac{1}{4\cdot 5\cdot 6} + \frac{1}{6\cdot 7\cdot 8} - \cdots$$

converges.

(b) Show that the sum of the series in (a) is $(\pi - 3)/4$. (Hint: use partial fractions and the series in Exercise 14.)

(c) Use the series

$$3 + 4\sum_{k=1}^{\infty} \frac{(-1)^{k+1}}{2k(2k+1)(2k+2)}$$

to approximate π correct to three decimal places.

18. The result in Exercise 12 in Section 10.2.3 can be written as

$$\frac{1}{1} + \frac{1}{1+2} + \frac{1}{1+2+3} + \cdots + \frac{1}{1+2+3+\cdots+n} + \cdots = 2.$$

Now show that

$$\frac{1}{1^2} + \frac{1}{1^2+2^2} + \frac{1}{1^2+2^2+3^2} + \cdots + \frac{1}{1^2+2^2+3^2+\cdots+n^2} + \cdots = 18 - 24\ln 2.$$

(Hint: show that the nth partial sum of this series is $18 - 24A_n + \frac{6}{n+1}$, where A_n is the nth partial sum of the alternating harmonic series.)

10.7 Explorations

10.7.1 The series of prime reciprocals

In Section 10.1 you learned that the harmonic series diverges. But if we discard sufficiently many terms in this series, it will converge. For example, if we discard all the terms that are not powers of $1/2$, the result is the convergent geometric series with $r = 1/2$. What if instead we discard all the terms that are not reciprocals of prime numbers? Will the resulting series—the *series of prime reciprocals*—converge or diverge?

In this exploration you will prove that the series of prime reciprocals diverges, even though its terms approach zero much faster than the terms in the harmonic series. You will show that $\sum_{p\,\text{prime}} 1/p$ diverges by showing that its nth partial sum, $P_n = \sum_{p\leq n} 1/p$, is larger than $(1/2)\ln H_n$ where H_n is the nth partial sum of the harmonic series. In addition to using the sigma notation for sums, we will employ the pi notation for products. For example, $\prod_{p\leq 11} p$

denotes the product $2 \cdot 3 \cdot 5 \cdot 7 \cdot 11$ of the primes less than or equal to 11 (in this Exploration p always denotes a prime).

Exercise 1. For a fixed integer $n \geq 2$, consider the set of all primes $p \leq n$ and the product

$$\prod_{p \leq n} \left(\frac{p}{p-1} \right).$$

Show that

$$\prod_{p \leq n} \left(\frac{p}{p-1} \right) = \prod_{p \leq n} \left(\frac{1}{1-1/p} \right) = \prod_{p \leq n} \left(1 + \frac{1}{p} + \frac{1}{p^2} + \cdots \right) \qquad (10.13)$$

and thus

$$\prod_{p \leq n} \left(\frac{p}{p-1} \right) > \sum_{k=1}^{n} \frac{1}{k} = H_n.$$

(Hint: each number $k \leq n$ is a product of powers of primes $p \leq n$, so for each $k \leq n$, $1/k$ must appear as one of the terms in the product on the right side in (10.13).)

Exercise 2. Show that

$$\sum_{p \leq n} [\ln p - \ln(p-1)] > \ln H_n. \qquad (10.14)$$

Exercise 3. Show that

$$\ln p - \ln(p-1) = \int_{p-1}^{p} \frac{1}{x}\, dx < \frac{1}{p-1} \leq \frac{2}{p},$$

and thus

$$\sum_{p \leq n} [\ln p - \ln(p-1)] \leq \sum_{p \leq n} \frac{2}{p}. \qquad (10.15)$$

Exercise 4. Combine (10.14) and (10.15) to show that

$$P_n > \frac{1}{2} \ln H_n.$$

Since H_n increases without bound as $n \to \infty$ so does P_n, and this completes the proof that $\sum_{p \text{ prime}} 1/p$ diverges.

Exercise 5. Use the divergence of the series of prime reciprocals to prove that there are infinitely many prime numbers.

10.7.2 An infinite series for e

In this Exploration you will derive the following series for the number e:

$$e = \sum_{k=0}^{\infty} \frac{1}{k!} = 1 + \frac{1}{1!} + \frac{1}{2!} + \cdots + \frac{1}{n!} + \cdots. \qquad (10.16)$$

(recall that $0!=1$). To do so, you will show that the limit of the nth partial sum $S_n = \sum_{k=0}^{n} 1/k!$ of the series in (10.16) is e. The tool to accomplish this is the sequence $\{x_n\}_{n=0}^{\infty}$ defined by $x_n = (1/n!) \int_0^1 t^n e^{-t} \, dt$.

Exercise 1. Show that $x_0 = -(1/e) + 1$ and for $n \geq 1$,

$$x_n = -\frac{1}{n!e} + x_{n-1}.$$

(Hint: use integration by parts.)

Exercise 2. Show that

$$x_n = -\frac{1}{n!e} - \frac{1}{(n-1)!e} - \cdots - \frac{1}{1!e} - \frac{1}{e} + 1 = 1 - \frac{1}{e}(S_n).$$

Exercise 3. Show that $0 \leq x_n \leq 1/n!$. (Hint: what are the bounds on the integrand in the integral defining x_n?)

Exercise 4. Show that $\lim x_n = 0$ to conclude that $\lim S_n = e$.

10.7.3 The number e is irrational

You probably already know that the number e, like $\sqrt{2}$ and π, is irrational. In this Exploration you will prove this fact using the infinite series (10.16) for e derived in Exploration 10.7.2. In general, a proof that a number is irrational is a proof by contradiction, since the definition of irrational is "not rational." So we begin the proof with the statement *Assume e is rational*, that is, there exist positive integers p and q such that $e = p/q$, and derive a contradiction (i.e., a false statement). The proof is in three steps.

Exercise 1. Show that

$$\frac{p}{q} = \sum_{n=0}^{q} \frac{1}{n!} + \sum_{n=q+1}^{\infty} \frac{1}{n!}. \tag{10.17}$$

Exercise 2. Show that

$$p(q-1)! - \sum_{n=0}^{q} \frac{q!}{n!} = \frac{1}{q+1} + \frac{1}{(q+1)(q+2)} + \frac{1}{(q+1)(q+2)(q+3)} + \cdots.$$

(Hint: transpose the first term on the right in (10.17) to the left, and multiply each side by $q!$.)

Exercise 3. Show that

$$\frac{1}{q+1} + \frac{1}{(q+1)(q+2)} + \frac{1}{(q+1)(q+2)(q+3)}$$
$$+ \cdots < \frac{1}{q+1} + \frac{1}{(q+1)^2} + \frac{1}{(q+1)^3} + \cdots = \frac{1}{q},$$

and hence

$$p(q - 1)! - \sum_{n=0}^{q} \frac{q!}{n!} < \frac{1}{q}. \tag{10.18}$$

Do you see the contradiction in (10.18)? (If not, what can you say about the number on the left of (10.18)? What can you say about the number on the right?) Since our assumption that $e = p/q$ leads to a contradiction, e must be irrational.

10.8 Acknowledgments

1. Part (a) of Figure 10.4 is from R. Mabry, Mathematics without words, *The College Mathematics Journal*, 32 (2001), p. 63; and part (b) is from S. A. Ajose, Proof without words: geometric series, *Mathematics Magazine*, **67** (1994), p. 230.
2. The illustration of the telescoping series in Example 10.11 is from J. H. Mathews, The sum is one, *College Mathematics Journal*, **22** (1991), p. 322.
3. Exercise 7 in Section 10.3.1 is adapted from R. F. Johnsonbaugh, Another proof of an estimate for e, *American Mathematical Monthly*, **81** (1974), pp. 1011–1012.
4. Exercise 8 in Section 10.3.1 is adapted from M. K. Kinyon, Another look at some p-series, *College Mathematics Journal*, **37** (2006), pp. 385–386.
5. Exercise 13 in Section 10.4.2 is from J. F. Fleron, Gabriel's wedding cake, *College Mathematics Journal*, **30** (1999), pp. 35–38.
6. Exercise 16 in Section 10.5.1 (the divergent p-series with $p = 1 + (1/n)$) is from R. M. Foster and M. S. Klamkin, On the convergence of the p-series, *American Mathematical Monthly*, **60** (1953), pp. 625–626.
7. The proof in Exploration 10.7.1 is from W. G. Leavitt, The sum of the reciprocals of the primes, *Two-Year College Mathematics Journal*, **10** (1979), pp. 198–199.
8. The derivation of the series for e in Exploration 10.7.2 is adapted from M. Chamberland, The series for e via integration, *College Mathematics Journal*, **30** (1999), p. 397.
9. The proof in Exploration 10.7.3 is from G. H. Hardy and E. M. Wright, *An Introduction to the Theory of Numbers*, fourth edition, Oxford University Press, London, 1960.

11

Infinite Series (Part II)

In the order of literature, as in others, there is no act that is not the coronation of an infinite series of causes and the source of an infinite series of effects.

Jorge Luis Borges

In this chapter we continue our study of infinite series. We begin by introducing two special forms of convergence, called absolute and conditional convergence. After developing tests for absolute and conditional convergence, we study power series representations of functions, and learn methods to find representations of several functions from your PCE such as exponential and logarithmic functions, and trigonometric functions and their inverses. These methods also allow us to represent and approximate many functions, useful in a variety of applied fields, that do not have simple forms like the functions from your PCE.

11.1 Absolute and conditional convergence

In Chapter 10 you learned that the alternating harmonic series converges, but the harmonic series diverges. However, when we replace $1/n$ in both series by $1/n^2$, the resulting series both converge. This raises the question: Does the convergence or divergence of $\sum_{n=1}^{\infty} |a_n|$ tell us anything about the convergence or divergence of $\sum_{n=1}^{\infty} a_n$? In this section we will answer that question.

In the preceding chapter we studied some methods for determining the convergence or divergence of a series $\sum_{n=1}^{\infty} a_n$ when a_n has a particular form. For example, when $a_n > 0$ for all n, we have the integral test and the comparison tests; and when the terms of $\sum_{n=1}^{\infty} a_n$ alternate in sign, we have the alternating series test. In this section we present tests that can be applied to $\sum_{n=1}^{\infty} a_n$ for general a_n. To do so, we first define two types of convergence, called *absolute convergence* and *conditional convergence*.

Definition 11.1 The series $\sum_{n=1}^{\infty} a_n$ is *absolutely convergent* (or *converges absolutely*) if the series $\sum_{n=1}^{\infty} |a_n|$ converges.

If $a_n \neq 0$ for all n, $\sum_{n=1}^{\infty} |a_n|$ will be a positive term series, and we can use tests from Chapter 10 to test $\sum_{n=1}^{\infty} a_n$ for absolute convergence, as in the following example.

Example 11.1 Consider the series $\sum_{n=1}^{\infty} \frac{\sin n}{n\sqrt{n}}$. The terms are both positive and negative, but do not alternate in sign, so let's consider the series $\sum_{n=1}^{\infty} |\frac{\sin n}{n\sqrt{n}}| = \sum_{n=1}^{\infty} \frac{|\sin n|}{n\sqrt{n}}$. Since $\frac{|\sin n|}{n\sqrt{n}} \leq \frac{1}{n\sqrt{n}}$ and $\sum_{n=1}^{\infty} \frac{1}{n\sqrt{n}}$ converges (why?), the series $\sum_{n=1}^{\infty} \frac{\sin n}{n\sqrt{n}}$ is absolutely convergent. ∎

Suppose $\sum_{n=1}^{\infty} a_n$ is absolutely convergent. Then the sequence of partial sums of the series $\sum_{n=1}^{\infty} |a_n|$ converges. But does $\sum_{n=1}^{\infty} a_n$ converge? That is, does the sequence of partial sums of $\sum_{n=1}^{\infty} a_n$ converge? The following theorem provides the answer.

Theorem 11.1 *If $\sum_{n=1}^{\infty} a_n$ is absolutely convergent, then $\sum_{n=1}^{\infty} a_n$ converges.*

Proof. Assume $\sum_{n=1}^{\infty} a_n$ converges absolutely, so that $\sum_{n=1}^{\infty} |a_n|$ converges. Every number a_n satisfies the double inequality $-|a_n| \le a_n \le |a_n|$, from which it follows that $0 \le a_n + |a_n| \le 2|a_n|$. Since $\sum_{n=1}^{\infty} |a_n|$ converges, so does $\sum_{n=1}^{\infty} 2|a_n|$, and therefore so does $\sum_{n=1}^{\infty} (a_n + |a_n|)$. But $a_n = (a_n + |a_n|) - |a_n|$, so from Theorem 10.4, $\sum_{n=1}^{\infty} a_n$ converges. $\qquad\square$

Example 11.2 In Example 10.26 we learned that the alternating harmonic series $\sum_{n=1}^{\infty} (-1)^{n+1}/n$ converges. But it is not absolutely convergent, since $\left|(-1)^{n+1}/n\right| = 1/n$ and the harmonic series $\sum_{n=1}^{\infty} 1/n$ diverges. ■

Convergent series that are not absolutely convergent are called *conditionally convergent*.

Definition 11.2 The series $\sum_{n=1}^{\infty} a_n$ is *conditionally convergent* (or *converges conditionally*) if it converges but $\sum_{n=1}^{\infty} |a_n|$ diverges.

Back to Example 11.2 The alternating harmonic series is conditionally convergent since it converges but the series of absolute values of the terms is the divergent harmonic series. ■

Definitions 11.1 and 11.2 lead to the classification scheme for infinite series illustrated in Figure 11.1. From now on, we will classify a series as absolutely convergent, conditionally convergent, or divergent. Every convergent positive term series is automatically absolutely convergent, and any conditionally convergent series must have infinitely many positive terms and infinitely many negative terms. (Do you see why?)

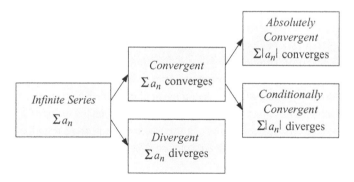

Figure 11.1. A classification scheme for infinite series

Back to Example 10.27 We used the alternating series test to show that $\sum_{n=1}^{\infty} (-1)^{n+1} \frac{n}{n^2+1}$ converges. Is the convergence absolute or conditional? To answer we consider the series $\sum_{n=1}^{\infty} \frac{n}{n^2+1}$. Since for large n, the term $n/(n^2 + 1)$ is approximately $1/n$, we suspect divergence. Using the limit comparison test, we have

$$\lim \frac{n/(n^2 + 1)}{1/n} = \lim \frac{n^2}{n^2 + 1} = 1,$$

and since the harmonic series diverges, so does $\sum_{n=1}^{\infty} \frac{n}{n^2+1}$. Hence $\sum_{n=1}^{\infty}(-1)^{n+1}\frac{n}{n^2+1}$ is conditionally convergent. ∎

The difference between absolute and conditional convergence

It is important to distinguish between the two types of convergence because they are *profoundly different*. Absolutely convergent series are in a sense very much like finite sums, where conditionally convergent series behave very differently. Here are some of the differences.

1. The *introduction of parentheses* (grouping finitely many consecutive terms) into a convergent series (absolute or conditional) produces a new series with the same sum, since the sequence of partial sums of the new series is a subsequence of the sequence of partial sums of the original series. Doing this with an absolutely convergence series always yields a new absolutely convergent series, but doing so with a conditionally convergent series may yield an absolutely convergent series. For example, let's group adjacent pairs of terms in the alternating harmonic series:

$$1 - \frac{1}{2} + \frac{1}{3} - \frac{1}{4} + \frac{1}{6} + \cdots = \left(1 - \frac{1}{2}\right) + \left(\frac{1}{3} - \frac{1}{4}\right) + \left(\frac{1}{5} - \frac{1}{6}\right) + \cdots$$

$$= \frac{1}{1 \cdot 2} + \frac{1}{3 \cdot 4} + \frac{1}{5 \cdot 6} + \cdots,$$

and the new series is absolutely convergent since it is a convergent positive term series.

2. A *rearrangement* of a series is formed by adding the terms of a series in a different order. For example $a_2 + a_1 + a_4 + a_3 + a_6 + a_5 + \cdots$ is a rearrangement of $\sum_{n=1}^{\infty} a_n$. Although we won't do so, it can be proved that any rearrangement of an absolutely convergent series with sum S is another absolutely convergent series with sum S. In this regard, absolutely convergent series behave exactly like finite sums.

   Is the same true for conditionally convergent series? Consider the following rearrangement of the alternating harmonic series (with sum $\ln 2$), where the positive terms appear in order and the negative terms appear in order, but each positive term is followed by two negative terms:

$$1 - \frac{1}{2} - \frac{1}{4} + \frac{1}{3} - \frac{1}{6} - \frac{1}{8} + \frac{1}{5} - \frac{1}{10} - \frac{1}{12} + \cdots.$$

To show that this series has a sum different from the alternating harmonic series, we introduce parentheses as follows (since doing so leaves its sum unaltered):

$$\left(1 - \frac{1}{2}\right) - \frac{1}{4} + \left(\frac{1}{3} - \frac{1}{6}\right) - \frac{1}{8} + \left(\frac{1}{5} - \frac{1}{10}\right) - \frac{1}{12} + \cdots$$

$$= \frac{1}{2} - \frac{1}{4} + \frac{1}{6} - \frac{1}{8} + \frac{1}{10} - \frac{1}{12} + \cdots,$$

resulting in one half the alternating harmonic series, so the sum of the rearranged series is $(1/2)\ln 2$. Consequently conditionally convergent series most definitely do not behave like finite sums!

Indeed, it can be proved that if $\sum_{n=1}^{\infty} a_n$ is a conditionally convergent series and r is any real number, then there exists a rearrangement of $\sum_{n=1}^{\infty} a_n$ that converges to r. In fact, this is also true for $r = \infty$ and $r = -\infty$. This remarkable result is known as *Riemann's theorem on rearrangements*.

While we can use the tests in Chapter 10 to determine whether or not a series converges absolutely, two other tests, the ratio test and the root test, are often more efficient. We introduce and study these tests in the next section.

11.1.1 Exercises

In Exercises 1 through 6, classify each series as absolutely convergent, conditionally convergent, or divergent.

1. $1 + \dfrac{1}{3} + \dfrac{1}{5} + \dfrac{1}{7} + \cdots$
   2. $1 - \dfrac{1}{3} + \dfrac{1}{5} - \dfrac{1}{7} + \cdots$
   3. $1 + \dfrac{1}{4} + \dfrac{1}{9} + \dfrac{1}{16} + \cdots$

4. $1 - \dfrac{1}{4} + \dfrac{1}{9} - \dfrac{1}{16} + \cdots$
   5. $1 - \dfrac{1}{3} + \dfrac{1}{9} - \dfrac{1}{27} + \cdots$
   6. $1 - \dfrac{1}{\sqrt{2}} + \dfrac{1}{\sqrt{3}} - \dfrac{1}{\sqrt{4}} + \cdots$

In Exercises 7 through 9, use a comparison test to show that each series is absolutely convergent.

7. $\displaystyle\sum_{n=1}^{\infty} \frac{\sin n}{n^2}$
   8. $\displaystyle\sum_{n=1}^{\infty} \frac{(-1)^{n(n+1)/2}}{n(n+1)}$
   9. $\displaystyle\sum_{n=1}^{\infty} \frac{\cos(n\pi/3)}{n\sqrt{n}}$

10. Show that a convergent geometric series must be absolutely convergent.

11. In Exercise 15 in Section 10.5, you proved that if $\sum_{n=1}^{\infty} a_n$ and $\sum_{n=1}^{\infty} b_n$ are convergent series of positive terms, then $\sum_{n=1}^{\infty} \sqrt{a_n^2 + b_n^2}$ converges. Now prove: if $\sum_{n=1}^{\infty} \sqrt{a_n^2 + b_n^2}$ converges, then $\sum_{n=1}^{\infty} a_n$ and $\sum_{n=1}^{\infty} b_n$ converge absolutely.

12. Prove: If $\sum_{n=1}^{\infty} a_n$ and $\sum_{n=1}^{\infty} b_n$ converge absolutely, then $\sum_{n=1}^{\infty} a_n b_n$ converges absolutely. (Hint: for sufficiently large n, $|a_n| < 1$.)

13. Prove that if $\sum_{n=1}^{\infty} a_n$ converges absolutely, then $\sum_{n=1}^{\infty} a_n^2$ converges. Is the converse true? If it is, prove it; if not, give a counterexample.

14. Suppose that $\sum_{n=1}^{\infty} a_n$ converges, but that $\sum_{n=1}^{\infty} a_n^2$ diverges. Prove that $\sum_{n=1}^{\infty} a_n$ converges conditionally. (Hint: show that the assumption that $\sum_{n=1}^{\infty} a_n$ converges absolutely leads to a contradiction.)

15. Given a series $\sum a_n$, let $a_n^+ = \max\{a_n, 0\}$ and $a_n^- = \min\{a_n, 0\}$ for each n.
    (a) Show that the nonzero terms of the series $\sum a_n^+$ are the positive terms of $\sum a_n$, and that the nonero terms of the series $\sum a_n^-$ are the negative terms of $\sum a_n$.
    (b) Show that $a_n^+ = (a_n + |a_n|)/2$ and $a_n^- = (a_n - |a_n|)/2$.
    (c) Show that if $\sum a_n$ is absolutely convergent, then both $\sum a_n^+$ and $\sum a_n^-$ converge, and if $\sum a_n$ is conditionally convergent, then both $\sum a_n^+$ and $\sum a_n^-$ diverge.

11.2 The ratio and root tests

In this section we study two tests for absolute convergence of a series.

Theorem 11.2 (The ratio test) *Let $\sum_{n=1}^{\infty} a_n$ be a series with nonzero terms.*

(i) If $\lim \left| \dfrac{a_{n+1}}{a_n} \right| = L < 1$, *then* $\sum_{n=1}^{\infty} a_n$ *converges absolutely.*

(ii) If $\lim \left| \dfrac{a_{n+1}}{a_n} \right| = L > 1$ *or* $\lim \left| \dfrac{a_{n+1}}{a_n} \right| = +\infty$, *then* $\sum_{n=1}^{\infty} a_n$ *diverges.*

(iii) If $\lim \left| \dfrac{a_{n+1}}{a_n} \right| = 1$, *then the test fails, that is,* $\sum_{n=1}^{\infty} a_n$ *may converge absolutely, converge conditionally, or diverge.*

Of course, the test also fails if $\lim \left| \frac{a_{n+1}}{a_n} \right|$ fails to exist (other than $+\infty$).

Before proving Theorem 11.2, here are two examples of its use.

Example 11.3 Discuss the convergence of $\sum_{n=1}^{\infty} n^2(-1/2)^n$. Here $a_n = n^2(-1/2)^n$ and so

$$\left| \frac{a_{n+1}}{a_n} \right| = \left| \frac{(n+1)^2(-1/2)^{n+1}}{n^2(-1/2)^n} \right| = \left| -\frac{1}{2} \right| \frac{(n+1)^2}{n^2} = \frac{1}{2} \left(1 + \frac{1}{n} \right)^2.$$

Thus,

$$\lim \left| \frac{a_{n+1}}{a_n} \right| = \lim \frac{1}{2} \left(1 + \frac{1}{n} \right)^2 = \frac{1}{2} < 1$$

and thus the series converges absolutely (and by Theorem 11.1, it converges). In Section 11.5 we will find the sum of this series. ∎

Back to Example 10.22 To use a comparison test on the series $\sum_{n=1}^{\infty} n!/n^n$, we had to suspect convergence, and then compare our series to a larger convergent series. The ratio test is simpler, in that we need only look at the ratio of consecutive terms in the given series:

$$\left| \frac{a_{n+1}}{a_n} \right| = \left| \frac{(n+1)!/(n+1)^{(n+1)}}{n!/n^n} \right| = \frac{(n+1)!}{(n+1)^{(n+1)}} \cdot \frac{n^n}{n!} = \frac{n^n}{(n+1)^n} = \frac{1}{\left(1 + \frac{1}{n} \right)^n}.$$

Hence

$$\lim \left| \frac{a_{n+1}}{a_n} \right| = \lim \frac{1}{\left(1 + \frac{1}{n} \right)^n} = \frac{1}{e} < 1 \tag{11.1}$$

and so the series converges (absolutely). ∎

☺ In using the ratio test, we recommend the following two-step process we used in the preceding examples:

1. Simplify the ratio $|a_{n+1}/a_n|$ as much as possible. If the terms are fractions, use the "invert and multiply" technique to divide a_{n+1} by a_n. Recall properties of absolute value such as $|ab| = |a| \cdot |b|$, $|a/b| = |a| / |b|$, $|a| = a$ for $a > 0$, and $|(-1)^n| = 1$ for all n. The fewer occurrence of n in $|a_{n+1}/a_n|$ the easier the next step will be.
2. Take the limit as $n \to \infty$ and compare it to 1.

Proof of Theorem 11.2. (i) Since $L < 1$ we can find a number r between L and 1, for example $r = (L + 1)/2$, and thus $L < r < 1$. Because the limit of the sequence $\{|a_{n+1}/a_n|\}$ is L, eventually its terms must be smaller than r. To be precise, there exists an integer N such that if $n \geq N$, then $|a_{n+1}/a_n| < r$, or equivalently $|a_{n+1}| < |a_n|r$. So we have

$$|a_{N+1}| < |a_N|r,$$

$$|a_{N+2}| < |a_{N+1}|r < |a_N|r^2,$$

$$|a_{N+3}| < |a_{N+1}|r < |a_{N+1}|r^2 < |a_N|r^3,$$

and in general $|a_{N+k}| < |a_N|r^k$ for all $k \geq 1$. Thus the series $\sum_{k=1}^{\infty} |a_{N+k}|$ converges by direct comparison to the convergent geometric series $\sum_{k=1}^{\infty} |a_N|r^k$. But $\sum_{k=1}^{\infty} |a_{N+k}|$ is simply the tail of $\sum_{n=1}^{\infty} |a_n|$, so $\sum_{n=1}^{\infty} |a_n|$ converges and thus $\sum_{n=1}^{\infty} a_n$ is absolutely convergent. (ii) If the limit of the sequence $\{|a_{n+1}/a_n|\}$ is $L > 1$ or $+\infty$, then there exists an integer N such that if $n \geq N$, then $|a_{n+1}/a_n| > 1$, or equivalently $|a_{n+1}| > |a_n|$. It follows that $|a_{N+k}| > |a_N|$ for all $k \geq 1$, and thus $\lim a_n \neq 0$ so that $\sum_{n=1}^{\infty} |a_n|$ diverges by the divergence test. In Exercise 13 you will prove part (iii). $\qquad\square$

Example 11.4 Let x be a real number, and consider the series

$$\sum_{n=0}^{\infty} \frac{x^n}{n!} = 1 + x + \frac{x^2}{2!} + \frac{x^3}{3!} + \cdots + \frac{x^n}{n!} + \cdots . \qquad (11.2)$$

For which values of x does the series converge absolutely? Here $a_n = x^n/n!$ so that

$$\left|\frac{a_{n+1}}{a_n}\right| = \left|\frac{x^{n+1}/(n+1)!}{x^n/n!}\right| = \left|\frac{x^{n+1}}{(n+1)!} \cdot \frac{n!}{x^n}\right| = \frac{|x|}{n+1}.$$

Hence

$$\lim \left|\frac{a_{n+1}}{a_n}\right| = \lim \frac{|x|}{n+1} = 0 \text{ for all } x$$

and so the series in (11.4) converges absolutely for all x. This is an important power series (recall Section 10.2) and so (11.4) defines a function $f(x)$ with domain $(-\infty, \infty)$. Clearly $f(0) = 1$, and in Exploration 10.7.2 you learned that $f(1) = e$. We will learn more about this function in Section 11.5. ∎

The preceding example establishes the following result, which will be useful in Section 11.6.

Theorem 11.3 *For any real number x, $\lim x^n/n! = 0$.*

Example 11.5 For what values of x does the series

$$1 - \frac{1}{2}x + \frac{1 \cdot 3}{2 \cdot 4}x^2 - \frac{1 \cdot 3 \cdot 5}{2 \cdot 4 \cdot 6}x^3 + \frac{1 \cdot 3 \cdot 5 \cdot 7}{2 \cdot 4 \cdot 6 \cdot 8}x^4 - \cdots + (-1)^n \frac{1 \cdot 3 \cdot 5 \cdots (2n-1)}{2 \cdot 4 \cdot 6 \cdots (2n)}x^n + \cdots$$

converge absolutely? For what values of x does the series diverge? Since

$$a_{n+1} = -a_n \frac{(2n+1)x}{2n+2}$$

we have

$$\lim \left| \frac{a_{n+1}}{a_n} \right| = \lim \frac{2n+1}{2n+2} |x| = |x|,$$

so that the series converges absolutely for $|x| < 1$ and diverges for $|x| > 1$. ∎

A second test or absolute convergence, similar to the ratio test, is the *root test*.

Theorem 11.4 (The root test) *Let $\sum_{n=1}^{\infty} a_n$ be a series with nonzero terms.*

(i) *If $\lim \sqrt[n]{|a_n|} = L < 1$, then $\sum_{n=1}^{\infty} a_n$ converges absolutely.*

(ii) *If $\lim \sqrt[n]{|a_n|} = L > 1$ or $\sqrt[n]{|a_n|} = +\infty$, then $\sum_{n=1}^{\infty} a_n$ diverges.*

(iii) *If $\lim \sqrt[n]{|a_n|} = 1$, then the test fails, that is, $\sum_{n=1}^{\infty} a_n$ may converge absolutely, converge conditionally, or diverge.*

The proof of the root test is almost the same as the proof of the ratio test. See Exercise 14.

Since the root test is based on the nth root of $|a_n|$, it is often easier to use than the ratio test when a_n involves functions of n raised to powers, as in the following example.

Example 11.6 Discuss the convergence of the series $\sum_{n=2}^{\infty} \frac{(-1)^n}{(\ln n)^n}$.

The presence of nth powers suggests using the root test: Since

$$\sqrt[n]{|a_n|} = \sqrt[n]{\left| \frac{(-1)^n}{(\ln n)^n} \right|} = \frac{1}{\ln n},$$

we have

$$\lim \sqrt[n]{|a_n|} = \lim \frac{1}{\ln n} = 0 < 1,$$

and so the series is absolutely convergent.

If the terms of a series have factorials, we usually recommend using the ratio test, since ratios of factorials simplify. But the limit in (9.4), $\lim(\sqrt[n]{n!}/n) = 1/e$, often makes it possible to use the root test when a_n involves both nth powers and factorials.

Example 11.7 Discuss the convergence of the series $\sum_{n=1}^{\infty} (-2)^n n!/n^n$.

Since $\sqrt[n]{|a_n|} = 2\sqrt[n]{n!}/n$, we have

$$\lim \sqrt[n]{|a_n|} = \lim \frac{2\sqrt[n]{n!}}{n} = \frac{2}{e} < 1.$$

Thus, the series $\sum_{n=1}^{\infty} (-2)^n n!/n^n$ converges. ∎

In an advanced calculus course you will prove that if the ratio test succeeds in showing absolute convergence or divergence, so does the root test. However, there are series for which the ratio test fails but the root test succeeds—see, for example, Exercise 15.

11.2.1 Exercises

In Exercises 1 through 12, classify each series as absolutely convergent, conditionally convergent, or divergent.

1. $\displaystyle\sum_{n=1}^{\infty} \frac{n^2}{2^n}$ *alt series* *limit comp*

2. $\displaystyle\sum_{n=0}^{\infty} \frac{2^n}{n!}$

3. $\displaystyle\sum_{n=1}^{\infty} \frac{\sin(2n)}{2^n}$

4. $\displaystyle\sum_{n=1}^{\infty}(-1)^n \frac{n}{n^2+1}$

5. $\displaystyle\sum_{n=2}^{\infty} \frac{(-1)^n}{(\ln n)^n}$

6. $\displaystyle\sum_{n=0}^{\infty} n! e^{-n}$

7. $\displaystyle\sum_{n=1}^{\infty}(-1)^n \frac{(n!)^2}{(2n)!}$

8. $\displaystyle\sum_{n=1}^{\infty}(-1)^n \frac{\ln n}{n}$

9. $\displaystyle\sum_{n=1}^{\infty} \frac{\ln n}{n^2}$

10. $\displaystyle\sum_{n=1}^{\infty} \frac{(-1)^n}{(\arctan n)^n}$

11. $\displaystyle\sum_{n=1}^{\infty}(-1)^n \frac{n+1}{n^2}$

12. $\displaystyle\sum_{n=2}^{\infty} \frac{(-1)^n}{n(\ln n)^2}$

13. Show that the ratio test yields $L = 1$ for the harmonic series, the alternating harmonic series, and the p-series with $p = 2$, proving part (iii) of Theorem 11.2.

14. Prove Theorem 11.3. (Hint: the proof parallels the proof of the ratio test (Theorem 11.2). Observe $\sqrt[n]{|a_n|} < r$ implies $|a_n| < r^n$.)

15. Consider the series

$$\sum_{n=1}^{\infty} a_n = \frac{1}{4} + \frac{1}{2} + \frac{1}{16} + \frac{1}{8} + \frac{1}{64} + \frac{1}{32} + \cdots$$

where $a_n = 1/2^{n-1}$ for n even and $a_n = 1/2^{n+1}$ for n odd. Show (a) that the ratio test fails but (b) the root test succeeds.

16. Consider the series

$$\sum_{n=1}^{\infty} a_n = \frac{1}{2} + \frac{1}{9} + \frac{1}{8} + \frac{1}{81} + \frac{1}{32} + \frac{1}{729} + \cdots$$

where $a_n = 1/2^n$ for n odd and $a_n = 1/3^n$ for n even. Show (a) that both the ratio and root tests fail yet (b) the series converges to 19/24. (Hint: recall Theorem 10.4(ii)).

17. Prove the following relationship between the ratio and root tests: Let $\sum_{n=1}^{\infty} a_n$ be a series with nonzero terms, and suppose $\lim |a_{n+1}/a_n| = L$ and $\lim \sqrt[n]{|a_n|} = M$, with L and M finite. Then $L = M$. (Hint: construct a proof by contradiction, i.e., show that it is impossible to have $L \neq M$. Assume $L < M$ (the case $L > M$ is similar), and let k satisfy $L < k < M$ (for example, let $k = (L + M)/2$). Note that $k > 0$ since $L \geq 0$. Now consider the series $\sum_{n=1}^{\infty} b_n$, where $b_n = a_n/k^n$, and test it for convergence by the ratio test and the root test.)

11.3 Convergence tests: a summary and strategy

Table 11.1 summarizes the convergence tests you have seen so far.

Table 11.1. A summary of convergence tests for infinite series

Test	Series	Converge/Diverge
Divergence test (§10.1)	$\sum a_n$	Diverges if $\lim a_n \neq 0$, test fails if $\lim a_n = 0$.
Geometric series (§10.2)	$\sum_{n=0}^{\infty} ar^n$, $(a \neq 0)$	Converges to $a/(1-r)$ if $\lvert r \rvert < 1$, diverges if $\lvert r \rvert \geq 1$.
Telescoping series (§10.2)	$\sum a_n$, $a_n = b_n - b_{n+1}$	Converges to $b_1 - L$ if $\lim b_n = L$.
p-series (§10.4)	$\sum 1/n^p$	Converges if $p > 1$, diverges if $p \leq 1$.
Integral test (§10.4)	$\sum a_n, a_n = f(n)$, f positive, continuous and decreasing	$\sum a_n$ and $\int_k^{\infty} f(x)\,dx$ both converge or both diverge.
Direct comparison test (§10.5)	$\sum a_n$ and $\sum b_n$, $0 < a_n \leq b_n$	$\sum a_n$ converges if $\sum b_n$ converges, $\sum b_n$ diverges if $\sum a_n$ diverges.
Limit comparison test (§10.5)	$\sum a_n$ and $\sum b_n$, $a_n > 0, b_n > 0$, $\lim a_n/b_n = L > 0$	$\sum a_n$ and $\sum b_n$ both converge or both diverge.
Alternating series test (§10.6)	$\sum (-1)^{n+1} b_n$ or $\sum (-1)^n b_n, b_n > 0$	Converges if $b_n \to 0$ and $b_{n+1} \leq b_n$.
Ratio test (§11.2)	$\sum a_n$, $\lim \lvert a_{n+1}/a_n \rvert = L$	Converges absolutely if $L < 1$, diverges if $L > 1$, test fails if $L = 1$.
Root test (§11.2)	$\sum a_n$, $\lim \sqrt[n]{\lvert a_n \rvert} = L$	Converges absolutely if $L < 1$, diverges if $L > 1$, test fails if $L = 1$.

Here are some hints for a strategic use of the tests. Of course, the best strategy is to become familiar with the tests through practice.

1. Perform the divergence test first. If $\lim a_n \neq 0$, the series diverges, and you are done. If $\lim a_n = 0$, the series may converge or diverge, so try another test.
2. Does the series have a special form, like geometric, telescoping, p-series, or alternating?
3. If $a_n = f(n)$ where f is positive, continuous, and decreasing, try the integral test.
4. If the ratio a_{n+1}/a_n of successive terms will simplify, try the ratio test.
5. If a_n is an nth power, try the root test.
6. If a_n is similar to a term of a geometric series or a p-series, try one of the comparison tests.

11.3.1 Exercises

In Exercises 1 through 12, classify each series as absolutely convergent, conditionally convergent, or divergent.

1. $\displaystyle\sum_{n=0}^{\infty}\left(\frac{3}{\pi}\right)^n$

2. $\displaystyle\sum_{n=1}^{\infty}2^{-1/n}$

3. $\displaystyle\sum_{n=1}^{\infty}(-1)^{n+1}\frac{n^2}{3^n}$

4. $\displaystyle\sum_{n=1}^{\infty}\frac{(-1)^{n+1}}{1+\ln n}$

5. $\displaystyle\sum_{n=1}^{\infty}\ln\left(\frac{2n-1}{2n+1}\right)$

6. $\displaystyle\sum_{n=1}^{\infty}(-1)^{n+1}\frac{e^n}{n^n}$

7. $\displaystyle\sum_{n=1}^{\infty}\frac{\sin n}{n^e}$

8. $\displaystyle\sum_{n=1}^{\infty}(-1)^{n+1}\tan(1/n)$

9. $\displaystyle\sum_{n=2}^{\infty}\frac{(-1)^n}{n^{1/\ln n}}$

10. $\displaystyle\sum_{n=0}^{\infty}\frac{(-1)^n}{1\cdot3\cdot5\cdots(2n+1)}$

11. $\displaystyle\sum_{n=0}^{\infty}\frac{(-3)^n}{(n+1)2^n}$

12. $\displaystyle\sum_{n=1}^{\infty}\left(1+\frac{1}{n}\right)^{-n}$

11.4 Power series

We first encountered power series informally in Section 10.2. We can now give a proper definition.

Definition 11.3 A *power series in x* is a series of the form

$$\sum_{k=0}^{\infty}c_kx^k = c_0 + c_1x + c_2x^2 + \cdots + c_nx^n + \cdots \tag{11.3}$$

where $\{c_k\}_{k=0}^{\infty}$ is a sequence of constants called the *coefficients* of the series and x is a variable.

In general a power series converges for some values of x and diverges for others. If the series converges for a particular x, its sum depends on x, and so a power series defines a *function* $f(x) = c_0 + c_1x + c_2x^2 + \cdots + c_nx^n + \cdots$ whose *domain* is the set of all x values for which the series converges. The partial sums of a power series are polynomials.

Example 11.8 Let the sequence $\{c_k\}$ of coefficients in (11.3) be given by $c_k = 1/2^k$ so that the power series is

$$1+\frac{1}{2}x+\frac{1}{4}x^2+\frac{1}{8}x^3+\cdots+\frac{1}{2^n}x^n+\cdots.$$

Since it is a geometric series with first term 1 and common ratio $x/2$, it converges to $1/[1-(x/2)] = 2/(2-x)$ for $|x/2| < 1$, or $|x| < 2$, and diverges elsewhere. (The ratio test shows that the convergence is absolute for $|x| < 2$.) We write the sum and domain as

$$1+\frac{1}{2}x+\frac{1}{4}x^2+\frac{1}{8}x^3+\cdots+\frac{1}{2^n}x^n+\cdots=\frac{2}{2-x},\quad -2 < x < 2. \blacksquare$$

Of course, power series take many forms other than geometric series, and we use the convergence tests we've studied to find the values of x for which they converge (i.e., to find the domain of the sum function).

Back to Example 11.4 In this example we used the ratio test to learn that the series

$$\sum_{n=0}^{\infty} \frac{x^n}{n!} = 1 + x + \frac{x^2}{2!} + \frac{x^3}{3!} + \cdots + \frac{x^n}{n!} + \cdots .$$

converges absolutely for all x. However, we do not (yet) know if there is a nice expression for its sum. ∎

Example 11.9 For what values of x does the series

$$\sum_{n=1}^{\infty} \frac{(-1)^{n+1}x^n}{n \cdot 3^n} = \frac{x}{3} - \frac{x^2}{2 \cdot 9} + \frac{x^3}{3 \cdot 27} - \frac{x^4}{4 \cdot 81} + \cdots$$

converge? Since $a_n = (-1)^{n+1}x^n/n3^n$, we have

$$\left| \frac{a_{n+1}}{a_n} \right| = \left| \frac{(-1)^{n+2}x^{n+1}}{(n+1)3^{n+1}} \cdot \frac{n3^n}{(-1)^{n+1}x^n} \right| = \frac{n|x|}{3(n+1)} \rightarrow \frac{|x|}{3},$$

so by the ratio test the series converges absolutely when $|x|/3 < 1$, or $|x| < 3$, and diverges when $|x|/3 > 1$, or $|x| > 3$. But the ratio test is inconclusive when $|x|/3 = 1$, or when $x = 3$ or $x = -3$. We test for convergence at these two values of x individually by examining the series at each one.

When $x = 3$ the series becomes the alternating harmonic series

$$1 - (1/2) + (1/3) - (1/4) + \cdots ,$$

which converges conditionally, and when $x = -3$ we have

$$-1 - (1/2) - (1/3) - (1/4) - \cdots ,$$

the negative of the harmonic series, which diverges. Hence the domain of the function defined by this series is the interval $(-3, 3]$. ∎

Example 11.10 For what values of x does the series

$$\sum_{k=1}^{\infty} k^k x^k = x + 4x^2 + 27x^3 + 64x^4 + \cdots + n^n x^n + \cdots$$

converge? Using the root test with $a_n = n^n x^n$ we have $\sqrt[n]{|n^n x^n|} = n|x|$, and

$$\lim n|x| = \begin{cases} 0, & x = 0, \\ \infty, & x \neq 0 \end{cases} .$$

Hence this power series converges (to 0) at $x = 0$ and diverges for all $x \neq 0$. ∎

So far in this section we've seen examples of power series with domains of three forms: all real numbers, an interval of finite length, and just the single number 0. Are there other forms of domains for a power series? Before answering the question, we first consider other types of power series in addition to the power series in x given by (11.3). For any real number a, *a power series in $x - a$*, or *power series centered at a*, or *power series about a*, is a series of the form given in (11.3) but with x replaced by $(x - a)$:

$$\sum_{k=1}^{\infty} c_k(x - a)^k = c_0 + c_1(x - a) + c_2(x - a)^2 + \cdots + c_n(x - a)^n + \cdots. \qquad (11.4)$$

The procedure for finding the values of x for which a power series in $(x - a)$ converges are the same as for a power series in x since the function in (11.4) is simply the one in (11.3) shifted horizontally by a units.

Example 11.11 For what values of x does the series

$$\sum_{n=0}^{\infty} \frac{(x + 1)^n}{2^n \sqrt{n + 1}} = 1 + \frac{x + 1}{2\sqrt{2}} + \frac{(x + 1)^2}{4\sqrt{3}} + \cdots$$

converge? Using the ratio test with $a_n = (x + 1)^n / (2^n \sqrt{n + 1})$ yields

$$\left| \frac{a_{n+1}}{a_n} \right| = \left| \frac{(x + 1)^{n+1}}{2^{n+1}\sqrt{n + 2}} \cdot \frac{2^n \sqrt{n + 1}}{(x + 1)^n} \right| = \frac{|x + 1|}{2} \sqrt{\frac{n + 1}{n + 2}} \to \frac{|x + 1|}{2}.$$

Thus the series converges absolutely for $|x + 1|/2 < 1$, or equivalently $|x + 1| < 2$, that is, $-2 < x + 1 < 2$ or $-3 < x < 1$. Similarly, the series diverges for $x < -3$ and for $x > 1$, and the ratio test fails for $x = -3$ and $x = 1$.

When $x = 1$ the series is

$$1 + (1/\sqrt{2}) + (1/\sqrt{3}) + \cdots,$$

the p-series with $p = 1/2$, which diverges. When $x = -3$ we have

$$1 - (1/\sqrt{2}) + (1/\sqrt{3}) + \cdots,$$

an alternating series. Since $1/\sqrt{n + 1}$ decreases to zero, this series converges by the alternating series test. The convergence at $x = -3$ is conditional since the corresponding series of absolute values is the same as the series with $x = 1$. Thus the given series converges in the interval $[-3, 1)$ and diverges elsewhere. ∎

In the preceding examples we've seen that the set of numbers for which a power series converges has one of three forms: all real numbers, an interval of finite length, or a single real number. These are the only possibilities, as we now prove.

Theorem 11.5 *Let $\sum_{k=0}^{\infty} c_k(x - a)^k$ be a power series. Then exactly one of the following three conditions is true:*

(i) *the series converges only for $x = a$,*

(ii) *the series converges absolutely for all x,*

(iii) *there exists a number $R > 0$ such that the series converges absolutely for all x for which $|x - a| < R$ and diverges for all x for which $|x - a| > R$.*

Condition (iii) can also be stated this way: *there exists a number $R > 0$ such that the series converges absolutely for all x in the interval $(a - R, a + R)$ and diverges for all $x < a - R$ and all $x > a + R$.* The number R is called the *radius of convergence* of the series, $R = 0$ when condition (i) holds, and $R = \infty$ when condition (ii) holds. Condition (iii) says nothing about whether the series converges or diverges at the endpoints of $(a - R, a + R)$; you have to test these values of x as in Examples 11.9 and 11.11.

When $R > 0$ the set of x values for which the series converges is called the *interval of convergence* of the series; when condition (ii) holds it is $(-\infty, \infty)$, and when condition (iii) holds it has one of four forms: $(a - R, a + R)$, $[a - R, a + R)$, $(a - R, a + R]$, or $[a - R, a + R]$. In each case, the radius of convergence R is one-half the length of the interval of convergence. See Figure 11.2 for an illustration of condition (iii).

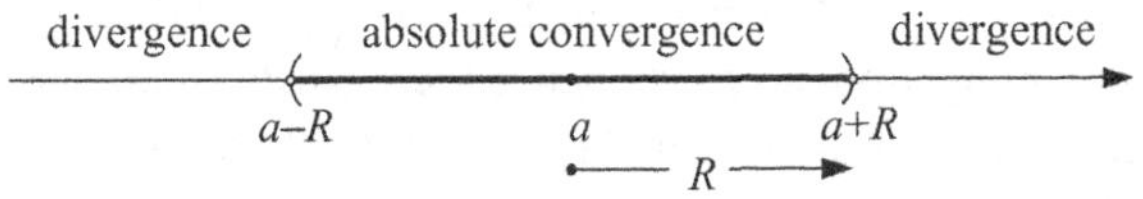

Figure 11.2. The interval of convergence

To prove Theorem 11.5, we will prove the special case when $a = 0$, given below as Theorem 11.6. This case implies the general situation in Theorem 11.5 since all we need to do to obtain Theorem 11.5 from Theorem 11.6 is replace each occurrence of x by $x - a$.

Theorem 11.6 *Let $\sum_{k=1}^{\infty} c_k x^k$ be a power series. Then exactly one of the following three conditions is true:*

  (i) the series converges only for $x = 0$,
  (ii) the series converges absolutely for all x,
  (iii) there exists a number $R > 0$ such that the series converges absolutely for all x for which $|x| < R$ and diverges for all x for which $|x| > R$.

The proof of Theorem 11.6 requires a preliminary lemma.

Lemma 11.7 *Let $\sum_{k=0}^{\infty} c_k x^k$ be a power series.*

  (i) If $\sum_{k=0}^{\infty} c_k x^k$ converges for $x = x_c$, then it converges absolutely for all x such that $|x| < |x_c|$.
  (ii) If $\sum_{k=0}^{\infty} c_k x^k$ diverges for $x = x_d$, then it diverges for all x such that $|x| > |x_d|$.

Proof. (i) Assume $\sum_{k=0}^{\infty} c_k x_c^k$ converges. Then $\lim c_k x_c^k = 0$ by Theorem 10.2, so there exists a positive integer N such that $\left| c_k x_c^k \right| < 1$ whenever $k \geq N$. So if $|x| < |x_c|$ and $k \geq N$, we have

$$\left| c_k x^k \right| = \left| c_k x_c^k \cdot \frac{x^k}{x_c^k} \right| = \left| c_k x_c^k \right| \cdot \left| \frac{x}{x_c} \right|^k < \left| \frac{x}{x_c} \right|^k.$$

The series $\sum_{k=N}^{\infty} |x/x_c|^k$ converges (it's geometric with $|r| = |x/x_c| < 1$), hence by the direct comparison test so does $\sum_{k=N}^{\infty} \left| c_k x^k \right|$ for $|x| < |x_c|$. Thus $\sum_{k=0}^{\infty} c_k x^k$ is absolutely convergent for $|x| < |x_c|$.

(ii) Let $\sum_{k=1}^{\infty} c_k x_d^k$ diverge, and suppose the series converges for some x for which $|x| > |x_d|$. Then by part (i) of this lemma the series converges for x_d, a contradiction. Thus $\sum_{k=0}^{\infty} c_k x^k$ diverges for $|x| > |x_d|$. $\qquad \Box$

Proof of Theorem 11.6. Clearly every series of the form $\sum_{k=0}^{\infty} c_k x^k$ converges for $x = 0$, and we've seen series that converge for all x, such as the one in Example 11.4. So to prove

Theorem 11.6, we need to show that if conditions (i) and (ii) do not hold, then (iii) must hold. To do so, assume that (i) and (ii) do not hold. Then there exist nonzero numbers x_c and x_d such that the series converges when $x = x_c$ and diverges when $x = x_d$. Let S be the set of all x for which $\sum_{k=0}^{\infty} c_k x^k$ converges absolutely. Then S is not empty since both 0 and x_c are in S. From Lemma 11.7 the series diverges for all x for which $|x| > |x_d|$, hence $|x| \leq |x_d|$ for all x in S. Since S is nonempty and bounded above, it has a least upper bound $R > 0$ by the completeness property of real numbers (which we encountered in the proof of Theorem 10.6). To finish the proof we must show that if $|x| < R$ then x is in S and if $|x| > R$ then $\sum_{k=0}^{\infty} c_k x^k$ diverges. If $|x| < R$ then $|x|$ is not an upper bound for S, hence there exists a number y in S with $|x| < y$. So by part (i) of Lemma 11.7, x is in S. If $|x| > R$, then there exists a number z such that $|x| > z > R$. Since z is not in S and $z = |z|$, $\sum_{k=1}^{\infty} c_k z^k$ diverges, so by part (ii) of Lemma 11.7 so does $\sum_{k=1}^{\infty} c_k x^k$. $\square$

Hypergeometric functions. Many important functions in applied mathematics are defined as power series, such as the hypergeometric function $F(a, b; c; x)$, which finds applications in differential equations and mathematical physics. It is defined for real values of a, b, c (c not 0 or a negative integer) by the series

$$F(a, b; c; x) = 1 + \frac{ab}{c}x + \frac{a(a+1)b(b+1)}{c(c+1)2!}x^2 + \frac{a(a+1)(a+2)b(b+1)(b+2)}{c(c+1)(c+2)3!}x^3 + \cdots.$$

To find its domain we use the ratio test to find the radius of convergence. If we denote the nth term by t_n, then when neither a nor b is 0 or a negative integer,

$$\left| \frac{t_{n+1}}{t_n} \right| = \frac{|a+n|\,|b+n|}{|c+n|\,(n+1)}|x| = \frac{|(a/n)+1|\,|(b/n)+1|}{|(c/n)+1|\,(1+(1/n))}|x| \to |x| < 1,$$

so that the series converges for $|x| < 1$ and for all a, b, c (none 0 or negative integer). The series converges at the enpoints 1 and -1 if a or b is 0 or a negative integer (in these cases $F(a, b; c; x)$ is a polynomial, defined for all x).

11.4.1 Exercises

In Exercises 1 through 12, find the interval of convergence of the power series.

1. $\sum_{n=0}^{\infty} \frac{x^n}{2n+1}$

2. $\sum_{n=0}^{\infty} (-1)^{n+1} \frac{x^n}{n^2+1}$

3. $\sum_{n=0}^{\infty} \frac{x^n}{(n+1)2^n}$

4. $\sum_{n=0}^{\infty} \frac{(x-2)^n}{\sqrt{n+2}}$

5. $\sum_{n=0}^{\infty} \frac{(2x+1)^n}{3^n}$

6. $\sum_{n=0}^{\infty} \frac{(3x+1)^n}{2^n}$

7. $\sum_{n=0}^{\infty} \frac{(-1)^n x^n}{\ln(n+3)}$

8. $\sum_{n=1}^{\infty} \frac{n^2 x^n}{2^n}$

9. $\sum_{n=0}^{\infty} \frac{(2x)^n}{n!}$

10. $\sum_{n=1}^{\infty} \frac{x^{2n}}{n3^n}$

11. $\sum_{n=1}^{\infty} (-1)^n [x \sin(1/n)]^n$

12. $\sum_{n=1}^{\infty} \frac{n! x^n}{n^n}$

13. Find the radius of convergence of the series $\sum_{n=1}^{\infty} H_n x^n$ where H_n is the nth harmonic number (defined in Back to Example 10.4 in Section 10.1).

14. Prove that if a power series is absolutely convergent at one endpoint of its interval of convergence, then it is absolutely convergent at both endpoints.

15. Prove that if a power series converges at one endpoint of its interval of convergence and diverges at the other endpoint, then it is conditionally convergent at the endpoint where it converges.

11.5 The calculus of power series

You have no doubt noticed the resemblance of power series to polynomials, since the partial sums of a power series are polynomials. In your PCE you learned that polynomials are the easiest functions to differentiate and integrate. In this section we explore the calculus, i.e., differentiation and integration, of functions defined by power series.

In the next theorem we express the derivative of a function defined by a power series as a power series with the same radius of convergence, and similarly for the indefinite and definite integrals of a power series. Just as polynomials are differentiated and integrated term by term, the same is true for functions defined by power series. The proof of the theorem requires concepts and techniques not encountered in a first course in calculus, and is therefore omitted.

Theorem 11.8 (Differentiation and integration of power series) *Let $\sum_{n=1}^{\infty} c_n x^n$ be a power series with radius of convergence $R > 0$, and let f be the function given by*

$$f(x) = \sum_{n=0}^{\infty} c_n x^n = c_0 + c_1 x + c_2 x^2 + c_3 x^3 + \cdots + c_n x^n + \cdots .$$

Then f is continuous and differentiable on $(-R, R)$, and for x in $(-R, R)$

$$f'(x) = c_1 + 2c_2 x + 3c_3 x^2 + 4c_4 x^3 + \cdots + nc_n x^{n-1} + \cdots = \sum_{n=1}^{\infty} nc_n x^{n-1},$$

$$\int f(x)\, dx = C + c_0 x + c_1 \frac{x^2}{2} + c_2 \frac{x^3}{3} + \cdots + c_n \frac{x^{n+1}}{n+1} + \cdots = C + \sum_{n=0}^{\infty} c_n \frac{x^{n+1}}{n+1},$$

and

$$\int_0^x f(t)\, dt = c_0 x + c_1 \frac{x^2}{2} + c_2 \frac{x^3}{3} + \cdots + c_n \frac{x^{n+1}}{n+1} + \cdots = \sum_{n=0}^{\infty} c_n \frac{x^{n+1}}{n+1},$$

and each series has radius of convergence R.

⊘ Note 1: While the radius of convergence R remains the same when a series is differentiated or integrated, the interval of convergence may be different. See Example 11.12.

☺ Note 2: Theorem 11.8 also holds for series of the form $\sum_{n=1}^{\infty} c_n(x - x_0)^n$ with the interval replaced by $(a - R, a + R)$ and with x replaced by $(x - a)$ in the series for $f'(x)$, $\int f(x)\,dx$, and $\int_0^x f(t)\,dt$.

Example 11.12 Consider the function $f(x) = \sum_{n=1}^{\infty} \frac{x^n}{n^2} = x + \frac{x^2}{4} + \frac{x^3}{9} + \frac{x^4}{16} + \cdots$. To find the radius of convergence, we use the ratio test:

$$\left| \frac{a_{n+1}}{a_n} \right| = \left| \frac{x^{n+1}}{(n+1)^2} \cdot \frac{n^2}{x^n} \right| = \left(\frac{n}{n+1} \right)^2 |x| \to |x| < 1,$$

and hence $R = 1$. To find the interval of convergence, we test the endpoints 1 and -1. At $x = 1$ we have the convergent p-series with $p = 2$, and at $x = -1$ the series is absolutely convergent since its series of absolute values is the same as the series at $x = 1$. Thus the interval of convergence is $[-1, 1]$.

By Theorem 11.8, f is continuous and differentiable on $(-1, 1)$, and

$$f'(x) = 1 + \frac{2x}{4} + \frac{3x^2}{9} + \frac{4x^3}{16} + \cdots = 1 + \frac{x}{2} + \frac{x^2}{3} + \frac{x^3}{4} + \cdots = \sum_{n=0}^{\infty} \frac{x^n}{n+1},$$

with radius of convergence $R = 1$. At $x = 1$ the series is the divergent harmonic series while at $x = -1$ the series is the conditionally convergence alternating harmonic series. Thus the interval of convergence of f' is $[-1, 1)$. While f and f' have the same radius of convergence, the intervals of convergence are different. ∎

Back to Examples 10.10 and 11.3 In Example 10.10 we hypothesized that power series might be differentiated term by term. By Theorem 11.8 we can differentiate the geometric series $1 + x + x^2 + x^3 + \cdots = 1/(1 - x)$ term by term for $|x| < 1$ to obtain

$$1 + 2x + 3x^2 + 4x^3 + \cdots = 1/(1 - x)^2 \quad \text{for} \quad |x| < 1.$$

By Theorem 10.4 we can multiply both sides by x (where $|x| < 1$) yielding

$$x + 2x^2 + 3x^3 + 4x^4 + \cdots = x/(1 - x)^2 \quad \text{for} \quad |x| < 1.$$

Differentiating once more yields

$$1 + 4x + 9x^2 + 16x^3 + \cdots = (1 + x)/(1 - x)^3 \quad \text{for} \quad |x| < 1.$$

Once again multiply both sides by x to obtain

$$x + 4x^2 + 9x^3 + 16x^4 + \cdots + n^2 x^n + \cdots = x(1 + x)/(1 - x)^3 \quad \text{for} \quad |x| < 1. \quad (11.5)$$

In Example 11.3 we showed that the series $\sum_{n=1}^{\infty} n^2(-1/2)^n$ converges. But this series is the series in (11.5) with x replaced by $-1/2$ (which is in the interval of convergence), and hence we have

$$\sum_{n=1}^{\infty} n^2(-1/2)^n = \frac{(-1/2)(1/2)}{(3/2)^3} = -\frac{2}{27} = -0.074074074\ldots. \quad ∎$$

In the next three examples we make use of the fact that we can express the sum of a geometric series as a rational function.

Example 11.13 Recalling that the alternating harmonic series sums to $\ln 2$ (see Example 10.26), let's construct a power series in x to evaluate other natural logarithms. But we can't represent $\ln x$ as a power series in x since its domain is $(0, \infty)$, which does not contain 0 (see Theorem 11.6). However, the domain of $\ln(1 + x)$ is $(-1, \infty)$, which does contain 0. Furthermore, the derivative of $\ln(1 + x)$ is $1/(1 + x)$, which is the sum of the geometric series with first term 1 and common ratio $-x$:

$$\frac{d}{dx} \ln(1 + x) = \frac{1}{1 + x} = 1 - x + x^2 - \cdots + (-1)^n x^n + \cdots .$$

The radius of convergence is 1. Within the interval $(-1, 1)$ we can integrate term by term:

$$\ln(1 + x) = \left[\ln(1 + t) \right]_{t=0}^{x} = \int_0^x \frac{d}{dt} \ln(1 + t)\, dt = \int_0^x \frac{1}{1 + t}\, dt$$

$$= \int_0^x (1 - t + t^2 - t^3 + \cdots)\, dt = t - \frac{t^2}{2} + \frac{t^3}{3} - \frac{t^4}{4} + \cdots \Big|_{t=0}^{x}$$

$$= x - \frac{x^2}{2} + \frac{x^3}{3} - \frac{x^4}{4} + \cdots = \sum_{n=1}^{\infty} (-1)^{n+1} \frac{x^n}{n}$$

with a radius of convergence 1. At $x = -1$ we have the negative of the harmonic series, which diverges. At $x = 1$ we have the alternating harmonic series, which converges to $\ln 2$ as shown in Example 10.26, the value of $\ln(1 + x)$ at $x = 1$. The result is an important and useful series so we put it in a box, as we will for other important series we derive in this section:

$$\ln(1 + x) = x - \frac{x^2}{2} + \frac{x^3}{3} - \frac{x^4}{4} + \cdots = \sum_{n=1}^{\infty} (-1)^{n+1} \frac{x^n}{n}, \text{ for } x \text{ in } (-1, 1]. \qquad \blacksquare \qquad (11.6)$$

The series for $\ln(1 + x)$ in (11.6) is sometimes called *Mercator's series* in honor of the German mathematician Nicolaus Mercator (1620–1687) who first published it in 1668.

Example 11.14 (The dilogarithm) The dilogarithm is a member of a class of functions known as *polylogarithms*, all defined in terms of integrals of a natural logarithm. The dilogarithm $\text{Li}_2(x)$ (the "Li" stands for Logarithmic integral) is

$$\text{Li}_2(x) = \int_0^x \frac{-\ln(1 - t)}{t}\, dt \quad \text{for} \quad x \text{ in } (-\infty, 1).$$

To find a series representation for $\text{Li}_2(x)$, we first find a series for the integrand by replacing x by $-x$ in (11.6) and dividing both sides by $-x$:

$$\frac{-\ln(1 - x)}{x} = 1 + \frac{x}{2} + \frac{x^2}{3} + \frac{x^3}{4} + \cdots = \sum_{n=0}^{\infty} \frac{x^n}{n + 1} \quad \text{for} \quad -1 \le x < 1.$$

(for $x = 0$ we interpret 0^0 as 1, the limit of $-\ln(1 - x)/x$ as $x \to 0$).

Integrating term by term yields

$$\text{Li}_2(x) = \int_0^x \left(1 + \frac{t}{2} + \frac{t^2}{3} + \frac{t^3}{4} + \cdots\right) dt = x + \frac{x^2}{4} + \frac{x^3}{9} + \frac{x^4}{16} + \cdots = \sum_{n=1}^{\infty} \frac{x^n}{n^2}.$$

Thus the power series for $\text{Li}_2(x)$ coincides with the function $f(x)$ we studied in Example 11.12. In particular, its interval of convergence is $[-1, 1]$, and with this series representation we can extend the domain of $\text{Li}_2(x)$ to $(-\infty, 1]$. In fact $\text{Li}_2(1)$ is the p-series with $p = 2$. ■

Example 11.15 Using (i) $\frac{d}{dx} \arctan x = 1/(1 + x^2)$, (ii) the sum of the geometric series with first term 1 and common ratio $-x^2$, and (iii) Exercise 14 in Section 10.6, you can follow the method in Example 11.13 to show (see Exercise 2) that

$$\arctan x = x - \frac{x^3}{3} + \frac{x^5}{5} - \frac{x^7}{7} + \cdots = \sum_{n=0}^{\infty} (-1)^n \frac{x^{2n+1}}{2n + 1} \text{ for } x \text{ in } [-1, 1]. \qquad (11.7)$$

The arctangent series in (11.7), often erroneously referred to as Gregory's series, was discovered by the Indian mathematician Mādhava of Sañgamāgrama (circa 1350–1425) and rediscovered by the Scottish mathematician James Gregory (1638–1675), who published it in 1668 (the same year that Mercator published his series).

With the aid of a graphing calculator or computer graphics software, it is easy to see the convergence of the series in (11.7) to $\arctan x$ for x in $[-1, 1]$ and its divergence outside that interval. In Figure 11.3 we have the graph of $y = \arctan x$ and the partial sums (polynomials) of degrees 1, 3, 5, and 7. ■

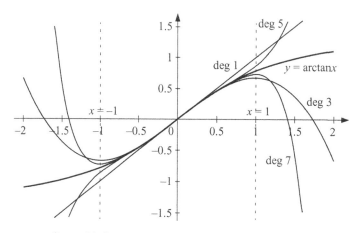

Figure 11.3. $y = \arctan x$ and the first four partial sums

Example 11.16 (Approximating π) Since $\arctan x = \pi/4$, (11.7) can be used to approximate $\pi/4$, and consequently to approximate π. However, the series for $\arctan 1$ converges much too slowly to be practical. With the aid of *Hutton's formula*

$$\frac{\pi}{4} = 2 \arctan \frac{1}{3} + \arctan \frac{1}{7},$$

we can use (11.7) to approximate π to as many decimal places as we wish rather easily. In Figure 11.4 we see a visual proof of Hutton's formula. The acute angles in the lower left corner are arctan(1/3) for the two light gray right triangles, arctan(1/7) for the dark gray right triangle, and they sum to $\pi/4$.

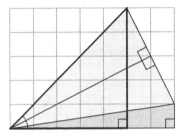

Figure 11.4. A visual proof of Hutton's formula

Using (11.7) with $x = 1/3$ and $x = 1/7$ in Hutton's formula yields the following alternating series for π, where we have used Theorem 10.4 to combine the series for arctan(1/3) and arctan(1/7) into a single series:

$$\pi = 4 \sum_{n=0}^{\infty} \frac{(-1)^n}{2n+1} \left[2 \left(\frac{1}{3} \right)^{2n+1} + \left(\frac{1}{7} \right)^{2n+1} \right].$$

Let's use this series to approximate π correct to eight decimal places. Since it's an alternating series, π lies between successive partial sums (which we denote by π_n). Computing the first few partial sums yields

$$\pi_0 = 3.328095238\ldots$$
$$\pi_1 = 3.135442536\ldots$$
$$\pi_2 = 3.142074498\ldots$$

$$\vdots$$

$$\pi_7 = 3.141592650\ldots$$
$$\pi_8 = 3.141592653\ldots$$

and we quit here since every number in the interval $(3.141592650, 3.141592653\ldots)$ has the same first eight decimals. Thus to eight decimals $\pi \approx 3.14159265$. ■

Back to Example 11.4 In the previous section we learned that the domain of the function

$$f(x) = \sum_{n=0}^{\infty} \frac{x^n}{n!} = 1 + x + \frac{x^2}{2!} + \frac{x^3}{3!} + \cdots + \frac{x^n}{n!} + \cdots$$

is $(-\infty, \infty)$. Since the derivative $f'(x)$ has the same radius of convergence, it too has domain $(-\infty, \infty)$ and is given by

$$f'(x) = \frac{d}{dx}\left(1 + x + \frac{x^2}{2!} + \frac{x^3}{3!} + \cdots + \frac{x^n}{n!} + \cdots\right)$$

$$= 0 + 1 + \frac{2x}{2!} + \frac{3x^2}{3!} + \cdots + \frac{nx^{n-1}}{n!} + \cdots$$

$$= 1 + x + \frac{x^2}{2!} + \cdots + \frac{x^{n-1}}{(n-1)!} + \cdots = f(x).$$

So if we let $y = f(x)$, then y satisfies the differential equation $dy/dx = y$ which is easily solved by separation of variables and the solution is $f(x) = Ce^x$ for some constant C. But since $1 = f(0) = C$, the function is $f(x) = e^x$:

$$\boxed{e^x = 1 + x + \frac{x^2}{2!} + \cdots + \frac{x^n}{n!} + \cdots = \sum_{n=0}^{\infty} \frac{x^n}{n!} \quad \text{for all } x.} \quad \blacksquare \qquad (11.8)$$

Back to Example 10.28 In this example we showed that the sum of the series $\frac{1}{2!} - \frac{1}{3!} + \frac{1}{4!} - \cdots$ was approximately .367879. But since this series can be written as

$$\frac{1}{2!} - \frac{1}{3!} + \frac{1}{4!} - \cdots = 1 - \frac{1}{1!} + \frac{1}{2!} - \frac{1}{3!} + \frac{1}{4!} - \cdots = \sum_{n=0}^{\infty} \frac{(-1)^n}{n!} = e^{-1},$$

the exact sum is $1/e = 0.367879441171\ldots$. $\blacksquare$

Example 11.17 (The standard normal probability distribution function.) In Example 9.32 we derived the normal density ϕ given by $\phi(z) = \frac{1}{\sqrt{2\pi}} e^{-z^2/2}$ for all z. If a measurement Z has this density, then the probability that $Z \leq z$ is given by the area $\Phi(z)$ under the graph of ϕ to the left of z, as illustrated in Figure 11.5.

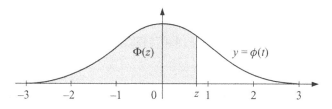

Figure 11.5. $\Pr(Z \leq z) = \Phi(z)$

Writing "Pr" for "probability," we have

$$\Pr(Z \leq z) = \Phi(z) = \int_{-\infty}^{z} \phi(t)\, dt = \frac{1}{\sqrt{2\pi}} \int_{-\infty}^{z} e^{-t^2/2}\, dt.$$

We can't use the fundamental theorem to evaluate the integral since the integrand has no closed form antiderivative, and the accuracy of the numerical integration techniques from Chapter 7 depend on the accuracy with which we evaluate the integrand. But we can expand the integrand in an infinite series and integrate term by term. To do so we use (11.8) to construct a power series for $\phi(z)$. Replacing x by $-t^2/2$ yields

$$e^{-t^2/2} = \sum_{n=0}^{\infty} (-1)^n \frac{t^{2n}}{2^n n!}$$

for all t, and we have

$$\Phi(z) - \Phi(0) = \frac{1}{\sqrt{2\pi}} \int_0^z e^{-t^2/2}\, dt$$

$$= \frac{1}{\sqrt{2\pi}} \sum_{n=0}^{\infty} \frac{(-1)^n}{2^n n!} \int_0^z t^{2n}\, dt$$

$$= \frac{1}{\sqrt{2\pi}} \sum_{n=0}^{\infty} \frac{(-1)^n z^{2n+1}}{2^n (2n+1)n!}.$$

The symmetry of ϕ implies that $\Phi(0) = 1/2$, and hence

$$\Phi(z) = \frac{1}{2} + \frac{1}{\sqrt{2\pi}} \sum_{n=0}^{\infty} \frac{(-1)^n z^{2n+1}}{2^n (2n+1)n!}$$

$$= \frac{1}{2} + \frac{1}{\sqrt{2\pi}} \left(z - \frac{z^3}{6} + \frac{z^5}{40} - \frac{z^7}{336} + \frac{z^9}{3456} - \cdots \right).$$

The function $\Phi(z)$ is known as the *standard normal probability distribution function*, and its power series converges for all z since the series for $\phi(z)$ converges for all z.

To illustrate the use of this series, let's estimate $\Phi(1/2)$. For most applications it suffices to know $\Phi(z)$ correct to four decimal places. Since $\Phi(z)$ is an alternating series for any z, the value of $\Phi(z)$ lies between successive partial sums. To compute $\Phi(1/2)$ we have partial sums

$$\Phi_n\left(\frac{1}{2}\right) = \frac{1}{2} + \frac{1}{\sqrt{2\pi}} \sum_{k=0}^{n} \frac{(-1)^k}{2^{3k+1}(2k+1)k!},$$

which yields the following sequence of approximations to $\Phi(1/2)$:

$$\Phi_0(1/2) = 0.699471\cdots$$

$$\Phi_1(1/2) = 0.691159\cdots$$

$$\Phi_2(1/2) = 0.691471\cdots$$

$$\Phi_3(1/2) = 0.691462\cdots$$

and we quit here since every number in the interval $(0.691462, 0.691472)$ rounds to 0.6915. Thus rounded to four decimal places, $\Phi(1/2) = 0.6915$. ∎

Calculation before computers Before the advent of electronic computers around the middle of the 20th century, many mathematical calculations were performed by hand either with paper and pencil or mechanical "adding machines" such as the one pictured below. These machines could add, subtract, multiply, and divide numbers, so they could evaluate polynomial functions. Hence the partial sums of a power series, since they are polynomials, were well suited for calculating values of a sophisticated function such as the normal probability distribution function. That is still true today, though calculators and computers have replaced the adding machines.

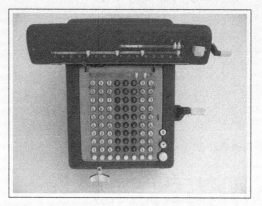

Nisa Model K adding machine, circa 1950.

11.5.1 Exercises

1. In Example 10.26 you learned that the alternating harmonic series converges to $\ln 2$. Now show that the series $\sum_{k=1}^{\infty} 1/(k2^k)$ also converges to $\ln 2$.

2. Derive the series in (11.7) for the arctangent function, and find its interval of convergence.

3. Find a power series representation and its interval of convergence for $\dfrac{x}{(1+x^2)^2}$.

4. Evaluate $\sum_{n=0}^{\infty} (n+1)^2/n!$. (Hint: $(n+1)^2 = n(n-1) + 3n + 1$.)

5. An *arithmetic-geometric series* is a power series whose coefficients are the terms of an arithmetic progression:

$$\sum_{n=0}^{\infty} (a+nb)x^n = a + (a+b)x + (a+2b)x^2 + (a+3b)x^3 + \cdots .$$

   Find (for nonzero a and b) its interval of convergence and sum function.

6. Show that $\sum_{n=0}^{\infty} \dfrac{1}{(n+2)n!} = 1$. (Hint: evaluate $\int_0^1 xe^x \, dx$ in two ways.)

7. Recall that $\sinh x = \frac{e^x - e^{-x}}{2}$ and $\cosh x = \frac{e^x + e^{-x}}{2}$. Use these definitions along with (11.8) to show that

$$\sinh x = x + \frac{x^3}{3!} + \frac{x^5}{5!} + \cdots + \frac{x^{2n+1}}{(2n+1)!} + \cdots = \sum_{n=0}^{\infty} \frac{x^{2n+1}}{(2n+1)!} \quad \text{for all } x,$$

$$\cosh x = 1 + \frac{x^2}{2!} + \frac{x^4}{4!} + \cdots + \frac{x^{2n}}{(2n)!} + \cdots = \sum_{n=0}^{\infty} \frac{x^{2n}}{(2n)!} \quad \text{for all } x.$$

8. Find a power series for $\tanh^{-1} x$ and its interval of convergence. (Hint: see Table 6.2.)

9. (a) Find a power series for $\ln\left(\frac{1+x}{1-x}\right)$ and its interval of convergence. (Hint: $\ln\left(\frac{1+x}{1-x}\right) = \ln(1+x) - \ln(1-x)$.)

   (b) Use the series in (a) to find a series that sums to $\ln 3$.

In Exercises 10 and 11, show that the power series is a solution to the DE.

10. $y = \sum_{n=2}^{\infty} \frac{x^n}{n!}, \quad \frac{dy}{dx} = x + y$

11. $y = \sum_{n=0}^{\infty} \frac{x^{2n}}{2^n n!}, \quad \frac{dy}{dx} = xy$

12. Show that $\sum_{n=1}^{\infty} n^2/2^n = 6$ and $\sum_{n=1}^{\infty} n^3/2^n = 26$. (Hint: see (11.5).)

13. Show that $\sum_{n=1}^{\infty} 2^n n^2/n! = 6e^2$. (Hint: find a series for $x^2 e^x + xe^x$.)

14. Evaluate $\int_0^{1/2} \arctan\left(x^2\right) dx$ correct to six decimal places.

15. In Exercise 13 of Section 10.6.1, we encountered *Catalan's constant G*, defined by the infinite series

$$G = \sum_{n=0}^{\infty} \frac{(-1)^n}{(2n+1)^2}.$$

   (a) Show that an integral expression for G is $G = \int_0^1 \frac{\arctan x}{x} dx$.

   (b) Show that other integral expressions for G are

$$G = -\int_0^1 \frac{\ln x}{1+x^2} dx = \frac{1}{2} \int_0^{\pi/2} \frac{x}{\sin x} dx = \frac{1}{2} \int_0^{\infty} \frac{x}{\cosh x} dx = -\int_0^{\pi/4} \ln \tan x \, dx.$$

   (c) Show that the area of the region bounded by the graph of the Gudermannian function $y = \mathrm{gd}x$, its asymptote $y = \pi/2$, and the y-axis (see Figure 6.6) is $2G$. (Hint: integrate with respect to y using $x = \mathrm{gd}^{-1}y = \ln(\sec y + \tan y) = \ln\left(\tan\left(\frac{y}{2} + \frac{\pi}{4}\right)\right)$.)

11.6 Taylor and Maclaurin series

In the preceding section we found power series representations for $\ln(1 + x)$ and $\arctan x$ by integrating geometric series, and the series for e^x by solving a differential equation. In this section we present a systematic method for finding power series representations for functions that are *infinitely differentiable* on an interval, that is, functions that have derivatives of all orders.

Suppose a function f is defined by a power series of the form in (11.4),

$$f(x) = c_0 + c_1(x - a) + c_2(x - a)^2 + \cdots + c_n(x - a)^n + \cdots,$$

with a positive radius of convergence R. The question we will now answer is: how are the coefficients $\{c_n\}$, the number a, and the function f related?

One relation is $f(a) = c_0$. By Theorem 11.8 f is infinitely differentiable on the interval $(a - R, a + R)$. Evaluating derivatives of the first few orders yields

$$f'(x) = c_1 + 2c_2(x - a) + 3c_3(x - a)^2 + 4c_4(x - a)^3 + \cdots + nc_n(x - a)^{n-1} + \cdots,$$

$$f''(x) = 2c_2 + 3 \cdot 2c_3(x - a) + 4 \cdot 3c_4(x - a)^2 + 5 \cdot 4c_5(x - a)^3 \cdots$$
$$+ n(n - 1)c_n(x - a)^{n-2} + \cdots,$$

$$f'''(x) = 3 \cdot 2c_3 + 4 \cdot 3 \cdot 2c_4(x - a) + 5 \cdot 4 \cdot 3c_5(x - a)^2$$
$$+ \cdots + n(n - 1)(n - 2)c_n(x - a)^{n-3} + \cdots,$$

$$f^{(4)}(x) = 4 \cdot 3 \cdot 2c_4 + 5 \cdot 4 \cdot 3 \cdot 2c_5(x - a)$$
$$+ \cdots + n(n - 1)(n - 2)(n - 3)c_n(x - a)^{n-4} + \cdots,$$

and so on, from which it follows that

$$f'(a) = c_1, \quad f''(a) = 2!c_2, \quad f'''(a) = 3!c_3, \quad f^{(4)}(a) = 4!c_4, \quad \cdots.$$

Continuing in this fashion, we have $f^{(n)}(a) = n!c_n$ for all positive integers n. Solving for the coefficient c_n yields $c_n = \frac{f^{(n)}(a)}{n!}$ (this also holds for $n = 0$ since $0! = 1$ and $f^{(0)}(a) = f(a)$). Thus we have proven

Theorem 11.9 *If f has a power series representation of the form*

$$f(x) = c_0 + c_1(x - a) + c_2(x - a)^2 + \cdots + c_n(x - a)^n + \cdots,$$

with a positive radius of convergence, then the coefficients are given by

$$c_n = \frac{f^{(n)}(a)}{n!},$$

that is,

$$f(x) = f(a) + \frac{f'(a)}{1!}(x - a) + \frac{f''(a)}{2!}(x - a)^2 + \cdots = \sum_{n=0}^{\infty} \frac{f^{(n)}(a)}{n!}(x - a)^n. \quad (11.9)$$

The series in (11.9) is called the *Taylor series of f at a* (or *about a* or *centered at a*). It is named for the English mathematician Brook Taylor (1685–1731). When $a = 0$ we have the special case known as a *Maclaurin series*

$$f(x) = f(0) + \frac{f'(0)}{1!}x + \frac{f''(0)}{2!}x^2 + \cdots = \sum_{n=0}^{\infty} \frac{f^{(n)}(0)}{n!}x^n, \quad (11.10)$$

named for the Scottish mathematician Colin Maclaurin (1698–1746).

⊘ Note 1. Read the statement of Theorem 11.9 carefully. It says that *if* f has a power series representation about a, then it is the Taylor series for f at a, i.e., the Taylor series converges to the function from which it was constructed. But functions exist that are not equal to the sum of their Taylor series—see Example 11.19 below.

Figure 11.6. Brook Taylor (left) and Colin Maclaurin (right)

☺ Note 2. If f has a power series representation about a, then it must be the Taylor series. That is, the power series representation of f about a is unique. For example, the series constructed in the preceding section for $\ln(1 + x)$, $\arctan x$, and e^x are Maclaurin series for those functions.

Example 11.18 Let's find the Maclaurin series for $f(x) = \cos x$ and its radius of convergence R. Differentiating the cosine yields

$$f'(x) = -\sin x, \quad f''(x) = -\cos x, \quad f'''(x) = \sin x, \quad f^{(4)}(x) = \cos x,$$

and the expressions for higher order derivatives repeat. Thus

$$f(0) = 1, \quad f'(0) = 0, \quad f''(0) = -1, \quad f'''(0) = 0, \quad f^{(4)}(0) = 1, \ldots$$

so that the derivatives of odd order are 0 and the derivatives of even order alternate between $+1$ and -1. So if the cosine has a power series representation centered at 0, then it is

$$1 - \frac{x^2}{2!} + \frac{x^4}{4!} - \frac{x^6}{6!} + \cdots = \sum_{n=0}^{\infty} (-1)^n \frac{x^{2n}}{(2n)!}. \tag{11.11}$$

To find R, we set $a_n = (-1)^n x^{2n}/(2n)!$ and use the ratio test:

$$\left| \frac{a_{n+1}}{a_n} \right| = \left| \frac{(-1)^{n+1} x^{2n+2}}{(2n+2)!} \cdot \frac{(2n)!}{(-1)^n x^{2n}} \right| = \frac{x^2}{(2n+2)(2n+1)} \to 0 < 1,$$

and thus $R = \infty$. We shall return to this example later in this section to show that the series in (11.11) converges to $\cos x$ for all x. ∎

Example 11.19 Let f be the function given by $f(0) = 0$ and $f(x) = e^{-1/x^2}$ for $x \neq 0$. In Figure 11.7 we see a graph of $y = f(x)$ and its horizontal asymptote.

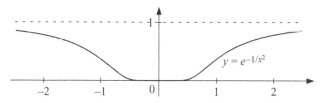

Figure 11.7. A graph of $y = e^{-1/x^2}$

Let's find the Maclaurin series for f. To find the derivatives of f at 0, we resort to the definition of the derivative and L'Hôpital's rule. For $f'(0)$ we have

$$f'(0) = \lim_{x \to 0} \frac{e^{-1/x^2} - 0}{x - 0} = \lim_{x \to 0} \frac{1/x}{e^{1/x^2}} \overset{\text{L'H}}{=} \lim_{x \to 0} \frac{-x^2}{e^{1/x^2}(-2x^3)} = \lim_{x \to 0} \frac{x}{2e^{1/x^2}} = 0.$$

Similarly it can be shown, using L'Hôpital's rule, that $f^{(n)}(0) = 0$ for all n (the graph of $y = f(x)$ is so flat near $(0, 0)$ that the derivatives of all f of all orders at 0 are equal to 0). Hence the Maclaurin series for f is $0 + 0x + 0x^2 + \cdots = 0$, which represents f at $x = 0$ but nowhere else. ■

So when is a function $f(x)$ equal to the sum of its Taylor series? More formally, we ask: if f is infinitely differentiable on some open interval centered at a, when is it true that for all x in this interval,

$$f(x) = \sum_{n=0}^{\infty} \frac{f^{(n)}(a)}{n!}(x - a)^n?$$

To answer the question we once again recall that the sum of a series is the limit of its sequence of partial sums. The nth partial sum $P_n(x)$ of a Taylor series is called the *nth degree Taylor polynomial of f at a*, and is given by

$$P_n(x) = f(a) + \frac{f'(a)}{1!}(x - a) + \frac{f''(a)}{2!}(x - a)^2 + \cdots + \frac{f^{(n)}(a)}{n!}(x - a)^n.$$

To show that $\lim P_n(x) = f(x)$, we will set $R_n(x) = f(x) - P_n(x)$ and show that $\lim R_n(x) = \lim [f(x) - P_n(x)] = 0$. The quantity $R_n(x)$ has a name—it is called the *Taylor remainder of f at a*—and we can write $f(x) = P_n(x) + R_n(x)$. There are several ways to express $R_n(x)$; a simple one (called the *Lagrange form of $R_n(x)$*, named for the French mathematician Joseph Louis Lagrange (1736–1813)) is given in the next theorem.

Theorem 11.10 (Taylor's theorem) *Let f be a function and I an interval containing the number a such that $f^{(n+1)}(x)$ exists for every x in I. Then for each x in I, there exists a number t between x and a such that $f(x) = P_n(x) + R_n(x)$ where*

$$R_n(x) = \frac{f^{(n+1)}(t)}{(n+1)!}(x - a)^{n+1}.$$

Figure 11.8. Joseph Louis Lagrange

In Taylor's theorem the remainder looks like the $(n + 1)$st term of the Taylor series for f at a, with one important difference: the $(n + 1)$st derivative of f is not evaluated at a, but at the unspecified number t between x and a.

Before proving Theorem 11.10, we consider two examples to illustrate its use.

Back to Example 11.18 To show that the series in (11.12) converges to $\cos x$ for all x, we need to evaluate $R_n(x)$ for $f(x) = \cos x$ and show that $\lim R_n(x) = 0$. Here $f^{(n+1)}(t)$ will be one of four numbers, depending on the value of n: $\sin t$, $\cos t$, $-\sin t$, or $-\cos t$. In each case $\left| f^{(n+1)}(t) \right| \leq 1$, so we can write

$$0 \leq |R_n(x)| = \frac{|f^{(n+1)}(t)|}{(n + 1)!}|x - a|^{n+1} \leq \frac{|x - a|^{n+1}}{(n + 1)!}.$$

In Theorem 11.3 we showed that $\lim x^n/n! = 0$ for any real number x. Since $|x - a|$ is a real number it follows that $\lim |x - a|^{n+1}/(n + 1)! = 0$, and hence by the squeeze theorem and Exercise 13 in Section 10.1, we have $\lim R_n(x) = 0$. Thus

$$\cos x = 1 - \frac{x^2}{2!} + \frac{x^4}{4!} - \frac{x^6}{6!} + \cdots = \sum_{n=0}^{\infty}(-1)^n \frac{x^{2n}}{(2n)!} \text{ for all } x. \qquad (11.12)$$

In Figure 11.9 we have the graph of $y = \cos x$ and the partial sums (polynomials) of degrees 2, 4, 6, and 8.

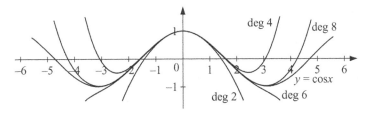

Figure 11.9. $y = \cos x$ and the first four partial sums

Since $\sin x = \int_0^x \cos t \, dt$, we replace x by t in (11.12) and integrate term by term to obtain

$$\sin x = x - \frac{x^3}{3!} + \frac{x^5}{5!} - \frac{x^7}{7!} + \cdots = \sum_{n=0}^{\infty} (-1)^n \frac{x^{2n+1}}{(2n+1)!} \text{ for all } x. \qquad (11.13)$$

In Figure 11.10 we have the graph of $y = \sin x$ and the partial sums (polynomials) of degrees 1, 3, 5, and 7 and 9.

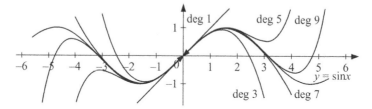

Figure 11.10. $y = \sin x$ and the first five partial sums

Example 11.20 Use the Maclaurin series for e^x to compute e correct to eight decimal places.

By virtue of Note 2 following Theorem 11.9, we know that the Maclaurin series for e^x is (11.8), so we will use $P_n(1)$, the nth degree Maclaurin polynomial for e^x evaluated at $x = 1$. But what value of n shall we use? To obtain eight decimal places of accuracy, we need the digit in the ninth place to be smaller than 5, that is, the remainder $R_n(1)$ must satisfy $|R_n(1)| < 5 \times 10^{-9}$. From Taylor's theorem 11.10, we have

$$|R_n(1)| = \frac{e^t}{(n+1)!}(1-0)^{n+1} < \frac{3}{(n+1)!} \le 5 \times 10^{-9},$$

since for $f(x) = e^x$, $f^{(n+1)}(t) = e^t$, and t is between 0 and 1. Hence n must satisfy $(n+1)! > 6 \times 10^8$, so $n = 12$ suffices. Using a calculator to sum the reciprocals of the first twelve factorials yields

$$e \approx P_{12}(1) = \sum_{n=0}^{12} 1/n! \approx 2.71828182829,$$

so rounded to eight places, we have $e \approx 2.71828183$. ∎

There are several ways to prove Taylor's Theorem 11.10, none very intuitive or simple. Our proof uses the mean value theorem from your PCE. Since you may not have thought about this theorem for a while, we state it here for reference:

The mean value theorem. *Let g be a function continuous on the interval $[a, b]$ and differentiable on the interval (a, b). Then there exists a number t in (a, b) such that*

$$g'(t) = \frac{g(b) - g(a)}{b - a}.$$

Proof of Theorem 11.10. Let x and a be two distinct points in I, and let $R_n(x) = f(x) - P_n(x)$. For any z in I let g be the function given by

$$g(z) = f(z) + \sum_{k=1}^{n} \frac{f^{(k)}(z)}{k!}(x - z)^k + R_n(x)\frac{(x-z)^{n+1}}{(x-a)^{n+1}}.$$

The existence of $f^{(n+1)}$ on I implies that g is a continuous and differentiable function of z on I, so g satisfies the hypotheses of the mean value theorem on the interval determined by x and a. Differentiating with respect to z yields

$$g'(z) = f'(z) + \sum_{k=1}^{n} \left[\frac{f^{(k+1)}(z)}{k!} (x-z)^k - \frac{f^{(k)}(z)}{(k-1)!} (x-z)^{k-1} \right] - (n+1)R_n(x) \frac{(x-z)^n}{(x-a)^{n+1}}.$$

The summation in the derivative telescopes, so that

$$g'(z) = f'(z) + \left[\frac{f^{(n+1)}(z)}{n!} (x-z)^n - f'(z) \right] - (n+1)R_n(x) \frac{(x-z)^n}{(x-a)^{n+1}}$$

$$= (x-z)^n \left[\frac{f^{(n+1)}(z)}{n!} - \frac{(n+1)R_n(x)}{(x-a)^{n+1}} \right].$$

Clearly $g(x) = f(x)$ and $g(a) = f(a) + (P_n(x) - f(a)) + R_n(x) = f(x)$, so by the mean value theorem there exists a t between x and a such that $g'(t) = (f(x) - f(x))/(x - a) = 0$. Hence

$$g'(t) = (x-t)^n \left[\frac{f^{(n+1)}(t)}{n!} - \frac{(n+1)R_n(x)}{(x-a)^{n+1}} \right] = 0.$$

But since t lies between x and a, $(x-t)^n \neq 0$ so that the term in brackets above is 0. Solving for $R_n(x)$ yields the desired conclusion.   □

Example 11.21 (The sine integral) The function known as the *sine integral* is given by

$$\mathrm{Si}(x) = \int_0^x \frac{\sin t}{t} \, dt \quad \text{for all } x,$$

and finds applications in signal processing. See Figure 11.11 for a graph of the sine integral for x in $[0, 20]$. Since the integrand has no simple antiderivative, we expand it in a power series and integrate term by term to express $\mathrm{Si}(x)$ as a series. Using (11.13) we have

$$\mathrm{Si}(x) = \int_0^x \left(1 - \frac{t^2}{3!} + \frac{t^4}{5!} - \frac{t^6}{7!} + \cdots \right) dt$$

$$= t - \frac{t^3}{3 \cdot 3!} + \frac{t^5}{5 \cdot 5!} - \frac{t^7}{7 \cdot 7!} + \cdots \Big]_{t=0}^{x}$$

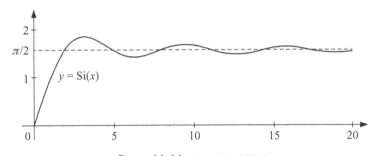

Figure 11.11. A graph of Si(x)

$$= x - \frac{x^3}{3 \cdot 3!} + \frac{x^5}{5 \cdot 5!} - \frac{x^7}{7 \cdot 7!} + \cdots$$

$$= \sum_{n=0}^{\infty} \frac{(-1)^n x^{2n+1}}{(2n+1)(2n+1)!}.$$

Let's use the series to approximate Si(1/2) rounded to eight decimal places. Since we don't have a closed form expression for the nth derivative of Si(x), we can't use Taylor's theorem. But Si(1/2) is represented by an alternating series, so the sum Si(1/2) lies between successive partial sums. If we let $\text{Si}_n(1/2)$ denote the nth partial sum, then

$$\text{Si}_0(1/2) = 0.5000000000 \cdots$$

$$\text{Si}_1(1/2) = 0.4930555555 \cdots$$

$$\text{Si}_2(1/2) = 0.4931076388 \cdots$$

$$\text{Si}_3(1/2) = 0.4931074174 \cdots$$

$$\text{Si}_4(1/2) = 0.4931074180 \cdots$$

and we stop here since every number in the interval $(0.4931074174, 0.4931074181)$ rounds to the same value. Thus $\text{Si}(1/2) \approx 0.49310742$. ∎

11.6.1 Exercises

1. Find the first three nonzero terms of the Maclaurin series for $\tan x$.

2. Use the result of the preceding exercise to find the first three nonzero terms of the Maclaurin series for $\sec^2 x$ and $\ln \sec x$.

3. Find the Maclaurin series for $\sin^2 x$ and $\cos^2 x$. (Hint: the formulas for $\cos 2x$ in terms of $\sin^2 x$ and $\cos^2 x$ might help.)

4. Find the Taylor series for $\ln x$ at $x = 1$.

5. FInd the Taylor series for $\sin x$ at $x = \pi/4$. (Hint: $\sin x = \sin[(x - \pi/4) + \pi/4]$.)

6. Observe that if you erase the exclamation marks (i.e., replace $n!$ with n) in the sine series (11.13), you obtain the arctangent series in (11.3). What functions are represented by the series obtained by erasing the exclamation marks in (a) the cosine series (11.12) and (b) the exponential function series (11.8)?

7. If we replace x by $\sqrt{x}$ in the cosine series (11.12), we obtain

$$f(x) = 1 - \frac{x}{2!} + \frac{x^2}{4!} - \cdots + (-1)^n \frac{x^n}{(2n)!} + \cdots.$$

   (a) Find the interval of convergence of the series for $f(x)$.
   (b) Clearly $f(x) = \cos \sqrt{x}$ for x positive. Is there a similar expression for $f(x)$ when x is negative?

In Exercises 8 through 10, use series to evaluate the integral rounded to five decimal places.

8. $\displaystyle\int_0^1 \sqrt{x} \sin x \, dx$ 9. $\displaystyle\int_0^1 \cos \sqrt{x} \, dx$ 10. $\displaystyle\int_0^{1/2} \ln(1 + x^3) \, dx$

11. In Exercise 11 of Section 8.3.4 we introduced *Fresnel integrals*: for t real,

$$S(t) = \int_0^t \sin(u^2)\, du \quad \text{and} \quad C(t) = \int_0^t \cos(u^2)\, du.$$

Find the Maclaurin series for $S(t)$ and $C(t)$.

11.7 Binomial series

We begin with an example that we shall return to throughout this section.

Example 11.22 (The Maclaurin series for the arcsine) To find this series, we need to find the derivatives of all orders of $f(x) = \arcsin(x)$ and evaluate them at 0. Thus $f(0) = 0$; $f'(x) = (1 - x^2)^{1/2}$, so $f'(0) = 1$; $f''(x) = x(1 - x^2)^{-3/2}$, so $f''(0) = 0$; after this point the derivatives become more difficult to generate. But the first derivative is rather simple. If we can find a series for the function $f'(x) = (1 - x^2)^{-1/2}$, we can integrate it term by term to find the series for $f(x) = \arcsin(x)$. Furthermore, we really only need a series for $(1 + x)^r$ for r real, beause then we could replace r by $-1/2$ and x by $-x^2$ to obtain a series for the derivative of the arcsine. We shall return to this example later in this section. ■

When r is a positive integer we use the *binomial theorem* to write $(1 + x)^r$ as a polynomial in x. Here's the binomial theorem applied to $(1 + x)^r$ for a positive integer r:

$$(1 + x)^r = 1 + \frac{r}{1!}x + \frac{r(r-1)}{2!}x^2 + \cdots + \frac{r(r-1)\cdots(r-k+1)}{k!}x^k + \cdots + x^r. \quad (11.14)$$

The expression in (11.14) for $(1 + x)^r$ can be written more compactly with the *binomial coefficients* $\binom{r}{k}$, defined for integers r and k with $0 \le k \le r$ as

$$\binom{r}{0} = 1 \quad \text{and} \quad \binom{r}{k} = \frac{r(r-1)(r-2)\cdots(r-k+1)}{k!} \quad \text{for } k = 1, 2, \cdots, r. \quad (11.15)$$

Then (11.14) becomes simply

$$(1 + x)^r = \sum_{n=0}^r \binom{r}{k}x^k. \quad (11.16)$$

Now consider finding the Maclaurin series for the function $f(x) = (1 + x)^r$ where r is a real number not a positive integer. Computing and evaluating derivatives yields

$$f(x) = (1 + x)^r \qquad\qquad f(0) = 1$$
$$f'(x) = r(1 + x)^{r-1} \qquad\qquad f'(0) = r$$
$$f''(x) = r(r-1)(1 + x)^{r-2} \qquad\qquad f''(0) = r(r-1)$$
$$f'''(x) = r(r-1)(r-2)(1 + x)^{r-3} \qquad\qquad f'''(0) = r(r-1)(r-2)$$

$$\vdots \qquad\qquad\qquad\qquad \vdots$$

$$f^{(k)}(x) = r(r-1)\cdots(r-k+1)(1 + x)^{r-k} \qquad f^{(k)}(0) = r(r-1)\cdots(r-k+1)$$

Thus the Maclaurin series for $f(x) = (1 + x)^r$ is

$$1 + \frac{r}{1!}x + \frac{r(r-1)}{2!}x^2 + \frac{r(r-1)(r-2)}{3!}x^3 + \cdots + \frac{r(r-1)\cdots(r-k+1)}{k!}x^k + \cdots. \quad (11.17)$$

This series looks *exactly* like (11.14) with one notable difference: it does not terminate at x^r. Consequently the series in (11.17) is called a *binomial series*, and extending the definition of the binomial coefficients to include non-integer values of r allows us to write (11.17) as

$$\sum_{k=0}^{\infty} \binom{r}{k} x^k.$$

Example 11.23 Let's find the Maclaurin series for $g(x) = 1/\sqrt{1+x} = (1+x)^{-1/2}$. Using (11.17) we have

$$1 - \frac{1}{2}x + \frac{(-1/2)(-3/2)}{2!}x^2 + \frac{(-1/2)(-3/2)(-5/2)}{3!}x^3 + \cdots,$$

which simplifies to

$$1 - \frac{1}{2}x + \frac{1 \cdot 3}{2^2 \cdot 2!}x^2 - \frac{1 \cdot 3 \cdot 5}{2^3 \cdot 3!}x^3 + \frac{1 \cdot 3 \cdot 5 \cdot 7}{2^4 \cdot 4!}x^4 - \cdots$$

$$= 1 - \frac{1}{2}x + \frac{1 \cdot 3}{2 \cdot 4}x^2 - \frac{1 \cdot 3 \cdot 5}{2 \cdot 4 \cdot 6}x^3 + \frac{1 \cdot 3 \cdot 5 \cdot 7}{2 \cdot 4 \cdot 6 \cdot 8}x^4 - \cdots,$$

the same series we saw in Example 11.5. Its radius of convergence is 1. ∎

We do not yet know that the binomial series (11.17) converges to $f(x) = (1+x)^r$; we will establish that shortly. But first let's find the radius of convergence of the series in (11.17). Setting $a_n = r(r-1)(r-2)\cdots(r-n+1)x^n/n!$ in the ratio test yields

$$\left| \frac{a_{n+1}}{a_n} \right| = \left| \frac{r(r-1)(r-2)\cdots(r-n+1)(r-n)x^{n+1}}{(n+1)!} \cdot \frac{n!}{r(r-1)(r-2)\cdots(r-n+1)x^n} \right|$$

$$= \left| \frac{r-n}{n+1} \right| |x| = \left| \frac{(r/n)-1}{1+(1/n)} \right| |x| \to |x| < 1,$$

and thus the radius of convergence of every binomial series is 1.

To show that the series (11.17) converges to $(1+x)^r$, we could attempt to show that the limit of the Taylor remainder $R_n(x)$ is zero. But this proves to be rather difficult, so we resort a different approach. We set the series in (11.17) equal to an unknown function $f(x)$, and differentiate the series term by term for values of x such that $|x| < 1$ (the interval where the series is absolutely convergent) to create a differential equation that we can solve. With

$$f(x) = 1 + \sum_{k=1}^{\infty} \frac{r(r-1)\cdots(r-k+1)}{k!}x^k$$

we have

$$f'(x) = \sum_{k=1}^{\infty} \frac{r(r-1)\cdots(r-k+1)}{(k-1)!}x^{k-1}, \tag{11.18}$$

and multiplication by x yields

$$xf'(x) = \sum_{k=1}^{\infty} \frac{r(r-1)\cdots(r-k)}{(k-1)!}x^k = \sum_{k=1}^{\infty} k\frac{r(r-1)\cdots(r-k+1)}{k!}x^k. \tag{11.19}$$

Now we rewrite (11.18) by replacing the index k by $k+1$,

$$f'(x) = \sum_{k=0}^{\infty} \frac{r(r-1)\cdots(r-k)}{k!} x^k = r + \sum_{k=1}^{\infty}(r-k)\frac{r(r-1)\cdots(r-k+1)}{k!} x^k. \quad (11.20)$$

We now add (11.19) and (11.20):

$$(1+x)f'(x) = r + \sum_{k=1}^{\infty}[(r-k)+k]\frac{r(r-1)\cdots(r-k+1)}{k!} x^k = rf(x). \quad (11.21)$$

If we set $y = f(x)$, (11.21) yields the separable differential equation $(1+x)(dy/dx) = ry$, or $(1/y)dy = r[1/(1+x)]dx$. The solution is $\ln|y| = r\ln(1+x) + C$. Hence $|y| = e^C(1+x)^r$ or equivalently $y = C(1+x)^r$. But in (11.17) we see that $f(0) = 1$ so that $C = 1$ and we have $y = f(x) = (1+x)^r$ as desired.

Hence we have proved

Theorem 11.11 *If r is a real number and $|x| < 1$, then*

$$(1+x)^r = \sum_{k=0}^{\infty} \frac{r(r-1)\cdot(r-k+1)}{k!} x^k = \sum_{k=0}^{\infty}\binom{r}{k}x^k.$$

Back to Examples 11.22 and 11.23 When $r = -1/2$ Theorem 11.11 yields

$$\frac{1}{\sqrt{1+x}} = 1 - \frac{1}{2}x + \frac{1\cdot3}{2\cdot4}x^2 - \frac{1\cdot3\cdot5}{2\cdot4\cdot6}x^3 + \frac{1\cdot3\cdot5\cdot7}{2\cdot4\cdot6\cdot8}x^4 - \cdots, \quad (11.22)$$

for $|x| < 1$. To find a series for $\arcsin x$, we will integrate the series for $1/\sqrt{1-t^2}$. Replacing x by $-t^2$ in (11.22) gives us

$$\frac{1}{\sqrt{1-t^2}} = 1 + \frac{1}{2}t^2 + \frac{1\cdot3}{2\cdot4}t^4 + \frac{1\cdot3\cdot5}{2\cdot4\cdot6}t^6 + \frac{1\cdot3\cdot5\cdot7}{2\cdot4\cdot6\cdot8}t^8 + \cdots.$$

Next we integrate each side from 0 to x for $|x| < 1$:

$$\arcsin x = \int_0^x \frac{1}{\sqrt{1-t^2}}\, dt = \int_0^x \left(1 + \frac{1}{2}t^2 + \frac{1\cdot3}{2\cdot4}t^4 + \frac{1\cdot3\cdot5}{2\cdot4\cdot6}t^6 + \frac{1\cdot3\cdot5\cdot7}{2\cdot4\cdot6\cdot8}t^8 + \cdots\right) dt$$

$$= x + \frac{1}{2}\cdot\frac{x^3}{3} + \frac{1\cdot3}{2\cdot4}\cdot\frac{x^5}{5} + \frac{1\cdot3\cdot5}{2\cdot4\cdot6}\cdot\frac{x^7}{7} + \frac{1\cdot3\cdot5\cdot7}{2\cdot4\cdot6\cdot8}\cdot\frac{x^9}{9} + \cdots$$

$$= x + \sum_{n=1}^{\infty} \frac{1\cdot3\cdots(2n-1)}{2\cdot4\cdots2n}\cdot\frac{x^{2n+1}}{2n+1}.$$

Thus the above power series represents $\arcsin x$ for x in $(-1, 1)$. Does this series converge at the endpoint $x = 1$? If it does, is the sum $\arcsin(1) = \pi/2$? We need only investigate this endpoint since the series at $x = -1$ is simply -1 times the series at $x = 1$. At $x = 1$, the series is

$$1 + \sum_{n=1}^{\infty} \frac{1\cdot3\cdots(2n-1)}{2\cdot4\cdots2n}\cdot\frac{1}{2n+1} \quad (11.23)$$

which we will now show convergent by comparison with the p-series with $p = 3/2$. In Exploration D.1 you will establish the limit

$$\lim_{n \to \infty} \left[\frac{2}{1} \cdot \frac{4}{3} \cdot \cdots \cdot \frac{2n}{2n-1} \right]^2 \frac{1}{2n+1} = \frac{\pi}{2}.$$

This limit can be written as

$$\lim \frac{1 \cdot 3 \cdot 5 \cdots (2n-1)}{2 \cdot 4 \cdot 6 \cdots (2n)} \sqrt{2n+1} = \sqrt{2/\pi}.$$

Now if we let

$$a_n = \frac{1 \cdot 3 \cdots (2n-1)}{2 \cdot 4 \cdots 2n} \cdot \frac{1}{2n+1} \quad \text{and} \quad b_n = \frac{1}{n^{3/2}},$$

then

$$\frac{a_n}{b_n} = \frac{1 \cdot 3 \cdots (2n-1)}{2 \cdot 4 \cdots 2n} \cdot \frac{n^{3/2}}{2n+1} = \frac{1 \cdot 3 \cdots (2n-1)}{2 \cdot 4 \cdots 2n} \sqrt{2n+1} \cdot \left(\frac{n}{2n+1} \right)^{3/2}.$$

Thus

$$\lim \frac{a_n}{b_n} = \sqrt{\frac{2}{\pi}} \cdot \frac{1}{2\sqrt{2}} = \frac{1}{2\sqrt{\pi}} > 0,$$

and since $\sum b_n$ converges, so does the series $\sum a_n$ in (11.23) by the limit comparison test. In fact, since $\sum a_n$ is a positive term series, it converges absolutely.

To find the sum function at $x = 1$ we use *Abel's theorem* (which we will state but not prove), established in 1826 by the Norwegian mathematician Niels Henrik Abel (1802–1829). The theorem states that if a power series converges at an endpoint of its interval of convergence, it converges to just what you hope it would converge to.

Theorem 11.12 (Abel's theorem) *Let $f(x) = \sum_{n=0}^{\infty} c_n x^n$ be a power series that converges for x in $(-R, R)$. If $\sum_{n=0}^{\infty} c_n R^n$ converges, then*

$$\sum_{n=0}^{\infty} c_n R^n = \lim_{x \to R^-} f(x),$$

and if $\sum_{n=0}^{\infty} c_n(-R)^n$ converges, then

$$\sum_{n=0}^{\infty} c_n(-R)^n = \lim_{x \to -R^+} f(x).$$

Hence the series in (11.23) sums to $\lim_{x \to 1^-} \arcsin(x) = \arcsin(1) = \pi/2$; similarly the series at $x = -1$ converges to $\arcsin(-1) = -\pi/2$, which establishes

$$\boxed{\arcsin x = x + \sum_{n=1}^{\infty} \frac{1 \cdot 3 \cdots (2n-1)}{2 \cdot 4 \cdots 2n} \cdot \frac{x^{2n+1}}{2n+1} \quad \text{for } x \text{ in } [-1, 1].}$$

Note: If we set $x = 1/2$ and recall that $\arcsin(1/2) = \pi/6$, we have a series for π:

$$\pi = 6\left(\frac{1}{2} + \frac{1}{2 \cdot 3 \cdot 2^3} + \frac{1 \cdot 3}{2 \cdot 4 \cdot 5 \cdot 2^5} + \frac{1 \cdot 3 \cdot 5}{2 \cdot 4 \cdot 6 \cdot 7 \cdot 2^7} + \cdots\right). \blacksquare$$

In Table 11.2 we list some of the most important Maclaurin series.

Table 11.2. Important Maclaurin series and their intervals of convergence

$$e^x = 1 + x + \frac{x^2}{2!} + \frac{x^3}{3!} + \cdots = \sum_{n=0}^{\infty} \frac{x^n}{n!} \text{ for all } x$$

$$\ln(1+x) = x - \frac{x^2}{2} + \frac{x^3}{3} - \frac{x^4}{4} + \cdots = \sum_{n=1}^{\infty} (-1)^{n+1}\frac{x^n}{n} \text{ for } x \text{ in } (-1, 1]$$

$$\sin x = x - \frac{x^3}{3!} + \frac{x^5}{5!} - \frac{x^7}{7!} + \cdots = \sum_{n=0}^{\infty} (-1)^n \frac{x^{2n+1}}{(2n+1)!} \text{ for all } x$$

$$\cos x = 1 - \frac{x^2}{2!} + \frac{x^4}{4!} - \frac{x^6}{6!} + \cdots = \sum_{n=0}^{\infty} (-1)^n \frac{x^{2n}}{(2n)!} \text{ for all } x$$

$$\arctan x = x - \frac{x^3}{3} + \frac{x^5}{5} - \frac{x^7}{7} + \cdots = \sum_{n=0}^{\infty} (-1)^n \frac{x^{2n+1}}{2n+1} \text{ for } x \text{ in } [-1, 1]$$

$$\arcsin x = x + \sum_{n=1}^{\infty} \frac{1 \cdot 3 \cdots (2n-1)}{2 \cdot 4 \cdots (2n)} \cdot \frac{x^{2n+1}}{2n+1} \text{ for } x \text{ in } [-1, 1]$$

$$\sinh x = x + \frac{x^3}{3!} + \frac{x^5}{5!} + \frac{x^7}{7!} + \cdots = \sum_{n=0}^{\infty} \frac{x^{2n+1}}{(2n+1)!} \text{ for all } x$$

$$\cosh x = 1 + \frac{x^2}{2!} + \frac{x^4}{4!} + \frac{x^6}{6!} + \cdots = \sum_{n=0}^{\infty} \frac{x^{2n}}{(2n)!} \text{ for all } x$$

$$(1+x)^r = \sum_{k=0}^{\infty} \frac{r(r-1)\cdots(r-k+1)}{k!}x^k = \sum_{k=0}^{\infty} \binom{r}{k}x^k \text{ for } x \text{ in } (-1, 1).$$

11.7.1 Exercises

In Exercises 1 through 6, use binomial series to find a Maclaurin series for the function.

1. $f(x) = \sqrt{1-x}$ 2. $f(x) = (1+x)^{2/3}$ 3. $f(x) = \sqrt[3]{1+x^3}$

4. $f(x) = x/\sqrt{1+x^2}$ 5. $f(x) = 1/\sqrt{4+x^2}$ 6. $f(x) = (1-4x)^{-1/2}$

7. Find the Maclaurin series for $\sinh^{-1} x$ and its interval of convergence. (Hint: see (6.6) and (11.22).)

In Exercises 8 through 10, use series to evaluate the integral rounded to five decimal places.

8. $\displaystyle\int_0^{1/2} (1+x^2)^{2/3}\, dx$ 9. $\displaystyle\int_0^{1/3} \sqrt[3]{1+x^3}\, dx$ 10. $\displaystyle\int_0^{1/2} \sinh^{-1}(x)\, dx$

In (11.5), we found the power series

$$\frac{x(1+x)}{(1-x)^3} = x + 4x^2 + 9x^3 + 16x^4 + \cdots = \sum_{n=1}^{\infty} n^2 x^n.$$

We say that the function $g(x) = x(1+x)/(1-x)^3$ *generates* the sequence $\{0, 1, 4, 9, 16, \ldots\}$ of square numbers, since the coefficient of x^n in its Maclaurin series is n^2. Similarly, the function $\operatorname{Li}_2(x)$ in Example 11.14 generates the sequence $\{1, 1/4, 1/9, 1/16, \ldots\}$ of reciprocals of the nonzero squares. *Generating functions* (functions that generate particular sequences) are used in courses in discrete mathematics, combinatorics, and probability. In Exercises 11 through 14 find generating functions $g(x)$ for the given sequence (assume that each series converges in some open interval centered at 0).

11. The odd numbers, i.e., $g(x) = x + 3x^2 + 5x^3 + 7x^4 + \cdots + (2n-1)x^n + \cdots$ (Hint: see Back to Examples 10.10 and 11.3 in Section 11.5.)

12. The triangular numbers, i.e., $g(x) = x + 3x^2 + 6x^3 + 10x^4 + \cdots + t_n x^n + \cdots$. (Hint: see Exercise 12 in Section 10.2.3.)

13. The Fibonacci numbers, i.e., $g(x) = x + x^2 + 2x^3 + 3x^4 + 5x^5 + 8x^6 + \cdots + F_n x^n + \cdots$, where $F_1 = F_2 = 1$ and $F_n = F_{n-1} + F_{n-2}$ for $n \geq 3$. (Hint: show that we can define $F_0 = 0$ and then show that $x = g(x) - xg(x) - x^2 g(x)$.) This generating function can be used to derive *Binet's formula* for the Fibonacci numbers: $F_n = \left[\varphi^n - (-\varphi)^{-n}\right]/\sqrt{5}$, where φ is the golden ratio. See Exploration 11.8.3.

14. The harmonic numbers, i.e., $g(x) = \sum_{n=1}^{\infty} H_n x^n$ where $H_n = 1 + \frac{1}{2} + \frac{1}{3} + \cdots + \frac{1}{n}$. (Hint: rearrange the series so that all the powers of x with coefficient 1 come first, then all the powers of x with coefficient $1/2$, etc.).

15. The binomial coefficients $\binom{2n}{n}$ for $n \geq 0$ are called the *central binomial coefficients*. Show that the function $1/\sqrt{1-4x}$ generates the sequence of central binomial coefficients, i.e.,

$$\frac{1}{\sqrt{1-4x}} = \sum_{n=0}^{\infty} \binom{2n}{n} x^n.$$

(Hint: see (11.15) and Exercise 6 above.)

16. Show that for x in $[-1, 1]$,

$$\arcsin x = \sum_{n=0}^{\infty} 4^{-n} \binom{2n}{n} \frac{x^{2n+1}}{2n+1}.$$

(Hint: see (11.15).)

11.8 Explorations

11.8.1 The sophomore's dream

The remarkable formula

$$\int_0^1 x^{-x} \, dx = \sum_{n=1}^{\infty} n^{-n} \tag{11.24}$$

was discovered by the Swiss mathematician Johann Bernoulli (1667–1748) in about 1697. Many sites on the world wide web refer to the formula in (11.24) as the sophomore's dream, observing that it seems almost too good to be true. But it is true, as you shall now show in this exploration. See Figure 11.12 for area interpretations of both sides of (11.24).

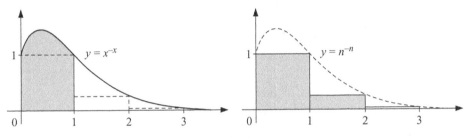

Figure 11.12. $\int_0^1 x^{-x} \, dx = \sum_{n=1}^{\infty} n^{-n}$

Exercise 1. Show that both the integral and the series in (11.24) converge.

Exercise 2. Show that $\int_0^1 x^{-x} \, dx = \int_0^1 \left(\sum_{n=0}^{\infty} \frac{(-x \ln x)^n}{n!} \right) \, dx$. (Hint: $x = e^{\ln x}$.)

Exercise 3. Show that $\int_0^1 \left(\sum_{n=0}^{\infty} \frac{(-x \ln x)^n}{n!} \right) \, dx = \sum_{n=0}^{\infty} \left(\frac{1}{n!} \int_0^1 (-x \ln x)^n \, dx \right)$.

Now let $I_{n,k} = \frac{1}{k!} \int_0^1 x^n \, (- \ln x)^k \, dx$ where n and k are integers such that $n \geq k \geq 0$. The integral expression in the sum on the right of the equals sign in Exercise 3 is $I_{n,n}$. Then

Exercise 4. Show that $I_{n,k}$ converges.

Exercise 5. Show that $I_{n,k} = \frac{1}{n+1} I_{n,k-1}$. (Hint: integrate by parts.)

Exercise 6. Show that $I_{n,0} = \frac{1}{n+1}$.

Exercise 7. Show that $I_{n,n} = (n + 1)^{-(n+1)}$.

Exercise 8. Put the results of Exercises 2, 3, and 7 together to conclude that $\int_0^1 x^{-x} \, dx = \sum_{n=1}^{\infty} n^{-n}$, which concludes the proof.

11.8.2 The Basel problem: the exact value of $\zeta(2)$.

One of the most celebrated accomplishments of the remarkable mathematician Leonard Euler (1707–1783) was an evaluation of the sum of the series of the reciprocals of the squares of the positive integers (i.e., $\zeta(2)$, the p series with $p = 2$). This problem is known as the *Basel problem*, named for Basel, Switzerland, Euler's hometown. In Section 10.4 we were able to show that the sum of this series lies in the interval $(1.64491, 1.64495)$, and now you will show that its exact value is $\pi^2/6$.

In this exploration you will use a variety of results from this course to establish Euler's result

$$\sum_{n=1}^{\infty} \frac{1}{n^2} = \frac{\pi^2}{6} \approx 1.64493. \tag{11.25}$$

Exercise 1. Show that $\frac{3}{4} \sum_{n=1}^{\infty} \frac{1}{n^2} = \sum_{n=0}^{\infty} \frac{1}{(2n+1)^2}$, and hence to establish 11.25, all we need to do is show that

$$\sum_{n=0}^{\infty} \frac{1}{(2n+1)^2} = \frac{\pi^2}{8} \tag{11.26}$$

and multiply each side by $4/3$. (Hint: $3/4 = 1 - 1/4$.)

Exercise 2. Show that

$$\int_0^1 \frac{\arcsin x}{\sqrt{1-x^2}}\, dx = \frac{\pi^2}{8}. \tag{11.27}$$

(Hint: although this integral is improper, a substitution will make it proper!)

In the remaining steps we will show that the integral in (11.27) can be expressed as the series in (11.26). We begin with the Maclaurin series for the arcsine that we derived after Theorem 11.11: for x in $[-1, 1]$,

$$\arcsin x = x + \sum_{n=1}^{\infty} \frac{1 \cdot 3 \cdots (2n-1)}{2 \cdot 4 \cdots 2n} \cdot \frac{x^{2n+1}}{2n+1},$$

so that

$$\int_0^1 \frac{\arcsin x}{\sqrt{1-x^2}}\, dx = \int_0^1 \frac{x + \sum_{n=1}^{\infty} \left(\frac{1 \cdot 3 \cdots (2n-1)}{2 \cdot 4 \cdots 2n} \cdot \frac{x^{2n+1}}{2n+1} \right)}{\sqrt{1-x^2}}\, dx. \tag{11.28}$$

Exercise 3. Use Theorem 11.8 to evaluate the integral in (11.28) term by term:

$$\int_0^1 \frac{\arcsin x}{\sqrt{1-x^2}}\, dx = \int_0^1 \frac{x}{\sqrt{1-x^2}}\, dx + \sum_{n=1}^{\infty} \frac{1 \cdot 3 \cdots (2n-1)}{2 \cdot 4 \cdots 2n(2n+1)} \int_0^1 \frac{x^{2n+1}}{\sqrt{1-x^2}}\, dx. \tag{11.29}$$

Exercise 4. Although both integrals on the right of the equals sign in (11.29) are improper, use a substitution to make them proper:

$$\int_0^1 \frac{\arcsin x}{\sqrt{1-x^2}}\, dx = \int_0^{\pi/2} \sin\theta\, d\theta + \sum_{n=1}^{\infty} \frac{1 \cdot 3 \cdots (2n-1)}{2 \cdot 4 \cdots 2n(2n+1)} \int_0^{\pi/2} \sin^{2n+1}\theta\, d\theta. \tag{11.30}$$

Exercise 5. From Exercise 35 in Section 1.2.1,

$$\int_0^{\pi/2} \sin\theta\, d\theta = 1 \quad \text{and} \quad \int_0^{\pi/2} \sin^{2n+1}\theta\, d\theta = \frac{2n}{2n+1}\cdot\frac{2n-2}{2n-1}\cdots\frac{2}{3}. \tag{11.31}$$

Combine this with (11.30) to yield

$$\int_0^1 \frac{\arcsin x}{\sqrt{1-x^2}}\, dx = 1 + \sum_{n=1}^\infty \frac{1}{(2n+1)^2} = \sum_{n=0}^\infty \frac{1}{(2n+1)^2},$$

which with (11.27) completes the proof.

11.8.3 Binet's formula for the Fibonacci numbers

In this exploration you will use the generating function derived in Exercise 13 in Section 11.7.1, along with partial fractions, geometric series, and some facts about the golden ratio φ to derive Binet's formula for the nth Fibonacci number: $F_n = \left[\varphi^n - (-\varphi)^{-n}\right]/\sqrt{5}$.

We begin with some useful facts about $\varphi = (1+\sqrt{5})/2$.

Exercise 1. Show that $-\varphi$ and $1/\varphi$ are the two roots of $x^2 + x - 1 = 0$ and consequently $\varphi - (1/\varphi) = 1$ and $\varphi + (1/\varphi) = \sqrt{5}$.

Exercise 2. In Exercise 13 in Section 11.7.1 you found the generating function for the Fibonacci numbers: $g(x) = x/(1 - x - x^2)$. Now use partial fractions to show that

$$g(x) = \frac{-x}{x^2 + x - 1} = \frac{-x}{(x - 1/\varphi)(x + \varphi)} = \frac{-1}{\sqrt{5}}\left[\frac{1/\varphi}{x - 1/\varphi} + \frac{\varphi}{x + \varphi}\right].$$

Exercise 3. Show that

$$g(x) = \frac{1}{\sqrt{5}}\left[\frac{1}{1 - \varphi x} - \frac{1}{1 + x/\varphi}\right].$$

Exercise 4. Interpret the two fractions in the last expression for $g(x)$ as sums of geometric series to show that

$$g(x) = \frac{1}{\sqrt{5}}\left[\sum_{n=0}^\infty (\varphi x)^n - \sum_{n=0}^\infty (-x/\varphi)^n\right] = \sum_{n=1}^\infty \frac{1}{\sqrt{5}}\left[\varphi^n - (-\varphi)^{-n}\right]x^n.$$

Exercise 5. Recall that $g(x) = \sum_{n=1}^\infty F_n x^n$ to conclude $F_n = \left[\varphi^n - (-\varphi)^{-n}\right]/\sqrt{5}$.

11.8.4 The perimeter of an ellipse

In Exercise 10 of Section 8.2.4 you set up an integral expression for the perimeter of the ellipse given parametrically by $x = a\cos t$, $y = b\sin t$, $0 \le t \le 2\pi$. In this exploration you will use a binomial series to convert that integral into an infinite series representing the perimeter. In this Exploration, we assume that $0 < b < a$; the case $0 < a < b$ is similar. See Figure 11.13.

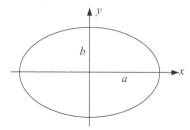

Figure 11.13. The ellipse $x = a \cos t, y = b \sin t, 0 \le t \le 2\pi$

Exercise 1. If you haven't done Exercise 10 in Section 8.2.4, do it now!

The integral you obtained in Exercise 1 (using the symmetry of the ellipse) for the perimeter P is equivalent to

$$P = \int_0^{\pi/2} \sqrt{a^2 \sin^2 t + b^2 \cos^2 t} \, dt.$$

Exercise 2. Show that the perimeter P can also be written as

$$P = 4a \int_0^{\pi/2} \sqrt{1 - \epsilon^2 \sin^2 \theta} \, d\theta$$

where $\epsilon = \sqrt{a^2 - b^2}/a$. (Hint: use $\sin^2 t = 1 - \cos^2 t$ and then the substitution $t = \pi/2 - \theta$.) The quantity ϵ, which satisfies $0 < \epsilon < 1$, is known as the *eccentricity* of the ellipse, as it is close to 0 for nearly circular ellipses and close to 1 for very elongated ellipses. The integral in P is the *elliptic integral* promised in Exercise 10 in Section 8.2.4.

Exercise 3. Show that the integrand for P can be written as

$$\sqrt{1 - \epsilon^2 \sin^2 \theta} = 1 - \frac{\epsilon^2 \sin^2 \theta}{2} - \sum_{k=2}^{\infty} \frac{1 \cdot 3 \cdot 5 \cdots (2n-3) \epsilon^{2k} \sin^{2k} \theta}{2^k k!}$$

and that the series converges for all θ. (Hint: see Theorem 11.11, and note that $0 \le \epsilon^2 \sin^2 \theta < 1$.)

Exercise 4. Use the results of Exercises 2 and 3 to show that

$$P = 2\pi a \left\{ 1 - \sum_{k=1}^{\infty} \left[\left(\frac{1 \cdot 3 \cdot 5 \cdots (2k-1)}{2 \cdot 4 \cdot 6 \cdots 2k} \right)^2 \cdot \frac{\epsilon^{2k}}{2k-1} \right] \right\},$$

i.e.,

$$P = 2\pi a \left[1 - \left(\frac{1}{2} \right)^2 \frac{\epsilon^2}{1} - \left(\frac{1 \cdot 3}{2 \cdot 4} \right)^2 \frac{\epsilon^4}{3} - \left(\frac{1 \cdot 3 \cdot 5}{2 \cdot 4 \cdot 6} \right)^2 \frac{\epsilon^6}{5} - \cdots \right].$$

(Hint: see Exercise 35 in Section 1.2.1.)

Exercise 5. The dwarf planet Pluto has an elliptical orbit with eccentricity $\epsilon \approx 0.2482$, and $a \approx 39.44$ AU (one AU, for *astronomical unit*, is approximately 93 million miles, the average distance between the centers of the Earth and the sun). In Figure 11.14 we see a drawing of the orbits of the planets and Pluto.

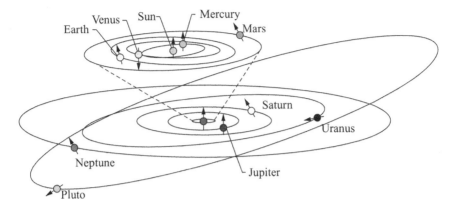

Figure 11.14. View of our solar system looking from the edge

Use the first four terms of the series in Exercise 4 to show that the perimeter of Pluto's orbit is approximately 244 AU.

11.9 Acknowledgments

1. The proof in Exercise 17 in Section 11.2.1 is adapted from P. N. Bajaj, Math Bite: Equality of limits in ratio and root test, *Mathematics Magazine*, **71** (1998), p. 299.
2. The proof in Exploration 11.8.2 is adapted from G. Kimble, Euler's other proof, *Mathematics Magazine*, **60** (1987), p. 282.

A Description of the AP Calculus AB Course

To be sure, many AP programs are first rate. Calculus, especially in the hands of a gifted teacher, is widely considered to be one of the best-thought-out AP programs. Two years ago, when the Center for Education at the National Academy of Sciences conducted one of the few serious studies of the AP curriculum ever done, it praised the AP Calculus program for achieving "an ap-propriate balance between breadth and depth."!

TIME Magazine (Nov. 1, 2004)

A description of the Advanced Placement Program Calculus AB course is given below. This description is intended to indicate the scope of the course, but not necessarily the order in which the topics are taught. Although the exam is based on the topics listed here, many teachers enrich their courses with additional topics. For the official topic outline provided by the College Board, visit media.collegeboard.com/digitalServices/pdf/ap/ap-calculus-course-description.pdf.

Many high school Calculus AB course syllabi consist of the following three units: Unit I: Functions and Limits; Unit II: Derivatives and their applications; and Unit III: Integrals and their applications.

I. Functions and Limits

- Students study functions from four points of view: Symbolic, graphical, numerical, and verbal. Limits of functions are presented intuitively, and students learn to calculate limits of functions using algebra and to estimate limits using graphs or from tables of data.
- One-sided limits and limits at infinity are used to study the asymptotic behavior of functions (for both horizontal and vertical asymptotes).
- Students study continuity from both an intuitive standpoint and in terms of limits. Important graphical properties of continuous functions are examined, such as those expressed in the intermediate value theorem and the extreme value theorem.

II. Derivatives and their applications

- The concept of the derivative is studied graphically, numerically, and analytically via the limit of a difference quotient.
- The connection between differentiable functions and continuous functions (differentiable functions are continuous) is presented both graphically and analytically.
- The rate of change of a function is studied from several viewpoints: approximate rates of change from graphs and tables of data; instantaneous rate of change as the limit of average rate of change; and derivatives as instantaneous rates of change.
- Students learn that the slope of a curve at a point is the slope of the line tangent to the curve at that point, and they learn about local linear approximations to functions.
- The relationship between the graph of a function and the graph of its derivative is studied, as well as the corresponding relationship between the increasing and decreasing behavior of a function and the sign of its derivative.
- Students learn how to differentiate basic functions, including power, exponential, logarithmic, trigonometric, and inverse trigonometric functions. They learn the rules for differentiating sums, differences, products, and quotients of functions, as well as the chain rule for differentiating composite functions and the implicit differentiation technique.
- Students learn about the mean value theorem and its geometric interpretation.
- The second derivative of a function is studied, including the relationship of its graph to the graphs of the function and the first derivative. Students learn to use the second derivative to study the concavity of a function, and to find inflection points of its graph.
- Common applications of derivatives in the course include the following: analysis of graphs of functions; optimization problems (absolute and relative extrema); related rates problems; rates of change in applied contexts, including velocity, acceleration, and speed.

III. Integrals and their applications

- The definite integral is defined as the limit of Riemann sums. Students learn basic properties of definite integrals such as linearity and additivity.
- A definite integral of the rate of change of a quantity over an interval is interpreted as the change of the quantity over the interval.
- Students use definite integrals in a variety of applications to model physical, biological, or economic situations. Common applications include finding the area of a region, the volume of a solid with known cross sections, the average value of a function, the distance traveled by a particle along a line, and accumulated change from a rate of change. The common emphasis in each application is on setting up an approximating Riemann sum and representing its limit as a definite integral.
- The fundamental theorem of calculus is undoubtedly the most import theorem encountered in the course. Students learn about its use to represent a particular antiderivative, and study the analytical and graphical analysis of functions so defined. Students also learn to use the fundamental theorem to evaluate definite integrals.
- Students learn two methods for finding antiderivatives: directly from known derivative formulas, and by the substitution technique.
- Students learn to translate appropriate word problems into differential equations. Students study differential equations via slope fields and the relationship between slope fields and solution curves for differential equations. Students learn how to solve separable

differential equations and how to use them in modeling (including the equation $y' = ky$ and exponential growth and decay).

- Numerical methods for approximating the value of a definite integral are studied, including the use of Riemann sums (with left, right, and midpoint sample points) and trapezoidal sums. These methods are used to approximate definite integrals of functions represented symbolically, graphically, and numerically.

B

Useful Formulas from Geometry and Trigonometry

B.1 Geometric formulas

(All angles are measured in radians)

Circle

Area $= \pi r^2$

Circumference $= 2\pi r$

Circular sector

Area $= \frac{1}{2}r^2\theta$

$L = r\theta$

Ellipse

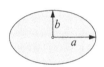

Area $= \pi ab$

Trapezoid

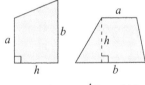

Area $= \frac{1}{2}(a+b)h$

Equilateral triangle

Altitude $h = s\frac{\sqrt{3}}{2}$

Area $= \frac{1}{2}sh = \frac{1}{4}s^2\sqrt{3}$

Sphere	Right circular cylinder	Right circular cone
		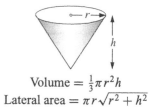

$$\text{Volume} = \tfrac{4}{3}\pi r^3 \qquad\qquad \text{Volume} = \pi r^2 h \qquad\qquad \text{Volume} = \tfrac{1}{3}\pi r^2 h$$
$$\text{Area} = 4\pi r^2 \qquad\qquad \text{Lateral area} = 2\pi r h \qquad\qquad \text{Lateral area} = \pi r\sqrt{r^2 + h^2}$$

B.2 Trigonometric formulas

Degrees and radians

$$180° = \pi \ (\text{rad}) \quad 1° = \frac{\pi}{180} \ (\text{rad}) \quad 1\ (\text{rad}) = \left(\frac{180}{\pi}\right)°$$

Right triangle and unit circle definitions (angles measured in radians)

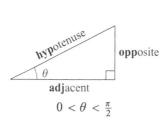

$$0 < \theta < \tfrac{\pi}{2}$$

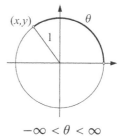

$$-\infty < \theta < \infty$$

$$\sin\theta = \frac{\text{opp}}{\text{hyp}} \qquad \csc\theta = \frac{\text{hyp}}{\text{opp}} \qquad\qquad \sin\theta = y \qquad \csc\theta = \frac{1}{y}$$

$$\cos\theta = \frac{\text{adj}}{\text{hyp}} \qquad \sec\theta = \frac{\text{hyp}}{\text{adj}} \qquad\qquad \cos\theta = x \qquad \sec\theta = \frac{1}{x}$$

$$\tan\theta = \frac{\text{opp}}{\text{adj}} \qquad \cot\theta = \frac{\text{adj}}{\text{opp}} \qquad\qquad \tan\theta = \frac{y}{x} \qquad \cot\theta = \frac{x}{y}$$

The six trigonometric functions as lengths of line segments (for $0 < \theta < \frac{\pi}{2}$)

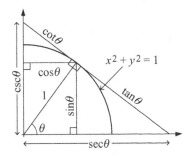

Reciprocal and quotient identities

$$\sin x = \frac{1}{\csc x} = \frac{\tan x}{\sec x} \qquad \cos x = \frac{1}{\sec x} = \frac{\cot x}{\csc x} \qquad \tan x = \frac{1}{\cot x} = \frac{\sin x}{\cos x}$$

$$\csc x = \frac{1}{\sin x} = \frac{\sec x}{\tan x} \qquad \sec x = \frac{1}{\cos x} = \frac{\csc x}{\cot x} \qquad \cot x = \frac{1}{\tan x} = \frac{\cos x}{\sin x}$$

Pythagorean identities

$$\sin^2 x + \cos^2 x = 1 \qquad \tan^2 x + 1 = \sec^2 x \qquad 1 + \cot^2 x = \csc^2 x$$

Symmetry identities

$$\sin(-x) = -\sin x \qquad \cos(-x) = \cos x \qquad \tan(-x) = -\tan x$$

$$\csc(-x) = -\csc x \qquad \sec(-x) = \sec x \qquad \cot(-x) = -\cot x$$

Cofunction identities

$$\sin\left(\frac{\pi}{2} - x\right) = \cos x \qquad \cos\left(\frac{\pi}{2} - x\right) = \sin x \qquad \tan\left(\frac{\pi}{2} - x\right) = \cot x$$

$$\csc\left(\frac{\pi}{2} - x\right) = \sec x \qquad \sec\left(\frac{\pi}{2} - x\right) = \csc x \qquad \cot\left(\frac{\pi}{2} - x\right) = \tan x$$

Sum and difference formulas

$$\sin(x \pm y) = \sin x \cos y \pm \cos x \sin y \qquad \cos(x \pm y) = \cos x \cos y \mp \sin x \sin y$$

$$\tan(x \pm y) = \frac{\tan x \pm \tan y}{1 \mp \tan x \tan y}$$

Double-angle formulas

$$\sin 2x = 2 \sin x \cos x \qquad \cos 2x = \begin{cases} \cos^2 x - \sin^2 x \\ 2 \cos^2 x - 1 \\ 1 - 2 \sin^2 x \end{cases} \qquad \tan 2x = \frac{2 \tan x}{1 - \tan^2 x}$$

Half-angle formulas

$$\sin^2 x = \frac{1 - \cos 2x}{2} \qquad \cos^2 x = \frac{1 + \cos 2x}{2}$$

Sum to product and product to sum formulas

$$\sin x \pm \sin y = 2 \sin \frac{x \pm y}{2} \cos \frac{x \mp y}{2} \qquad \sin x \sin y = \frac{1}{2}[\cos(x - y) - \cos(x + y)]$$

$$\cos x + \cos y = 2 \cos \frac{x + y}{2} \cos \frac{x - y}{2} \qquad \cos x \cos y = \frac{1}{2}[\cos(x - y) + \cos(x + y)]$$

$$\cos x - \cos y = -2 \sin \frac{x + y}{2} \sin \frac{x - y}{2} \qquad \sin x \cos y = \frac{1}{2}[\sin(x - y) + \sin(x + y)]$$

C

Supplemental Topics in Single Variable Calculus

The AP Calculus AB course taught in most high schools generally follows the Topic Out-line in the Course Description for AP Calculus published by the College Board and re-produced in Appendix A. A close examination of that Outline will reveal that there are several topics often encountered in the first term of a college course in calculus that do not appear in it. In this Appendix we present two of those topics: *Newton's method* and *mathematical induction*.

C.1 Newton's method

> *If I have seen farther it is by standing on the shoulders of giants.*
>
> Isaac Newton

Newton's method is a technique for approximating the roots of an equation of the form $f(x) = 0$ for a differentiable function f. It was developed in the early eighteenth century by Isaac Newton and others as an algebraic technique. However, differential calculus is generally employed today in implementing the method. Before presenting Newton's method, let's examine another method—the *bisection method*—based on the intermediate value theorem. We begin with an example.

Example C.1 Find a decimal approximation to $\sqrt{7}$. This would be easy with your graphing calculator, but unfortunately its batteries are dead. All you have is a simple calculator that will add, subtract, multiply, and divide numbers, such as the one illustrated in Figure 10.2 in Section 10.1. It does not have a $\boxed{\sqrt{\ }}$ button.

Since $\sqrt{7}$ is a root of the equation $x^2 - 7 = 0$ we can apply the intermediate value theorem to the continuous function $f(x) = x^2 - 7$ to find an interval (a, b) containing $\sqrt{7}$. One such interval is $(2, 3)$ since 0 lies between $f(2) = -3$ and $f(3) = 2$. We can find a smaller interval containing $\sqrt{7}$ by bisecting $(2, 3)$ at its midpoint 2.5. If $f(2.5)$ is positive, $\sqrt{7}$ lies in $(2, 2.5)$ and if $f(2.5)$ is negative, $\sqrt{7}$ lies in $(2.5, 3)$. If we continue bisecting, we will locate $\sqrt{7}$ in smaller and smaller intervals, as indicated in Table C.1. In going from one row to the next, we replace a by the midpoint m whenever $f(m) < 0$, and we replace b by m whenever $f(m) > 0$ (since f is an increasing function on $(2, 3)$). This method is known as the *bisection method*.

Table C.1. The bisection method for approximating $\sqrt{7}$.

a	b	$m = (a + b)/2$	$f(m)$
2.0	3.0	2.5	−
2.5	3.0	2.75	+
2.5	2.75	2.625	−
2.625	2.75	2.6875	+
2.625	2.6875	2.65625	+
2.625	2.65625	2.640625	−
2.640625	2.65625	2.6484375	+
2.640625	2.6484375		

After seven steps we have $\sqrt{7}$ trapped in the interval $(2.640625, 2.6484375)$, and hence we know the first two decimal places of $\sqrt{7}$, $\sqrt{7} \approx 2.64$. The bisection method does not use technology very efficiently, since with new batteries in your graphing calculator you could graph $y = x^2 - 7$ and zoom in on the root $\sqrt{7}$ much more quickly. ■

Like many concepts in calculus such as limits, derivatives, and definite integrals, Newton's method is based on successive approximations. The idea behind Newton's method can be illustrated graphically. In Figure C.1 we have the graph $y = f(x)$ with a root r that we wish to approximate. From the graph we chose an initial approximation x_0, compute $f(x_0)$ and $f'(x_0)$, and construct the tangent line to $y = f(x)$ at the point P_0 with coordinates $(x_0, f(x_0))$. If $f'(x_0) \neq 0$ this line will intersect the x-axis at a point x_1 closer (we hope) to r.

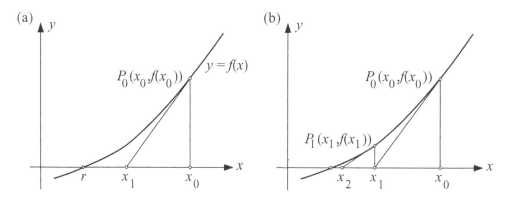

Figure C.1. Newton's method

The equation of the tangent line to $y = f(x)$ at P_0 is $y = f(x_0) + f'(x_0)(x - x_0)$, and if we call its x-intercept x_1, then $0 = f(x_0) + f'(x_0)(x_1 - x_0)$ so that

$$x_1 = x_0 - \frac{f(x_0)}{f'(x_0)}.$$

We continue in this manner, producing points closer and closer to r. The way in which we find a new approximation x_{new} from an old one x_{old} is the same at each step in Newton's method:

$$x_{new} = x_{old} - \frac{f(x_{old})}{f'(x_{old})}, \quad f'(x_{old}) \neq 0.$$

If we wish to evaluate r correct to d decimal places, we terminate the process when two successive approximations agree to d decimal places (called a *stopping rule* for the method). In an advanced mathematics course in numerical analysis you will learn that in general, the number of correct decimal places approximately doubles with each step in Newton's method. In the following examples we terminate the process when two successive approximations agree to the number of decimal places shown on a calculator.

Back to Example C.1 To illustrate Newton's method and compare it to the bisection method, let's use it to approximate the positive root of $x^2 - 7 = 0$. With $f(x) = x^2 - 7$ we have

$$x_{new} = x_{old} - \frac{f(x_{old})}{f'(x_{old})} = x_{old} - \frac{x_{old}^2 - 7}{2x_{old}} = \frac{x_{old}^2 + 7}{2x_{old}} = \frac{1}{2}\left(x_{old} + \frac{7}{x_{old}}\right).$$

With an initial approximation $x_0 = 3$ we have

$$x_1 = 2.66666666667$$
$$x_2 = 2.64583333333$$
$$x_3 = 2.64575131234$$
$$x_4 = 2.64575131106$$
$$x_5 = 2.64575131106$$

and we stop here with eleven decimal places after only five steps: $\sqrt{7} \approx 2.64575131106$. We can implement Newton's method to approximate $\sqrt{7}$ with a simple calculator that will only add, subtract, multiply and divide. ∎

Example C.2 Find the root of $x^3 - 2x - 5 = 0$. This is Newton's own example, from his work *De Analysi per Aequationes Numero Terminorum Infinitas* published in 1669. With $f(x) = x^3 - 2x - 5 = 0$ we have

$$x_{new} = x_{old} - \frac{f(x_{old})}{f'(x_{old})} = x_{old} - \frac{x_{old}^3 - 2x_{old} - 5}{3x_{old}^2 - 2} = \frac{2x_{old}^3 + 5}{3x_{old}^2 - 2}.$$

To choose a first approximation x_0, we look at a graph of $y = f(x)$, as shown in Figure C.2. Since the root r is close to 2, we begin with $x_0 = 2$.

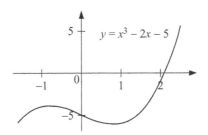

Figure C.2. A graph of $y = f(x) = x^3 - 2x - 5$

Using a calculator we obtain the following approximations to the root r:

$$x_1 = 2.1$$
$$x_2 = 2.0945681211$$
$$x_3 = 2.0945514817$$
$$x_4 = 2.0945514815$$
$$x_5 = 2.0945514815$$

and hence to ten decimal places, $r \approx 2.0945514815$. ■

☺ To check your work, now show that for the above value of r, $r^3 - 2r - 5 \approx 0$.

In the above example you may have been tempted to use a graphing calculator and its zoom-in feature to approximate the root. This is a good procedure to use when you are only interested in, say, two or three decimal places of accuracy. But if you wish to have ten or twelve places of accuracy, zooming in can be tedious and Newton's method is more efficient.

Solving polynomial equations

In your high school algebra course you learned a general method for solving quadratic equations: the quadratic formula. There is also a general method for solving cubic equations called *Cardano's formula,* named for the Italian mathematician Girolamo Cardano (1501–1576) who published it in 1545. When applied to Newton's cubic equation in the preceding example, Cardano's formula yields the solution

$$r = \left(\frac{\sqrt{643}}{6\sqrt{3}} + \frac{5}{2} \right)^{1/3} + \frac{2}{3} \left(\frac{\sqrt{643}}{6\sqrt{3}} + \frac{5}{2} \right)^{-1/3} .$$

Such an exact solution using square and cube roots is called a *solution by radicals* (the word *radical* comes from the Latin *radix* meaning "root", and our symbol $\sqrt{}$ stands for the initial letter r of radix). Cardano's student, the Italian mathematician Lodovico Ferrari (1522–1565), discovered a solution in radicals for the general quartic (degree 4) equation, which was published along with Cardano's formula in 1545.

Inspired by the success of Cardano and Ferrari, renaissance mathematicians began the search for solutions in radicals for the quintic (degree 5) equation. However, the work of the French mathematician Évariste Galois (1811–1832) and the Norwegian mathematician Niels Henrik Abel (1802–1829) ended that search by proving that *for general quintic and all higher degree polynomial equations, no general solutions in radicals exist.*

Newton's method can also be used to approximate numbers other than square roots. In the next example we use Newton's method to approximate π.

Example C.3 To use Newton's method to approximate π, we first need a function f with π as a root. One such function is $f(x) = \sin x$. With this function we have

$$x_{\text{new}} = x_{\text{old}} - \frac{f(x_{\text{old}})}{f'(x_{\text{old}})} = x_{\text{old}} - \frac{\sin(x_{\text{old}})}{\cos(x_{\text{old}})} = x_{\text{old}} - \tan(x_{\text{old}}).$$

A reasonable initial approximation is $x_0 = 3$. Then a calculator (one that can evaluate trigonometric functions) yields

$$x_1 = 3.14254654307$$

$$x_2 = 3.14159265330$$

$$x_3 = 3.14159265359$$

$$x_4 = 3.14159265359$$

and so to eleven decimal places, $\pi \approx 3.14159265359$. While you may be surprised to obtain such accuracy so easily, remember that there is a lot of mathematics hiding behind that $\boxed{\text{TAN}}$ button on your calculator. ∎

There are instances where Newton's method fails to approximate a root of an equation. For example, the method may lead to an approximation x_n for which $f'(x_n) = 0$ and $f(x_n) \neq 0$, so that the tangent line at $(x_n, f(x_n))$ is horizontal and doesn't intersect the x-axis. In other cases, the approximations may oscillate between two values, or become successively farther and farther away from the root. See Exercises 12, 13, and 14 for examples of this behavior.

Our advice is to use technology to locate an initial approximation x_0, and if possible x_0 should be in an interval where neither f' nor f'' changes sign. As you generate approximations, use technology to verify that they are indeed getting closer to the root.

Our final example shows how Newton's method is used to find a point of intersection of two graphs.

Example C.4 In Exercise 2 of Exploration 6.4.1 we noted that the graphs of $y = \cosh x$ and $y = x^2 + 1$ intersect at approximately $x = 2.9828$. Let's use Newton's method to find the x-value correct to ten places. To solve $\cosh x = x^2 + 1$ we set $f(x) = \cosh x - x^2 - 1$ and solve $f(x) = 0$ with

$$x_{\text{new}} = x_{\text{old}} - \frac{f(x_{\text{old}})}{f'(x_{\text{old}})} = x_{\text{old}} - \frac{\cosh(x_{\text{old}}) - x_{\text{old}}^2 - 1}{\sinh(x_{\text{old}}) - 2x_{\text{old}}}.$$

An examination of the graph (see Figure 6.10 in Exploration 6.4.1) indicates a starting value of $x_0 = 3$, which leads to

$$x_1 = 2.98315975559$$

$$x_2 = 2.98286722283$$

$$x_3 = 2.98286713574$$

$$x_4 = 2.98286713575$$

so that to ten places, $x = 2.9828671357$. ∎

Newton's method in the complex plane

Newton's method can be applied to functions of complex numbers, The equation $z^3 - 1 = 0$ has three roots, one real and two non-real. The set of starting values x_0 in the complex plane that lead to a root is called the *basin of attraction* for that root. In the figure below we have used different shades of gray for the three basins. The boundaries of the basins form a fractal called a *Newton fractal*.

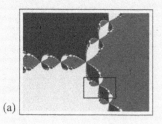 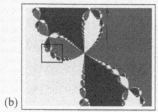 

(a)   (b)   (c)

We can see the intricate structure of the basins and the Newton fractal by zooming in on the picture. Zooming in on the small rectangular region outlined in black in (a) yields the region in (b), and zooming in on the region outlined in black in (b) yields the region in (c).

C.1.1 Exercises

In Exercises 1 through 6 use Newton's method to approximate (correct to three decimal places) the zero(s) of the following functions. Use a graphing utility to find an appropriate value for x_0.

1. $f(x) = 3 - x^3$ 2. $f(x) = x^3 - 3x^2 - 1$ 3. $f(x) = x^5 - x - 1$

4. $f(x) = e^x + x$ 5. $f(x) = \sin x - 2x + 1$ 6. $f(x) = \arctan x + 1 - x$

In Exercises 7 through 9 use Newton's method to approximate (correct to three decimal places) the solution(s) to the following equations.

7. $e^x = x^3$ 8. $\sin x = x^2 - 1$ 9. $x \ln x = 1$

10. (a) Show that the number $\sqrt{1/3}$ approximates the Euler-Mascheroni constant γ (see Section D.2) correct to three decimal places.
    (b) Show that the number $(7/83)^{2/9}$ approximates the Euler-Mascheroni constant γ correct to six decimal places.

11. Show that the number $(9/10)^{5/6}$ approximates Catalan's constant $G \approx 0.9159655\ldots$ (see Exercise 13 in Section 10.6.1) correct to four decimal places.

12. The equation $x^3 - 3x^2 + 3x + 1 = 0$ has a root between 0 and -1. Show that Newton's method with $f(x) = x^3 - 3x^2 + 3x + 1$ and $x_0 = 2$ fails to approximate this root since one of the steps in the approximation leads to a horizontal tangent line.

13. The equation $x^3 - 2x - 2 = 0$ has a root between 1 and 2. Show that Newton's method with $f(x) = x^3 - 2x - 2$ and $x_0 = 0$ fails to approximate this root, since the approximations alternate between 0 and -1.

14. The number 2 is clearly the only root of the equation $\sqrt[3]{x-2} = 0$. Show that Newton's method fails to "approximate 2" with $f(x) = \sqrt[3]{x-2}$ and $x_0 = 2 + h$ for $h \neq 0$. Hence Newton's method fails in this example for any x_0 except 2.

15. Use the function $f(x) = \ln x - 1$ to approximate e as accurately as you can.

16. (a) Use your calculator and Newton's method with $x_{new} = x_{old} + \sin(x_{old})$ and $x_0 = 3$ to find x_1, x_2, and x_3. What number do you think is being approximated?
    (b) What is the function f in the equation $f(x) = 0$ being solved in (a)? (Hint: solve the differential equation $f(x)/f'(x) = -\sin x$.)

17. (a) Use Newton's method to approximate $\sqrt{5/7}$ correct to eight decimal places. Call the result z.
    (b) Show that $3 - (z/3)$ approximates a well-known mathematical constant (which one?) correct to six decimal places.

18. Use Newton's method to approximate $\sqrt[4]{2143/22}$ correct to eight decimal places. Is the result familiar?

A number x_{fix} is a *fixed point* of a function f if $f(x_{fix}) = x_{fix}$. Approximate, correct to three decimal places, fixed points of the functions in Exercises 19 through 21.

19. $f(x) = \cos x$

20. $f(x) = e^{-x}$

21. $f(x) = \ln(1/x)$

22. Why do the functions in Exercises 20 and 21 have *exactly* the same fixed point?

23. The tangent function has infinitely many fixed points (why?). Approximate correct to three decimal places the smallest positive one.

C.2 Mathematical induction

> *Induction makes you feel guilty for getting something out of nothing, and it is artificial, but it is one of the greatest ideas of civilization.*
>
> Herbert Wilf

Mathematical induction (or simply *induction*) is an important and powerful proof technique. Many students encounter it for the first time in a calculus course in college. We have not used induction in this text since it is not on the topic outline for AP® calculus, and hence not taught in many high school calculus courses. But if you intend to study additional mathematics, statistics, or computer science courses in college, you should become familiar with mathematical induction.

Induction is a method of proof for statements about the set $\mathbb{N}$ of natural numbers or counting numbers, $\mathbb{N} = \{1, 2, 3, \cdots\}$. Simple examples from your PCE, which you used when you first evaluated simple definite integrals using the definition of the integral as the limit of a Riemann sum, are:

$$1 + 2 + 3 + \cdots + n = \frac{n(n+1)}{2} \quad \text{for all } n \geq 1 \tag{C.1}$$

$$1^2 + 2^2 + 3^2 + \cdots + n^2 = \frac{n(n+1)(2n+1)}{6} \quad \text{for all } n \geq 1 \tag{C.2}$$

$$1^3 + 2^3 + 3^3 + \cdots + n^3 = \frac{n^2(n+1)^2}{4} \quad \text{for all } n \geq 1. \tag{C.3}$$

(We use the phrases "all n in $\mathbb{N}$" and "all $n \geq 1$" interchangeably.)

Each of the statements asserts that the equation is an identity, true for infinitely many values of n. To explain how induction works, it will be useful to have a name for the statement we wish to prove true. Since the statement usually concerns a natural number n, we give it the name $S(n)$. For example, in (C.1), $S(n)$ is the statement "$1 + 2 + 3 + \cdots + n = n(n+1)/2$." We prove that $S(n)$ is true for all n in $\mathbb{N}$ in two steps, traditionally denoted step I and step II (step I is called the *basis* step and step II is called the *induction* step):

I. Show that $S(1)$ is true
II. Show that for any n in $\mathbb{N}$, if $S(n)$ is true then $S(n+1)$ must also be true.

In step II we don't prove that either $S(n)$ or $S(n+1)$ is true, but only that if S is true for one number (n) in $\mathbb{N}$ it must be true for the next one ($n+1$). The fact that steps I and II together imply that $S(n)$ is true for all n in $\mathbb{N}$ is called the *principle of mathematical induction*. The logic behind the principle of mathematical induction is the same as that behind what's called the domino effect in everyday speech. If, on a very long table, we stand up on end infinitely many dominoes (each domino represents one of the statements) close enough together so that if one falls over (falling over means the statement has been proven true) so will the next one, and then knock over the first one, all the dominoes must fall (that is, all the statements are true). See Figure C.3.

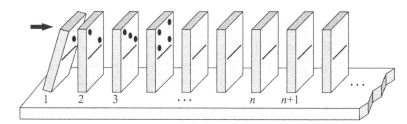

Figure C.3. Mathematical induction and the domino effect

Example C.5 Let's use induction to prove the statement $S(n)$ in (C.1). First observe that the natural number n is used two ways in $S(n)$, in the values of the right-hand side and the last term on the left-hand side, and the number of terms on the left-hand side. To begin, we examine the basis step $S(1)$: it is the statement $1 = (1 \cdot 2)/2$, which is clearly true. For the induction step, we need to show that the truth of the equation in $S(n+1)$

$$1 + 2 + 3 + \cdots + (n+1) = \frac{(n+1)(n+2)}{2}$$

follows from the truth of the statement of the equation in $S(n)$. That's done as follows:

$$1 + 2 + 3 + \cdots + (n+1) = (1 + 2 + 3 + \cdots + n) + (n+1)$$
$$= \frac{n(n+1)}{2} + (n+1)$$
$$= \frac{(n+1)}{2}(n+2)$$
$$= \frac{(n+1)(n+2)}{2}.$$

Thus, by the principle of mathematical induction, $1 + 2 + 3 + \cdots + n = n(n+1)/2$ for all $n \geq 1$. ■

Example C.6 Prove (C.2). This proof by induction is similar to the one in Example C.5. $S(1)$ is the statement $1^2 = (1 \cdot 2 \cdot 3)/6$, which is true. The statement $S(n+1)$ in the induction step is "$1^2 + 2^2 + 3^2 + \cdots + (n+1)^2 = (n+1)(n+2)(2n+3)/6$." To show that $S(n+1)$ follows from $S(n)$, we have:

$$1^2 + 2^2 + 3^2 + \cdots + (n+1)^2 = (1^2 + 2^2 + 3^2 + \cdots + n^2) + (n+1)^2$$
$$= \frac{n(n+1)(2n+1)}{6} + (n+1)^2$$
$$= \frac{(n+1)}{6}[(2n^2 + n) + 6(n+1)]$$
$$= \frac{(n+1)}{6}(n+2)(2n+3)$$
$$= \frac{(n+1)(n+2)(2n+3)}{6},$$

and hence by the principle of mathematical induction,

$$1^2 + 2^2 + 3^2 + \cdots + n^2 = \frac{n(n+1)(2n+1)}{6} \quad \text{for all } n \geq 1. \quad ■$$

We leave the proof of (C.3) as an exercise.

Example C.7 Here is an example where induction can be used early in calculus, just after learning that $\frac{d}{dx}x = 1$ and the product rule for differentiation. Prove that for all $n \geq 1$,

$$\frac{d}{dx}x^n = nx^{n-1}. \tag{C.4}$$

If $S(n)$ is the statement in C.4, then $S(1)$ is $\frac{d}{dx}x = 1$, which is true. To show that $S(n+1)$ follows from $S(n)$, we have:

$$\frac{d}{dx}x^{n+1} = \frac{d}{dx}(x \cdot x^n) = x \cdot \frac{d}{dx}x^n + x^n \cdot \frac{d}{dx}x$$
$$= x \cdot nx^{n-1} + x^n \cdot 1 = (n+1)x^n,$$

and hence by the principle of mathematical induction, $\frac{d}{dx}x^n = nx^{n-1}$ for all $n \geq 1$. ■

Example C.8 Prove that for all $n \geq 4$, $n! > 2^n$. Here the statement $S(n)$ is $n! > 2^n$. This example differs from the earlier ones in two ways: (i) $S(n)$ is an inequality (which only changes the algebra in the induction step), and (ii) we are to prove it true for $n \geq 4$ rather than for all of $\mathbb{N}$ (which simply changes the basis step to $S(4)$ rather than $S(1)$).

$S(4)$ is the statement $4! = 24 > 16 = 2^4$, which is clearly true. To show that the truth of $S(n)$ implies the truth of $S(n + 1)$, we have

$$(n + 1)! = (n + 1)n! > (n + 1)2^n > 2 \cdot 2^n = 2^{n+1},$$

as desired. Thus $n! > 2^n$ for all $n \geq 4$. ∎

As a proof technique, mathematical induction is a formal process for establishing the truth of a statement about the natural numbers. It sheds no light on the origin or context of the statement being proved. As the Italian-American mathematician Gian-Carlo Rota (1932–1999) put it, "If we have no idea why a statement is true, we can still prove it by induction." For example, in Exercise 2 we ask you to prove that for all $n \geq 1$, the sum of the first n odd numbers is the square of n. The induction proof won't tell you why this is true, but perhaps this picture does (hint: count the balls by L-shaped regions):

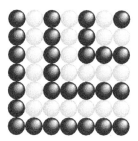

Figure C.4. A visual argument for $1 + 3 + 5 + \cdots + (2n - 1) = n^2$ for $n = 7$

C.2.1 Exercises

In Exercises 1–8 prove the statement true for all $n \geq 1$.

1. $1^3 + 2^3 + 3^3 + \cdots + n^3 = \dfrac{n^2(n + 1)^2}{4}$ for all $n \geq 1$ (this is (C.3)).

2. $1 + 3 + 5 + \cdots + (2n - 1) = n^2$.

3. $1^2 + 3^2 + 5^2 + \cdots + (2n - 1)^2 = \dfrac{n(2n - 1)(2n + 1)}{3}$.

4. $1^3 + 3^3 + 5^3 + \cdots + (2n - 1)^3 = n^2(2n^2 - 1)$.

5. $1 + 3 + 3^2 + 3^3 + \cdots + 3^{n-1} = \dfrac{3^n - 1}{2}$.

6. $\dfrac{1}{1 \cdot 2} + \dfrac{1}{2 \cdot 3} + \dfrac{1}{3 \cdot 4} + \cdots + \dfrac{1}{n(n + 1)} = \dfrac{n}{n + 1}$.

7. $t_1 + t_2 + t_3 + \cdots + t_n = \dfrac{n(n + 1)(n + 2)}{6}$ where $t_n = n(n + 1)/2$ is the nth triangular number (see Exercise 12 in Section 10.2.3).

8. $\cos a + \cos 3a + \cos 5a + \cdots + \cos(2n - 1)a = \dfrac{\sin 2na}{2 \sin a}.$

9. Let $f(x) = xe^x$. In this exercise you will find (and prove) a formula for $f^{(n)}(x)$. Evaluate $f'(x)$, $f''(x)$, $f'''(x)$, and $f^{(4)}(x)$. Do you see a pattern? Formulate a hypothesis for $f^{(n)}$ and prove your hypothesis true using mathematical induction.

10. *Bernoulli's inequality* (named for Jacob Bernoulli (1654–1705)) states that for all $n \geq 0$ and $x \geq -1$, $(1 + x)^n \geq 1 + nx$.
    (a) Use mathematical induction to prove Bernoulli's inequality.
    (b) The inequality is also true when n is replaced by any real number $r \geq 1$. To see why (and how to prove it), graph $y = (1 + x)^r$ and $y = 1 + rx$ on the same set of axes. What happens if r satisfies $0 < r < 1$? If $r < 0$?

11. In Exercise 10 in Section 9.5.1 you showed that for all $\alpha > 0$ the gamma function defined by

$$\Gamma(\alpha) = \int_0^\infty x^{\alpha-1} e^{-x} \, dx$$

satisfies $\Gamma(\alpha + 1) = \alpha \Gamma(\alpha)$. Prove that for a positive integer n, $\Gamma(n) = (n - 1)!$.

D
Supplemental Explorations

D.1 The Wallis product for $\pi/2$

In 1656 John Wallis (1616–1703) published the following curious but remarkable formula expressing $\pi/2$ as an infinite product:

$$\frac{\pi}{2} = \frac{2}{1} \cdot \frac{2}{3} \cdot \frac{4}{3} \cdot \frac{4}{5} \cdot \frac{6}{5} \cdot \frac{6}{7} \cdot \frac{8}{7} \cdot \frac{8}{9} \cdots . \tag{D.1}$$

To prove this, you will use calculus, a tool that Wallis did not possess. You will show that

$$\lim_{n\to\infty} \left[\frac{2}{1} \cdot \frac{4}{3} \cdots \frac{2n}{2n-1} \right]^2 \frac{1}{2n+1} = \frac{\pi}{2} . \tag{D.2}$$

Exercise 1. Let $I_n = \int_0^{\pi/2} \sin^n x \, dx$. Show that $I_{n+1} < I_n$ for n a nonnegative integer.

Exercise 2. In Exercise 35 in Section 1.2.1 you established

$$I_{2n} = \frac{2n-1}{2n} \cdot \frac{2n-3}{2n-2} \cdots \frac{3}{4} \cdot \frac{1}{2} \cdot \frac{\pi}{2} \quad \text{and} \quad I_{2n+1} = \frac{2n}{2n+1} \cdot \frac{2n-2}{2n-1} \cdots \frac{2}{3} \cdot 1.$$

Show that $I_{2n+1} < I_{2n}$ implies

$$\left[\frac{2}{1} \cdot \frac{4}{3} \cdots \frac{2n}{2n-1} \right]^2 \frac{1}{2n+1} < \frac{\pi}{2},$$

and that $I_{2n} < I_{2n-1}$ implies

$$\frac{2n}{2n+1} \cdot \frac{\pi}{2} < \left[\frac{2}{1} \cdot \frac{4}{3} \cdots \frac{2n}{2n-1} \right]^2 \frac{1}{2n+1} .$$

Exercise 3. Combine the inequalities from Exercise 2 and take the limit as $n \to \infty$ to obtain (D.2). (Hint: use the squeeze theorem.)

As a practical tool for calculating π, Wallis's product is useless, since the product on the right in (D.1) approaches $\pi/2$ much too slowly. Computing the product in (D.1) for $n = 500$ gives only $\pi \approx 3.13989$.

D.2 The Euler-Mascheroni constant

In Section 10.1 we showed that the harmonic series $1 + \frac{1}{2} + \frac{1}{3} + \cdots + \frac{1}{n} + \cdots$ diverges by showing that for each $n \geq 1$, the partial sum $H_n = 1 + \frac{1}{2} + \frac{1}{3} + \cdots + \frac{1}{n}$ is larger than $\ln(n+1)$.

345

But how much larger? The answer to this question leads to a number known as the *Euler-Mascheroni constant*.

Exercise 1. Add two rows to Table 10.1, exhibiting $\ln(n+1)$ and $H_n - \ln(n+1)$ for the selected value of n. Round the entries to four decimal places. You should obtain something like Table D.1.

Table D.1. Comparing H_n to $\ln(n+1)$

n	10	100	1,000	10,000	100,000	1,000,000
H_n	2.9290	5.1874	7.4855	9.7876	12.0901	14.3927
$\ln(n+1)$	2.3979	4.6151	6.9088	9.2104	11.5129	13.8155
$H_n - \ln(n+1)$	0.5311	0.5723	0.5767	0.5772	0.5772	0.5772

It should appear to you that the sequence $\{\gamma_n\}$ of terms $\gamma_n = H_n - \ln(n+1)$ (with γ_0 equal to 0) in the last row is converging. You will now prove that $\{\gamma_n\}$ converges by showing that it is increasing and bounded above. Observe that γ_n is the sum of the areas of the portions of the rectangles representing H_n that lie above the curve $y = 1/x$ over the interval $[1, n+1]$. See Figure D.1 where γ_n appears in gray.

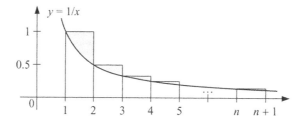

Figure D.1. A visual representation of γ_n

Exercise 2. Show that $\{\gamma_n\}$ is increasing. (Hint: show that $\gamma_n - \gamma_{n-1}$ represents the area of the region shaded dark gray in Figure D.2 and thus $\gamma_n - \gamma_{n-1} \geq 0$.)

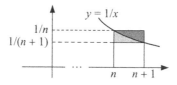

Figure D.2. A visual representation of $\gamma_n - \gamma_{n-1}$

Exercise 3. Show that $\{\gamma_n\}$ is bounded above. (Hint: Figure D.2 suggests that $\gamma_n - \gamma_{n-1}$ is less than the area of the rectangle in two shades of gray or

$$\gamma_n - \gamma_{n-1} < \frac{1}{n} - \frac{1}{n+1}. \tag{D.3}$$

Now replace n by k and sum both sides of (D.3) from 1 to n to obtain $\gamma_n < 1 - \frac{1}{n+1} < 1$.

Thus the sequence $\{y_n\}$ converges, and its limit is traditionally denoted by the Greek letter γ:

$$\gamma = \lim \gamma_n = \lim[H_n - \ln(n+1)]. \tag{D.4}$$

This limit is known as the *Euler-Mascheroni constant* after the Swiss mathematician Leonhard Euler (1707–1783) and the Italian mathematician Lorenzo Mascheroni (1750–1800). Evaluated to twenty decimal places, $\gamma \approx 0.57721566490153286060$. It is still unknown whether γ is rational or irrational.

In Example 10.26 you learned that the alternating harmonic series converges to $\ln 2$. Using (D.4) you can now construct a second proof. Let $A_n = \sum_{k=1}^{n}(-1)^{k+1}\frac{1}{k}$ denote the nth partial sum of the alternating harmonic series.

Exercise 4. Show that $A_{2n+1} = H_{2n+1} - H_n$ for $n \geq 1$.

Exercise 5. Use (D.4) to show that $\lim A_{2n+1} = \ln 2$ and then show $\lim A_{2n} = \ln 2$ to conclude $\lim A_n = \ln 2$.

Consider the series $1 + \frac{1}{2} - \frac{1}{3} + \frac{1}{4} + \frac{1}{5} - \frac{1}{6} + \cdots$ in which the signs follow the pattern two positive followed by one negative. Let S_n denote its nth partial sum.

Exercise 6. Show that $S_{3n} = H_{3n} - (2/3)H_n$ for $n \geq 1$.

Exercise 7. Evaluate $\lim S_{3n}$. Does the series converge or diverge?

Consider the series $1 + \frac{1}{2} - \frac{2}{3} + \frac{1}{4} + \frac{1}{5} - \frac{2}{6} + \cdots$ in which the signs follow the same pattern, but now the numerator of each negative term is 2. Let S_n denote its nth partial sum.

Exercise 8. Show that $S_{3n} = H_{3n} - H_n$ for $n \geq 1$.

Exercise 9. Evaluate $\lim S_{3n}$. Does the series converge or diverge?

Exercise 10. Show that

$$\gamma = \lim \left(1 + \frac{1}{2} + \cdots + \frac{1}{n} - \frac{1}{n+1} - \frac{1}{n+2} - \cdots - \frac{1}{n^2}\right).$$

D.3 The Euler-Mascheroni constant, continued

In Exploration D.2, you showed that the Euler-Mascheroni constant γ is given by $\gamma = \lim[H_n - \ln(n+1)]$, where H_n is the nth partial sum of the harmonic series. H_n is also called the nth *harmonic number*, and in Exercise 12 of 10.1.1 you encountered the fractional harmonic numbers: for x in $[0, 1]$, $H_x = \int_0^1 \frac{1-t^x}{1-t}\, dt$. In this exploration you will prove the surprising result that $\int_0^1 H_x\, dx = \gamma$. See Figure D.3.

Exercise 1. Show that $H_x = \sum_{k=1}^{\infty}\left(\frac{1}{k} - \frac{1}{x+k}\right)$. (Hint: express the integrand in the defining integral for H_x as a geometric series and integrate term by term.)

Exercise 2. Integrate the result of Exercise 1, recalling that an infinite series is the limit of its sequence of partial sums.

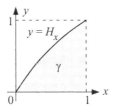

Figure D.3. The relationship between γ and H_x

D.4 Stirling's formula

Stirling's formula, named for the Scottish mathematician James Stirling (1692–1770), is an approximation to $n!$, usually written as

$$n! \sim \sqrt{2\pi} \cdot n^{n+(1/2)} e^{-n}$$

where for two sequences $\{a_n\}$ and $\{b_n\}$, $a_n \sim b_n$ means $\lim_{n\to\infty} a_n/b_n = 1$ (the symbol $\sim$ is pronounced "is asymptotic to"). You will now derive Stirling's formula using some of the calculus you have learned in this course. We begin with a trapezoidal rule approximation to a definite integral.

Exercise 1. Show that $\int_1^n \ln x \, dx = n \ln n - n + 1$ for a positive integer n.

Exercise 2. Let T_n denote the trapezoidal rule approximation to $\int_1^n \ln x \, dx$ using $n - 1$ intervals of width 1. Show that

$$T_n = \sum_{k=2}^{n} \ln k - \frac{1}{2} \ln n = \ln n! - \frac{1}{2} \ln n.$$

Now let c_n denote the error in approximating $\int_1^n \ln x \, dx$ by T_n, i.e.,

$$c_n = \int_1^n \ln x \, dx - T_n = \left(n + \frac{1}{2}\right) \ln n - n + 1 - \ln n!.$$

Exercise 3. Show that $n! = C_n n^{n+(1/2)} e^{-n}$ where $C_n = e^{1-c_n}$.

The final step is to evaluate $\lim C_n$, which you will do by examining the sequence $\{c_n\}$.

Exercise 4. Show that $\{c_n\}$ is an increasing sequence.

You will now show that $\{c_n\}$ is bounded. To do so, begin with the error formula for the trapezoidal rule approximation from Theorem 7.3.

Exercise 5. Show that the error in the approximating $\int_k^{k+1} \ln x \, dx$ with a single trapezoid is less than $1/(12k^2)$, and consequently that

$$c_n \leq \frac{1}{12}\left(1 + \frac{1}{4} + \frac{1}{9} + \cdots + \frac{1}{(n-1)^2}\right).$$

Exercise 6. Why does the result of Exercise 5 imply that $\{c_n\}$ converges?

Since $\{c_n\}$ converges, so does $\{C_n\}$. Let $C = \lim\limits_{n \to \infty} C_n$. To evaluate C, proceed as follows.

Exercise 7. Show that

$$\frac{C_n^2}{C_{2n}} = \frac{(n!)^2 e^{2n}}{n^{2n+1}} \cdot \frac{(2n)^{2n+(1/2)}}{(2n)! e^{2n}} = \frac{2 \cdot 4 \cdot 6 \cdots (2n)}{1 \cdot 3 \cdot 5 \cdots (2n-1)} \sqrt{\frac{2}{n}}$$

$$= \frac{2 \cdot 4 \cdot 6 \cdots (2n)}{1 \cdot 3 \cdot 5 \cdots (2n-1)} \cdot \frac{1}{\sqrt{2n+1}} \sqrt{\frac{2(2n+1)}{n}}.$$

Exercise 8. Take the limit as $n \to \infty$ of both sides of the above expression to show that $C = \sqrt{2\pi}$, which completes the derivation of Stirling's formula. (Hint: Use the Wallis formula from Exploration D.1.)

Exercise 9. A typical exercise in a probability course is, show that the probability p_n of obtaining exactly n heads and n tails in $2n$ flips of a fair coin is

$$p_n = \binom{2n}{n} \left(\frac{1}{2}\right)^{2n}.$$

($\binom{2n}{n}$ is the central binomial coefficient discussed in Exercise 15 in Section 11.7.1.) Use Stirling's formula to show that $p_n \approx 1/\sqrt{\pi n}$.

Another formula for $n!$ is Burnside's formula, which can be derived from Stirling's formula and is more accurate. Burnside's formula is

$$n! \sim \sqrt{2\pi} \left(\frac{n + 1/2}{e}\right)^{n+1/2}.$$

Exercise 10. Use Stirling's formula to show that

$$n! \sim \sqrt{2\pi}(n + 1/2)^{n+1/2} e^{-n} \left(\frac{n}{n + 1/2}\right)^{n+1/2}.$$

Exercise 11. Show that

$$\lim_{n \to \infty} \left(\frac{n}{n + 1/2}\right)^{n+1/2} = e^{-1/2}$$

which completes the derivation of Burnside's formula.

D.5 Newton's approximation of π

In 1671 Isaac Newton wrote the book *Methodus Fluxionum et Serierum Infinitarum*, which contains an approximation to π correct to sixteen decimal places based on what we now call the Maclaurin series for $\sqrt{1-x}$. In the exploration you will recreate his approximation using modern terminology. It is based on Figure D.4.

Exercise 1. Show that the graph of $y = \sqrt{x - x^2}$ is the semicircle with radius $1/2$ centered at $(1/2, 0)$, as shown in Figure D.4. Compute the area of the shaded region in two ways to show that

$$\frac{\pi}{24} = \frac{\sqrt{3}}{32} + \int_0^{1/4} \sqrt{x - x^2}\, dx$$

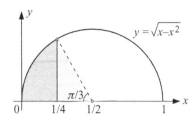

Figure D.4. The semicircle in Newton's approximation of π

and hence

$$\pi = \frac{3\sqrt{3}}{4} + 24 \int_0^{1/4} \sqrt{x - x^2}\, dx. \tag{D.5}$$

Exercise 2. Write the integrand in D.5 as $\sqrt{x}\sqrt{1-x}$, use Theorem 11.11 to expand $\sqrt{1-x}$ into its Maclaurin series, multiply by $\sqrt{x}$ and integrate to establish

$$\int_0^{1/4} \sqrt{x - x^2}\, dx = \frac{1}{12} - \frac{1}{5 \cdot 2^5} - \frac{1}{4 \cdot 7 \cdot 2^7} - \frac{1 \cdot 3}{4 \cdot 6 \cdot 9 \cdot 2^9} - \frac{1 \cdot 3 \cdot 5}{4 \cdot 6 \cdot 8 \cdot 11 \cdot 2^{11}} - \cdots.$$

Exercise 3. Using the first nine terms of the series in Exercise 2, show that D.5 yields the approximation $3.1415926683\ldots$, which is accurate to seven decimal places.

You may be wondering how Newton computed $\sqrt{3}$ in his approximation. He used the Maclaurin series for $\sqrt{1-x}$ after writing $\sqrt{3}$ as $2\sqrt{1 - (1/4)}$. At the time he wrote "I am ashamed to tell you to how many places of figures I carried these computations, having no other business at the time."

D.6 Estimating a value of a function from its derivatives

Let f be a function that is twice differentiable for $x > 0$ with $f(1) = 1$ and $f''(1) = 5.3146$. In Table D.2 we have values of $f'(x)$ for selected values of x.

Table D.2. Values of $f'(x)$

x	1	1.25	1.5	1.75	2
$f'(x)$	2.4645	3.9925	5.5452	6.2483	5.5452

In this exploration you will approximate the value of $f(2)$ using several techniques from calculus. You should be able to obtain the approximations in brackets [ ].

Exercise 1. Write an equation for the line tangent to the graph of $y = f(x)$ at $x = 1$ and use this line to estimate $f(2)$. [3.4645]

Exercise 2. Show that $f(2) = f(1) + \int_1^2 f'(x)\, dx$, and estimate $f(2)$ by using Simpson's rule with four subintervals to approximate the integral. [6.0019]

Exercise 3. Use Euler's method with four steps of size 0.25 to estimate $f(2)$. [5.5601]
Exercise 4. Write the second degree Taylor polynomial for f about $x = 1$, and use the polynomial to estimate $f(2)$. [6.1218]

Which of the four methods in the four exercises is likely to be a poor estimate of $f(2)$? Which is likely to be a better estimate? Why? The data in this exploration were generated from the function $f(x) = 9/(1 + 2^{7-4x})$, rounded to four decimal places. Thus $f(2) = 6$. Are you surprised by the accuracy or lack of accuracy of any of the estimation methods?

D.7 Acknowledgments

1. Exploration D.1 is adapted from A. E. Taylor and W. R. Mann, *Advanced Calculus*, 2nd ed., Xerox College Publishing, Lexington, MA, 1972.
2. The proof in Exploration D.2 that the alternating harmonic series converges to ln 2 is adapted from L. Gillman, The alternating harmonic series, *College Mathematics Journal*, **33** (2002), pp. 143–145.
3. Exercise 10 in Exploration D.2 is from J. J. Mačys, A new problem, *American Mathematical Monthly*, **119** (2012), p. 82.
4. Exploration D.5 is adapted from Chapter 7 in W. Dunham, *Journey Through Genius: The Great Theorems of Mathematics*, John Wiley & Sons, Inc., New York, 1990.

E

Answers to Odd-Numbered Exercises

1.1.1

1. 8π 3. $96\pi/5$ 5. $16\pi/3$ 7. 8π

9. $\pi/5$ 11. $2\pi/5$ 13. $5\pi/14$ 15. $7\pi/15$

17. $\pi/10$ 19. $10\pi/21$ 21. (a) $8\pi/3$ (b) 8π 23. $35\pi/2$

25. (a) $V = \pi \int_{-r}^{r} \left(\left(R + \sqrt{r^2 - y^2} \right)^2 - \left(R - \sqrt{r^2 - y^2} \right)^2 \right) dy$

(b) $V = 2\pi \int_{-r}^{r} (R - x) \cdot 2\sqrt{r^2 - x^2}\, dx$ (c) $V = 2\pi R \cdot \pi r^2$

1.2.1

1. $\sin x - x \cos x + C$

3. $\theta \tan \theta - \ln |\sec \theta| + C$

5. $\frac{1}{4} t^2 (2 \ln t - 1) + C$

7. $-(\ln y)/y + C$

9. $-\left(x^3 + 3x^2 + 6x + 6 \right) e^{-x} + C$

11. $\frac{1}{4}\left(x\sqrt{1 - x^2} + (2x^2 - 1) \arcsin x \right) + C$

13. $\frac{1}{13} e^{2x} (3 \sin(3x) + 2 \cos(3x)) + C$

15. $(x + 1) \arctan \sqrt{x} - \sqrt{x} + C$

17. $\frac{1}{5} (3 \sin(2x) \sin(3x) + 2 \cos(2x) \cos(3x)) + C$ 19. $\frac{1}{8} (\sin(2x) - 2x \cos(2x)) + C$

21. $\frac{1}{8} \left(2 \sec^3 x \tan x + 3 \sec x \tan x + 3 \ln |\sec x + \tan x| \right) + C$

23. $\pi \sqrt{2}/4 - \ln |\sqrt{2} + 1|$

25. $(\ln 2 - 1)/2$

27. 0

29. $2\pi^2$

31. (b) Different constants of integration.

(c) $\ln(\ln x) + C$

1.3.1

1. 2   3. $4\pi^2$ 5. 1440 m^3

2.1.1

1. 22/3   3. $\ln \left(\sqrt{2} + 1 \right)$ 5. $\frac{16}{3}\sqrt{2} - 2\sqrt{6}$ 7. $16\sqrt{2} - 5\sqrt{5}$

9. area $= \frac{4}{3}h^2$, perimeter $= h\left(2 + \sqrt{5} + \frac{1}{2}\ln\left(2 + \sqrt{5}\right)\right)$

11. (a) $(3 + 2\ln 2)/4$ (b) $17/12$ (c) $33/16$

    (d) yes. $f(x) = \dfrac{x^{r+1}}{2(r+1)} + \dfrac{x^{1-r}}{2(r-1)}, r \neq \pm 1$, and $f(x) = \pm\left(\dfrac{x^2}{4} - \dfrac{\ln x}{2}\right), r = \pm 1$

13. $\pi\sqrt{2}$ 15. 2π 17. (a) $L = \int_0^{\pi/2}\left(1 - e^{-2x}\right)^{-1/2} dx$

    (b) $L = \int_0^{\pi/3} \sec x \, dx$ (c) $\ln(2 + \sqrt{3})$ 19. $6a$

2.2.1

1. $\ln\left(\sqrt{1 + x^2} + x\right) + C$ 3. $\ln\left|t + \sqrt{t^2 - 4}\right| + C$

5. $\dfrac{1}{6}\operatorname{arcsec}\left(\dfrac{y}{3}\right) - \dfrac{\sqrt{y^2 - 9}}{2y^2} + C$ 7. $\dfrac{u}{2(1 - u^2)} + \dfrac{1}{4}\ln\dfrac{1 + u}{1 - u} + C$

9. $\frac{1}{2}\operatorname{arcsec}(x^2) + C$ 11. $2\arcsin\left(\dfrac{\theta - 2}{2}\right) + \dfrac{1}{2}(\theta - 2)\sqrt{4\theta - \theta^2} + C$

13. $3/10$ 15. $\left(4\sqrt{2} - 3\sqrt{3}\right)/6$

17. $\pi/3$ 19. (a)$(\ln 3)/4$ (b) $\ln(3/5)/4$

2.3.1

1. (a) πab (b) $a/b = \varphi$ 3. $2\sqrt{3} - \ln(2 + \sqrt{3})$

2.4.1

1. $A = \int_0^{\pi} 2\pi \sin x \sqrt{1 + \cos^2 x}\, dx$ 3. $A = \int_1^e 2\pi \ln x \sqrt{1 + (1/x)^2}\, dx$

5. $\frac{\pi}{27}\left(145^{3/2} - 1\right)$ 7. $\dfrac{1604}{45}\pi$

9. $\frac{\pi}{6}\left(e^3 + 12e - 13\right)$ 11. $\pi r \sqrt{r^2 + h^2}$

13. $2\pi R \cdot 2\pi r$ 15. $\pi(b^2 + h^2)$

3.1.1

1. $e^{2x}, -5e^{2x}$ are solutions   3. $y = t^2$ and $y = t^2 \ln t$ are solutions

5. $y = \cos x$ and $y = 2\cos x$ are solutions   7. $y = \frac{1}{4}x^4 - \dfrac{2}{3}x^3 + 3x + C$

9. $y = -\cos x + C$ 11. $y = \frac{1}{4}\ln(2x^2 - 1) + C$

13. $y = -\ln|\cos x| + C$ 15. $y = \dfrac{1}{2}x\sqrt{-x^2 + 36} + 18\arcsin\left(\dfrac{x}{6}\right)$

17. $y = -\dfrac{1}{2}\dfrac{\ln x}{x^2} - \dfrac{1}{4x^2} + C$ 19. $y = e^{-x} + C$

21. (a) $\dfrac{d^2 F}{dt^2} = \ln(4/3)\left(30 - \dfrac{dF}{dt}\right) = \ln(4/3)(30 - \ln(4/3)(30 - F))$

    (c) $F(0) = 150°$. $F(t) = 100$ when $t = \ln(7/12)\big/\ln(3/4) \approx 1.8736$.

23. The rate of absorption by the body is the negative of the rate of change of A with respect to time. Therefore, $\frac{dA}{dt} = -kA(t)$. With this equation, k is positive.

3.2.1

1. slope field for $\dfrac{dy}{dx} = x$ slope field for $\dfrac{dy}{dx} = y$

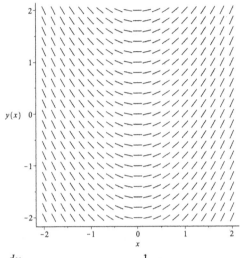 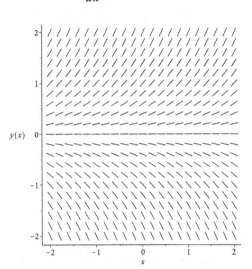

$\dfrac{dy}{dx} = x$ has solution $y = \dfrac{1}{2}x^2 + C$ $\dfrac{dy}{dx} = y$ has solution $y = e^x + C$

3. slope field for $\dfrac{dy}{dx} = y/x$ 5. slope field for $\dfrac{dy}{dx} = x/y$

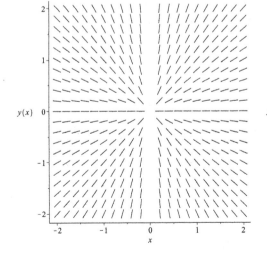 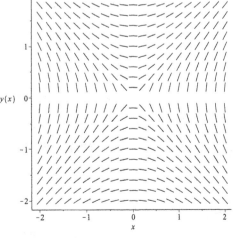

7. $\dfrac{d^2y}{dx^2} = 0$ 9. $\dfrac{d^2y}{dx^2} = \dfrac{y^2 - x^2}{y^2}$

11. slope field for $\dfrac{dy}{dx} = x^2/(1-y^2)$

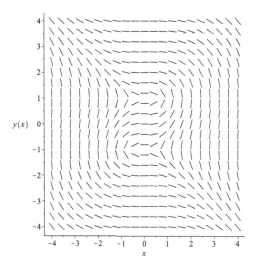

3.3.1

1. $y(3) = 6$ 3. $y(3) \approx \dfrac{278801}{82212} \approx 3.391$

5. (a) -4 (b) -4.678 (c) -5.384

7. (a) $\dfrac{dy}{dx}$ does not exist when either $y = 1$ or $y = -1$.

(b) The first step in Euler's method produces a y-value that is less than 1. But since the initial condition has a y-value greater than 1, this would force our solution to have $y = 1$ for some x-value. This cannot happen since the DE is undefined at $y = 1$.

3.4.1

1. $y = kx$ 3. $y = \pm\sqrt{x^2 + C}$ 5. $y = ke^{x^3/3}$

7. $y = ke^{x^2/2} - 1$ 9. $y = \arcsin\left(x^3/3 + C\right)$ 11. $y = \pm(2\arcsin(x) + C)^{1/2}$

13. $y = 3e^{2x-2}$ 15. $Q(t) = 200(1 - e^{x/2})$ 17. $y = 1 + \frac{8}{x}$

3.5.2

1. $y = \sin x + \frac{\cos x}{x} + \frac{k}{x}$ 3. $y = \dfrac{\ln(1 + x^2)}{2x} + \dfrac{C}{x}$

5. $y = \dfrac{e^{x^2}}{2x} + \dfrac{C}{x}$ 7. $y = 1 - 4e^{-x}$

9. $y = x^3 + 3x$

11. $y = 1 + 2e^{-\sin^{-1}(x)}$

4.1.1

1. (b) $t_{1/2} = (\ln A)/k$ 3. (a) $x = \left(1 + Ce^{-kt}\right)^{-1}$ (b) approximately 3 weeks

5. exponential growth in each one-year period. For 10 years, $r(t) = \dfrac{1}{10} \ln \dfrac{P(t + 10)}{P(t)}$.

4.2.2

1. $\dfrac{1}{4} \ln \left| \dfrac{x - 2}{x + 2} \right| + C$

3. $\dfrac{2}{3} \ln |x - 1| + \dfrac{1}{6} \ln \left(x^2 + x + 1\right) - \sqrt{3} \arctan \dfrac{2x + 1}{\sqrt{3}} + C$

5. $3x + 5 \ln |x - 3| - \ln |x + 1| + C$

7. $\ln \left| \dfrac{2x + 1}{x + 2} \right| + C$

9. $\dfrac{1}{2} \ln \dfrac{(x - 1)^2}{x^2 + x + 1} - \sqrt{3} \arctan \dfrac{2x + 1}{\sqrt{3}} + C$

13. $\dfrac{2}{3} \arctan \dfrac{x}{2} - \dfrac{1}{3} \arctan x + C$

15. $\dfrac{-1}{2(x - 1)^2} - \dfrac{5}{3(x - 1)^3} - \dfrac{5}{2(x - 1)^4} - \dfrac{2}{(x - 1)^5} + C$

17. $\ln x - \dfrac{1}{3} \ln |x^3 + 1| + C$

21. $x = \dfrac{ab(1 - e^{-kt})}{a - be^{-kt}}$

23. $\dfrac{1}{2} \ln \dfrac{x^2 - 2x + 2}{x^2 + 2x + 2} + \arctan(x - 1) + \arctan(x + 1) + C$

25. $y''' = \dfrac{-12}{x^4} + \dfrac{48}{(2x + 3)^4} + \dfrac{486}{(3x - 2)^4}$

27. $y^{(n)} = \dfrac{(-1)^n n!}{\sqrt{5}} \left((x - \varphi)^{-(n+1)} - \left(x + \dfrac{1}{\varphi}\right)^{-(n+1)} \right)$

4.3.1

1. $\dfrac{3}{2} \ln \left| x^{2/3} - 1 \right| + C$

3. $\dfrac{6}{7} x^{7/6} - \dfrac{6}{5} x^{5/6} + 2\sqrt{x} - 6 x^{1/6} + 6 \arctan x^{1/6} + C$

5. $-2\sqrt{x} - 4\sqrt[4]{x} + C$

7. $(9\sqrt{6} - 20)/12$

9. $2 \arctan 2 - \pi/2$

11. $\dfrac{2\pi}{1215}(247\sqrt{13} + 64)$

4.4.1

1. (b) The line is vertical when all the x_i are the same.

5. (a)   7(a)

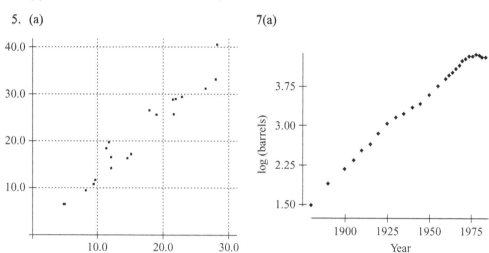

5. (b) $m \approx 1.289$, rather close to $4/\pi$

7. (b) Using technology, $b \approx -120.2733$, $A = e^b \approx 5.834 \cdot 10^{-53}$, $k = m \approx 0.06596$.

5.1.1

1. 160   3. $2e^{-1} - 4e^{-3}$ 5. $\frac{1}{4} \ln 3$

7. 40 foot-pounds.   9. $\frac{1}{2}k(x_2^2 - x_1^2)$ 11. 6750 foot-pounds

13. ≈ 6701.65 ft-lbs.   15. (a) $\approx 1.768 \times 10^{12}$ n   (b) $\approx 1.766 \times 10^{12}$ n

5.2.3

1. 4.8 feet.   3. (a) 2 (b) $\pi/2$ 5. (a) 1/4 (b) 8/15

7. (a) e^2 (b) $\frac{1}{4}(3e^2 - 1)$ 9. $\bar{x} = \dfrac{-1 + \pi\sqrt{2}/4}{\sqrt{2} - 1}, \bar{y} = \dfrac{1}{4(\sqrt{2} - 1)}$

11. $\bar{x} = \frac{1}{4}(1 + e^2), \bar{y} = -1 + \frac{e}{2}$.

13. $\bar{x} = \dfrac{2}{3} \dfrac{(1 - (1 - a^2)^{3/2})}{(\sin^{-1}(a) + a(1 - a^2)^{1/2})}, \bar{y} = \dfrac{1}{3} \dfrac{3a - a^3}{\sin^{-1}(a) + a(1 - a^2)^{1/2}}$

15. $\bar{x} = \dfrac{1 + \sqrt{5}}{2}, \bar{y} = 0$

5.3.1

1. $1404\pi \approx 4410.796$ lbs   3. 18,000 n   5. 5500/3 n   7. $2500/3 \times 62.4 = 52,000$ lbs

6.1.1

7. $f'(x) = \tanh x$

9. $h'(x) = -\operatorname{csch} x$

11. $g'(x) = x \sinh x$

13. $\frac{1}{2} (\ln(\cosh x))^2 + C$

15. $x \cosh x - \sinh x + C$

17. $x \tanh x - \ln(\cosh x) + C$

19. $\frac{1}{4} (\sinh(2x) - 2x) + C$

21. $\frac{1}{2} (\sin x \sinh x + \cos x \cosh x) + C$ 23. $\pi \tanh 2$

25. The two occurrences of $\int e^x \sinh x \, dx$ have different constants of integration.

27. (c) $f(x) = \cosh x, g(x) = \sinh x$

6.2.1

5. $g'(x) = \sec x$ 7. $g'(x) = \sinh^{-1} x$ 9. $g'(x) = \tanh^{-1} x$

11. See Exercises 7–9   15. $\frac{1}{2} \left((x^2 - 1) \tanh^{-1} x + x \right) + C$

17. $e^{\sinh^{-1}(1/2)} = \varphi$, the golden ratio

6.3.1

13. $e^{\operatorname{gd}^{-1}(\arctan(1/2))} = \varphi$, the golden ratio

7.1.2

1. The integral is exactly 0.2. The estimates and errors are

method	estimate	error
left endpoint	0.1427001953	−0.0572998047
right endpoint	0.2677001953125	+0.0677001953
trapezoid	0.2052001953125	+0.0052001953125
midpoint	0.1974029541015625	−0.0025970458984375

3. The integral is exactly $-\pi/4 + \arctan(3)$. The estimates and errors are

method	estimate	error
left endpoint	0.5728116711	+0.109164063
right endpoint	0.3728116711	−0.090835937
trapezoid	0.4728116711	+0.009164063
midpoint	0.4590672617	−0.0045803477.

5. The integral and all estimates are 0.

7. The integral and both estimates are 2.5 with no error. The region represented in the integral is a trapezoid with height 0 on left and 5 on right. Thus the trapezoid approximation

with one trapezoid is exact. The midpoint is exact because the area of a trapezoid is the same as the area of a rectangle with same base and height the average of the heights of the trapezoid. This average height occurs for this linear function at the midpoint of the interval.

9. For left and right: $n \geq 106{,}067$. For trapezoid: $n \geq 77{,}340$. For midpoint: $n \geq 38{,}670$.

11. $R_{100} = 0.7436573980$ (to ten decimal places) with error bound 0.008577638847.

13. $M_{100} = 0.1260499117$ (to ten decimal places) with error bound 0.00007284.

17. (a) $L_{10} = 9.359950394$, $R_{10} = 9.859950391$.
    (b) Both error bounds are 2.5.
    (c) $g'(x) = 2e^{-2x}/(1 + e^{-2x})^2 > 0$ for all x, g is an increasing function. Thus, the exact value of the integral is between L_{10} and R_{10}.
    (d) The error bound is greater than or equal to exact error, which in this case is much smaller than the error bound.

7.2.2

1. $S_8 = 0.2000325521$, error ≈ 0.0000325521

3. $S_4 = 0.4637488948$, error ≈ 0.0001012863 5. $S_6 = 0$ with 0 error.

7. $n \geq 10$ 9. $n \geq 44$

11. $S_{100} = 0.7468241331$, error bound $6.67 \cdot 10^{-10}$

13. $S_{50} = 2.644589422$, error bound 0.0006371555556

8.1.4

1. $y = \frac{2}{3}(x + 8)$, $-8 \leq x \leq 1$ 3. $y = x^3$, $-\sqrt{2} \leq x \leq \sqrt{2}$

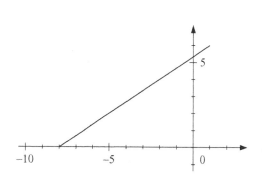

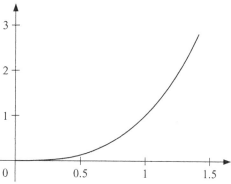

5. $y = x^4 - 4$, $\quad 0 \le x \le \sqrt{5}$

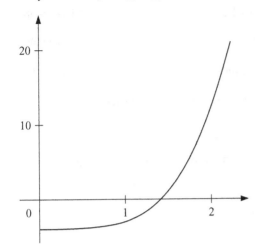

7. $y = 1/x$, $\quad x \ge 1$

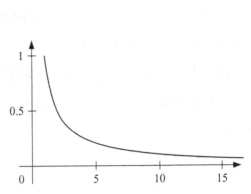

9. $4x^2 + y^2 = 1$

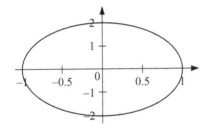

11. $y = 3x^2$, $\quad 0 \le x \le 1$

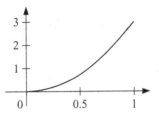

13. $x^2 - y^2 = 1$, $\quad \sinh(-2) \le y \le \sinh(2)$

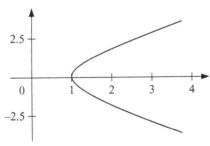

15. $y = x^2$, $\quad x \ge 0$

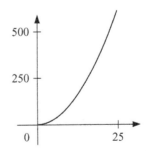

17. $\dfrac{dy}{dx} = -2\dfrac{\cos 2t}{\sin t}$

19. $\dfrac{dy}{dx} = \dfrac{1}{te^t}$

21. $y = \dfrac{\sqrt{3}}{2} + \dfrac{2}{\sqrt{3}}(x - 1/2)$

23. $y = 2 + \dfrac{3}{112}(x - 17)$

25. Tangent line is vertical. $x = 3/2$

27. Tangent line is horizontal. $y = 3$

29. Area $= 6a^2\pi$

8.2.4

1. (i) $(-2\sin 2t, \cos t)$ (ii) $(-4\cos 2t, -\sin t)$ (iii) ≈ 9.688

3. (i) $(3t^2 - 4, 2t - 3)$ (ii) $(6t, 2)$ (iii) ≈ 6.902

5. (i) $(\sinh t, \cosh t)$ (ii) $(\cosh t, \sinh t)$ (iii) ≈ 2.634

7. (i) $(\cos t - t\sin t, \sin t + t\cos t)$ (ii) $(-2\sin t - t\cos t, 2\cos t - t\sin t)$

   (iii) $\frac{1}{2}\left(\pi\sqrt{\pi^2+1} - \ln\left(-\pi + \sqrt{\pi^2+1}\right)\right)$

9. (i) $(t\cos t, t\sin t)$ (ii) $(\cos t - t\sin t, \sin t + t\cos t)$ (iii) $\dfrac{\pi^2}{2}$

11. t 13. $16a$

8.3.4

1. $(1, 0)$ 3. $(2, \pi/3)$ 5. $(2\sqrt{2}, 3\pi/4)$

7. $(\sqrt{10}, \tan^{-1} 3)$ 9. $(1, 0)$ 11. $(1, \sqrt{3})$

13. $(-1, -\sqrt{3})$ 15. $(\sqrt{2}, -\sqrt{2})$

17. (a) (b)

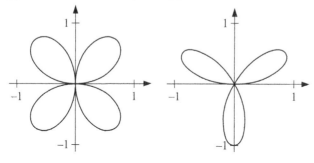

(c)

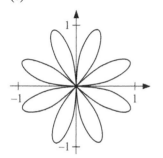

(d) When n is even, the area is $\pi/2$; when the area is odd the area is $\pi/4$.

19. length $= \sqrt{2}(e^{2\pi} - 1)$ 21. length $= 2\pi\sqrt{16\pi^2 + 1} - (1/2)\ln(-4*\pi + \sqrt{16\pi^2 + 1})$

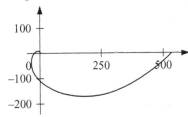
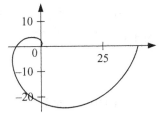

23. perimeter $= 8a$ 25. area $= 2a^2 - \frac{\pi}{2}a^2$

9.1.1

1. 2   3. 1/2   5. 1   7. diverges

9. diverges   11. 1/2   13. $\pi/2$ 15. 2

17. 1   19. $V = \pi$, $A = \infty$

21. Volume converges to $\pi/(p^2 - 3p + 2)$ for $p > 2$; diverges for $p \le 2$.

9.2.1

1. 4/3   3. 1   5. 1/2   7. 0

9. 1   11. 0   13. 1   15. 0

17. 3/4   19. diverges   21. $\ln(4/3)$ 23. $\frac{1}{2}\ln 2 + \pi/4$

25. $\dfrac{1}{s-a}, s > a$ 27. $\pi^2/4$

9.3.1

1. 4   3. $\pi/2$ 5. $-1/4$ 7. 9/2   9. π 11. π 13. $\pi + \ln(2 + \sqrt{3})$

15. Volume converges to $(2p\pi)/(2p - 1)$ for $p < 1/2$; diverges for $p \ge 1/2$.

17. (b) no 21. $A = \pi/4$, $L = -\ln x_0$, $V = \pi/3$, $S = 2\pi$

9.4.1

1. 1/2   3. 0   5. 1/2   7. e 9. 1   11. e

9.5.1

1. $K = a/\pi$ 3. $K = \dfrac{1}{2a}$ 5. $\mu = 1/\lambda$ 7. $\mu = 0$

9. $\mu = \begin{cases} a/(a-1), & a > 1 \\ \text{undefined}, & a \le 1 \end{cases}$ 13. $\mu = \alpha\beta$ 15. $M(t) = (1 - \beta t)^{-\alpha}$

10.1.1

1. (b) $1/3$, (c) $1/4$ 3. diverges   5. $1/2$ 7. diverges

15. No. Let $a_n = 1/n$ and $b_n = -1/n$.

10.2.3

1. converges to $2/3$ 3. converges to $1/12$ 5. converges to $e/(e-1)$

7. converges to $3/4$ 9. converges to $3/4$ 11. converges to 1

13. (a) converges to 1   (b) converges to 2

10.3.1

1. increasing and bounded above   3. decreasing and bounded below

5. decreasing and bounded below

7. (e) $s_n < s_{n+1}$ and $s_{2n-1} < s_{2n} < 4$ for all n, so $\{s_n\}$ is increasing and bounded above.

10.4.2

1. converges   3. diverges   5. converges   7. converges

9. converges   11. $\zeta(3) \approx 1.20206$

13. $V = \pi \sum_{n=1}^{\infty} 1/n^2$, $S = \pi + 2\pi \sum_{n=1}^{\infty} 1/n$

10.5.1

1. converges   3. diverges   5. converges   7. converges

9. converges   11. diverges

10.6.1

1. converges   3. converges   5. diverges   7. converges

9. converges   11. $S = 0.095\ldots$ 13. $C \approx 0.916$

15. $\sum_{n=1}^{\infty} 1/t_{2n} = 2 - 2\ln 2$

11.1.1

1. divergent   3. absolutely convergent   5. absolutely convergent

11.2.1

1. absolutely convergent   3. absolutely convergent   5. absolutely convergent

7. absolutely convergent   9. absolutely convergent   11. conditionally convergent

11.3.1

1. absolutely convergent   3. absolutely convergent   5. divergent

7. absolutely convergent   9. divergent   11. divergent

11.4.1

1. $[-1, 1)$ 3. $[-2, 2)$ 5. $(-2, 1)$ 7. $(-1, 1]$

9. $(-\infty, \infty)$ 11. $(-\infty, \infty)$ 13. $(-1, 1)$

11.5.1

3. $\sum_{n=1}^{\infty}(-1)^{n-1}nx^{2n-1}, x \in (-1, 1)$ 5. $x \in (-1, 1), \dfrac{a + (b - a)x}{(a - x)^2}$

9. (a) $2\sum_{n=0}^{\infty}x^{2n+1}/(2n + 1), x \in (-1, 1)$ (b) $\sum_{n=0}^{\infty}1/((2n + 1)4^n)$

11.6.1

1. $\tan x = x + \frac{1}{3}x^3 + \frac{2}{15}x^5 + \cdots$

3. $\sin^2 x = \sum_{n=1}^{\infty}(-1)^{n+1}\dfrac{2^{2n-1}x^{2n}}{(2n)!}, \cos^2 x = 1 + \sum_{n=1}^{\infty}(-1)^n\dfrac{2^{2n-1}x^{2n}}{(2n)!}$

5. $\sin x = \dfrac{\sqrt{2}}{2}\sum_{n=0}^{\infty}\dfrac{(-1)^{n(n-1)/2}}{n!}\left(x - \dfrac{\pi}{4}\right)^n$

7. (a) $(-\infty, \infty)$ (b) Yes, $\cosh(\sqrt{-x})$ 9. 0.763547

11. $S(t) = \sum_{n=0}^{\infty}(-1)^n\dfrac{t^{4n+3}}{(4n + 3)(2n + 1)!}, C(t) = \sum_{n=0}^{\infty}(-1)^n\dfrac{t^{4n+1}}{(4n + 1)(2n)!}$

11.7.1

1. $1 - \dfrac{x}{2} - \sum_{n=2}^{\infty}\dfrac{1 \cdot 3 \cdot 5 \cdots (2n - 3)}{2^n n!}x^n$

3. $1 + \dfrac{1}{3}x^3 + \sum_{n=2}^{\infty}(-1)^{n+1} \cdot \dfrac{2 \cdot 5 \cdot 8 \cdots (3n - 4)}{3^n n!}x^{3n}$

5. $\dfrac{1}{2} + \sum_{n=1}^{\infty}(-1)^n\dfrac{1 \cdot 3 \cdots (2n - 1)}{2^{3n+1}n!}x^{2n}$

7. $x + \sum_{n=1}^{\infty}(-1)^n\dfrac{1 \cdot 3 \cdots (2n - 1)}{2 \cdot 4 \cdots (2n)} \cdot \dfrac{x^{2n+1}}{2n + 1}$

9. 0.33435   11. $g(x) = \dfrac{x(1 + x)}{(1 - x)^2}$ 13. $g(x) = \dfrac{x}{1 - x - x^2}$

C.1.1

1. $x = 1.442\ldots$ 3. $x = 1.167\ldots$ 5. $x = 0.887\ldots$

7. $x = 1.857\ldots$ 9. $x = 1.763\ldots$ 11. $(9/10)^{5/6} = 0.9159\ldots$

13. $x_1 = -1$ and $x_2 = 0 = x_0$ so the approximations alternate between 0 and -1.

15. $x_4 = x_5 = e$ to ten decimal places   17. $(a)\, z = 0.84515425\ldots$ $(b)\, 3 - (z/3) \approx e$

19. $x_{fix} = 0.739\ldots$ 21. $x_{fix} = 0.567\ldots$

23. The line $y = x$ intersects $y = \tan x$ once in each interval of the form
    $(n\pi - (\pi/2),\ n\pi + (\pi/2))$ for n an integer. The smallest positive x_{fix} is $4.493\ldots$.

C.2.1

9. $f^{(n)} = (x + n)e^x$

Index